VOID

Library of
Davidson College

Undergraduate Texts in Mathematics

Editors
F. W. Gehring
P. R. Halmos

Advisory Board
C. DePrima
I. Herstein
J. Kiefer

William McGowen Priestley

Calculus:
An Historical Approach

Springer-Verlag
New York Heidelberg Berlin

W. M. Priestley
The University of the South
Department of Mathematics
Sewanee, Tennessee 37375
USA

Editorial Board
P. R. Halmos
Managing Editor
Indiana University
Department of Mathematics
Bloomington, Indiana 47401
USA

F. W. Gehring
University of Michigan
Department of Mathematics
Ann Arbor, Michigan 48104
USA

AMS Subject Classifications: 26-01, 26A06

With 335 Figures

Library of Congress Cataloging in Publication Data

Priestley, William McGowen, 1940-
   Calculus, an historical approach.

   Includes index.
   1. Calculus.   2. Mathematics—History.   3. Calculus—History.
   I. Title.
QA303.P92   515   78-13681

All rights reserved.

No part of this book may be translated or reproduced
in any form without written permission from Springer-Verlag.

© 1979 by Springer-Verlag New York Inc.

Printed in the United States of America.

9 8 7 6 5 4 3 2 1

ISBN 0-387-90349-6   Springer-Verlag New York
ISBN 3-540-90349-6   Springer-Verlag Berlin Heidelberg

*To William Montgomery
and Thomas Carter,
for their delight in
sandpiles, seesaws, and pebbles.*

Even now there is a very wavering grasp of the true position of mathematics as an element in the history of thought. I will not go so far as to say that to construct a history of thought without profound study of the mathematical ideas of successive epochs is like omitting Hamlet from the play which is named after him. That would be claiming too much. But it is certainly analogous to cutting out the part of Ophelia. This simile is singularly exact. For Ophelia is quite essential to the play, she is very charming—and a little mad. Let us grant that the pursuit of mathematics is a divine madness of the human spirit, a refuge from the goading urgency of contingent happenings.

Alfred North Whitehead
from <i>Mathematics as an
Element in the
History of Thought</i>

Preface

This book is for students being introduced to calculus, and it covers the usual topics, but its spirit is different from what might be expected. Though the approach is basically historical in nature, emphasis is put upon ideas and their place—not upon events and their dates. Its purpose is to have students to learn calculus first, and to learn incidentally something about the nature of mathematics.

Somewhat to the surprise of its author, the book soon became animated by a spirit of opposition to the darkness that separates the sciences from the humanities. To fight the spell of that darkness anything at hand is used, even a few low tricks or bad jokes that seemed to offer a slight promise of success. To lighten the darkness, to illuminate some of the common ground shared by the two cultures, is a goal that justifies almost any means. It is possible that this approach may make calculus more fun as well.

Whereas the close ties of mathematics to the sciences are well known, the ties binding mathematics to the humanities are rarely noticed. The result is a distorted view of mathematics, placing it outside the mainstream of liberal arts studies. This book tries to suggest gently, from time to time, where a kinship between mathematics and the humanities may be found.

There is a misconception today that mathematics has mainly to do with scientific technology or with computers, and is thereby unrelated to humanistic thought. One sees textbooks with such titles as *Mathematics for Liberal Arts Majors*, a curious phrase that seems to suggest that the liberal arts no longer include mathematics.

No discipline has been a part of liberal arts longer than mathematics. Three—logic, arithmetic, and geometry—of the original seven liberal arts are branches of mathematics. Plato's friend Archytas, who helped develop the whole idea of liberal education, was a distinguished mathematician. No true student of liberal arts can neglect mathematics.

How did it happen that mathematics, in the public eye, became dissociated from the humanities? In brief, the emergence and growth of scientific knowledge in the seventeenth century led to a polarization in academic circles. Science went one way, the humanities went another. Mathematics, at first in the middle, seems now to be more commonly identified with the sciences and with the technology they engendered.

Today in some academic institutions the state is not healthy. The ground between the sciences and the humanities is so dark that many well-meaning members on each side lack the education to see the most valuable contributions of the other. To the disadvantage of students, this is sometimes the case even among the faculties of so-called "liberal arts" colleges.

In the seventeenth century mathematics was a bridge between the two kinds of knowledge. Thus, for example, Isaac Newton's new physics could be read by Voltaire, who was at home both with Homer and with Archimedes. Voltaire even judged Archimedes to be superior, in imagination, to Homer.

The unity of knowledge which seemed attainable in the seventeenth century, and which has long been an ideal of liberal education, is still worth seeking. Today as in the time of Voltaire, and in the time of Plato, mathematics calls us to eye this goal.

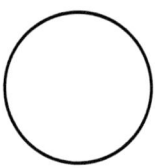

For Anyone Afraid of Mathematics

Maturity, it has been said, involves knowing when and how to delay succumbing to an urge, in order by doing so to attain a deeper satisfaction. To be immature is to demand, like a baby, the immediate gratification of every impulse.

Perhaps happily, none of us is mature in every respect. Mature readers of poetry may be immature readers of mathematics. Statemen mature in diplomacy may act immaturely in dealing with their own children. And mature mathematicians may on occasion act like babies when asked to listen to serious music, to study serious art, or to read serious poetry.

What is involved in many such cases is how we control our natural urge to get directly to the point. In mathematics, as in serious music or literature, the point sometimes simply cannot be attained immediately, but only by indirection or digression.

The major prerequisite for reading this book is a willingness to cultivate some measure of maturity in mathematics. If you get stuck, be willing to forge ahead, with suspended disbelief, to see where the road is leading. "Go forward, and faith will follow!" was d'Alembert's advice in the eighteenth century to those who would learn the calculus. Your puzzlement may vanish upon turning a page.

All that will be assumed at the outset is a nodding acquaintance with some elementary parts of arithmetic, algebra, and geometry, most of which was developed long before A.D. 1600. There will be some review in the early chapters, offering us as well a chance to outline the early history of mathematics.

To the Instructor

This book aspires to aid a student interested in either

(1) receiving an elementary introduction to the basic ideas of calculus; or
(2) learning "about" calculus, as a significant element in the history of thought.

At first these goals may seem incompatible. In fact, each tends to reinforce the other. A remark made by George Pólya suggests how both goals might be accomplished at once.

> If the learning of mathematics reflects to any degree the invention of mathematics, it must have a place for guessing, for plausible inference.

The reader will find plenty of opportunity here for guessing. The early chapters go at a gentle pace and invite the reader to enter into the spirit of the investigation.

For those whose backgrounds in mathematics are not especially strong, Chapters 1-6, together with Chapter 10, have been designed to form *a terminal course in mathematics*. (Chapter 10 can be read immediately after Chapter 6.) In these chapters algebraic manipulations have been kept at a simple level. Negative and fractional exponents can be avoided altogether here, and even the absolute value function is omitted. Trigonometry is not introduced until Chapter 7. The result, even with the omission of some of the harder problems at the ends of chapters, is not a watered-down course, but one that retains the full flavor of calculus.

This book is intended for use in a different way by students well prepared in mathematics. They should move rapidly through Chapters 1-6, omitting many of the routine exercises placed at the ends of sections and concentrating upon the more challenging problem sets located at the ends of chapters. After Chapter 6 the pace of the book picks up moderately. Chapter 7 develops the calculus of trigonometric functions, following a quick introduction to trigonometry. Chapter 8 discusses the integral in more detail than before, adding a little more rigor to a treatment that remains basically intuitive. Chapter 9 deals with the exponential and logarithmic functions. Chapter 10 is mainly historical in nature, but it sets the stage for the study of integration techniques through the use of formal manipulations with differentials.

The book ends with Chapter 10, where a leisurely two-semester calculus course will find itself near the end of the year. In a fast-moving course the instructor may have to look elsewhere for additional material to cover. Had there been a Chapter 11, it would have covered techniques of integration, a subject often touched upon in Chapters 6–10 but whose systematic study was postponed. The theory of infinite series is another important topic that has been touched upon but not developed in this volume.

In time, perhaps, a second volume will appear—to include these and other topics and to bring the story more up-to-date.

I wish to thank Mary Priestley for helping me in this enterprise and for sharing with me its ups and downs. I am grateful also to Paul Halmos for his interest and encouragement.

W. M. P.

May, 1978

Sewanee

Contents

Chapter 1. Tokens from the Gods ... 1
  Variables, Functions, and Limits

Chapter 2. The Spirit of Greece ... 28
  Pre-calculus Mathematics

Chapter 3. Sherlock Holmes Meets Pierre de Fermat ... 50
  Derivatives

Chapter 4. Optimistic Steps ... 78
  Techniques of Optimization

Chapter 5. Chains and Change ... 108
  Instantaneous rates

Chapter 6. The Integrity of Ancient and Modern Mathematics ... 148
  Integrals and Antiderivaties

Chapter 7. A Circle of Ideas ... 194
  Calculus of Trigonometric Functions

Chapter 8. House of Integrals ... 245
  Fundamental Principles Revisited

Chapter 9. The Central Height 299
Logarithmic and Exponential Functions

Chapter 10. Romance in Reason 352
Seventeenth-century Mathematics

Review Problems for Chapters 1–10 376

Appendices and Tables 385

Index 435

Acknowledgments

Grateful acknowledgment is given for permission to quote from copyrighted sources. Generally, full acknowledgment for a quotation may be found on the page where the quotation is found; however, to avoid awkwardness, sometimes only the publisher holding the U.S. rights is given in the case of short quotations whose rights are shared. More complete acknowledgments are given below (page numbers refer to the location of the quotation in the present text).

George Allen & Unwin, Ltd., for permission to quote from Tobias Dantzig, *Number, the Language of Science* (pp. 375, 427); from A. N. Whitehead, *The Aims of Education* (pp. 49, 413–414); and from Bertrand Russell, *Our Knowledge of the External World* (p. 368).

Cambridge University Press, for permission to quote from A. N. Whitehead, *Science and the Modern World* (p. v.); and from G. H. Hardy, *A Mathematician's Apology* (p. 384).

The Macmillan Publishing Company, for permission to quote from Tobias Dantzig, *Number, the Language of Science,* copyright 1930, 1933, 1939, and 1954 by Macmillan Publishing Co., Inc., and renewed 1958, 1961, and 1967 by Anna G. Dantzig (pp. 375, 427); from A. N. Whitehead, *The Aims of Education,* copyright 1929 by Macmillan Publishing Co., Inc., and renewed 1957 by Evelyn Whitehead (pp. 49, 413–414); and from A. N. Whitehead, *Science and the Modern World,* copyright 1925 by Macmillan Publishing Co., Inc., and Renewed 1953 by Evelyn Whitehead (p. v.).

Oxford University Press, for permission to quote from Leo Tolstoy, *War and Peace,* translated by Louise and Aylmer Maude (p. 371).

Rathbone Books Limited, for permission to quote from Bertrand Russell, *Wisdom of the West,* copyright 1959 (pp. 368, 433).

Simon and Schuster, for permission to quote from Bertrand Russel, *The Basic*

Writings of Bertrand Russell, edited by Robert E. Egner and Lester E. Denonn, copyright 1961 (pp. 110–111, 384).

Finally, the quotation by Richard Courant on page 353 is reprinted by kind permission of the author's son, who holds the copyright.

Anecdote of the Jar

Wallace Stevens

I placed a jar in Tennessee
And round it was, upon a hill.
It made the slovenly wilderness
Surround that hill.

The wilderness rose up to it,
And sprawled around, no longer wild.
The jar was round upon the ground
And tall and of a port in air.

It took dominion everywhere.
The jar was gray and bare.
It did not give of bird or bush,
Like nothing else in Tennessee.

Copyright 1923 and renewed 1951 by Wallace Stevens. Reprinted from *The Collected Poems of Wallace Stevens*, by Wallace Stevens, by permission of Alfred A. Knopf, Inc., and Faber and Faber Ltd.

Tokens from the Gods 1

A calculus is a pebble, or small stone.* Playing with pebbles, or "calculating", is a primitive form of arithmetic. *The calculus*, or *calculus*, refers to some mathematics that was developed principally in the seventeenth century.

Today the calculus can be seen as a natural result of a certain point of view. This point of view is reached in three steps. One begins by inventing the notion of a *variable* and trying to see situations, where possible, in terms of variables. The second step is to focus attention upon the relationship between the variables arising in a particular situation. This leads to the idea of a *function*. The third step involves the notion of the *limit* of a function. This simple yet subtle notion, which makes it all work, was recognized in the seventeenth century as being a key idea.

We shall discuss limits a little later. Right now, let us look at a couple of concrete situations where we can get hold of the idea of a variable and of a function relating one variable to another.

§1. A Calculus Problem

Let us become acquainted with a type of problem that calculus can handle. We shall not be able to solve this problem until certain tools are developed in a later chapter.

---

* Physicians still use the word *calculus* in this sense, to describe an unwelcome presence in the kidney or bladder. The success of a textbook on calculus is measured by the degree to which its contents are *not* described by the physician's usage of the word.

EXAMPLE 1. A small rectangular pen containing 12 square yards is to be fenced in. The front, to be made of stone, will cost $5 per yard of fencing, while each of the other three wooden sides will cost only $2 per yard. What is the least amount of money that will pay for the fencing?

In this example, the total *cost* of the fencing obviously will vary in terms of the design of the rectangle. Our job is to become familiar enough with *how the cost varies* in order to recognize the least possible cost. Toward this end we first pick at random a few possible designs and calculate their corresponding costs. There are lots of ways to enclose 12 square yards:

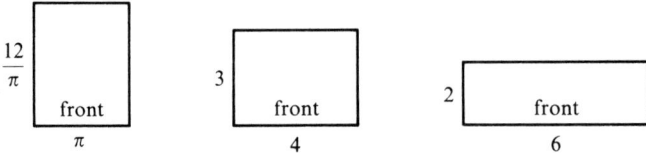

EXERCISES

1.1. Suppose the front is 1 yard in length. Find the cost. *Hint.* The cost is the sum of the costs of each of the four sides. First find the lengths of the sides, remembering that the area must be 12 square yards.

1.2. Suppose the front is 2 yards. Find the cost.

1.3. Suppose the front is 3 yards. Find the cost. *Answer:* $37.00.

1.4. Suppose the front is π yards. Find the cost. *Answer:* $7\pi + (48/\pi)$ dollars.

§2. Variables and Functions

The information obtained in the exercises above may be conveniently summarized in a table. Here, L is an abbreviation for the length in yards of the front, and C stands for the cost of the fencing in dollars.

L	C
1	55
2	38
3	37
π	$7\pi + (48/\pi)$

We have seen, in the exercises above, that the value of C is entirely determined by the value of L. In other words, there is a *rule* by which one gets

2. Variables and Functions

from L to C. This rule is simply given by

$$C = \text{cost in dollars of } \boxed{} \begin{array}{c} \frac{12}{L} \\ L \end{array} \qquad (1)$$

$$= \text{cost of front, plus cost of other sides}$$

$$= 5L + 2L + 2\left(\frac{12}{L}\right) + 2\left(\frac{12}{L}\right)$$

$$= 7L + \frac{48}{L}. \qquad (2)$$

Because the cost C varies in terms of the length L, it is natural to speak of C as a *variable* whose value is determined by the value of the variable L. In other words (and more explanation will be forthcoming below), C is a *function of* L, which we express succinctly by writing

$$C = f(L) \qquad (3)$$

(read "C equals f of L"). The symbol f denotes the function, or rule, by which C is given in terms of L. Putting lines (2) and (3) together shows that the rule f can be expressed by the equation

$$f(L) = 7L + \frac{48}{L}.$$

The notation $f(L)$ does not, of course, denote multiplication, but rather denotes the effect of the rule f acting upon the variable L. For example, by this rule,

$$f(\pi) = 7\pi + \frac{48}{\pi},$$

$$f(3) = 7 \cdot 3 + \frac{48}{3} = 21 + 16 = 37,$$

$$f(2) = 7 \cdot 2 + \frac{48}{2} = 14 + 24 = 38,$$

$$f(1) = 7 \cdot 1 + \frac{48}{1} = 7 + 48 = 55.$$

Since the equation $C = 7L + (48/L)$ says virtually the same thing as the equation $f(L) = 7L + (48/L)$, one might ask the reason for introducing this new symbol f. The reason is that we shall need to have a name for the mechanism, or rule, by which one gets from the left column above to the right column. It is, after all, this mechanism f that we want to study in order to recognize the least possible value of the cost C.

Note that f is *not* a variable, but stands for a fixed rule relating the two variables C and L.

EXERCISES

2.1. Use the rule given by $f(L) = 7L + (48/L)$ to find each of the following.
    (a) $f(4)$.
    (b) $f(5)$.
    (c) $f(x)$.
    (d) $f(\sqrt{2})$.
    (e) $f(3 + \sqrt{2})$.
    (f) $f(3 + h)$.
    (g) $f(x + h)$.
    (h) $f(\pi^2)$.
    (i) $f(x^2)$.
    (j) $f(4\pi)$.
    (k) $f(4t)$.
    (l) $f(8/7)$.
    (m) $f(6)$.
Answers: (c) $7x + (48/x)$. (g) $7(x + h) + (48/(x + h))$. (k) $28t + (12/t)$.

2.2. In the following table, fill in the question marks appropriately. (Your answers to the preceding exercise may be helpful here.)

L	C
4	?
2.5	?
x	?
x^2	?
$x + h$	?
?	50
?	50

§3. Three Ways of Looking at a Function

In this book, the word *function* will be used a little loosely and may have either of these three meanings:

(A) A function is a pair of columns of numbers. Not just any pair of columns, but a pair *whose first column has no number repeated*. We speak of the function as being *from* the first column *to* the second.

1	55
2	38
3	37
4	40
(etc.)	

(B) A function is a rule of correspondence. Not just any rule, but a rule which associates to each number *exactly one* second number. We picture the correspondence as going *from* a horizontal number line *to* a vertical number line.

3. Three Ways of Looking at a Function

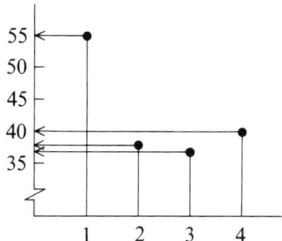

(C) A function is a curve in the plane. Not just any curve, but a curve that no vertical line crosses more than once. (Occasionally, instead of calling the curve a function, we call it the *graph* of a function.)

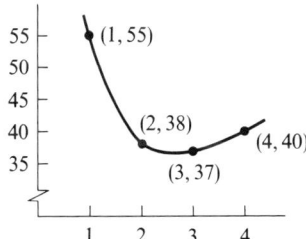

Do you agree that (A), (B), and (C) are, at heart, expressions of the same idea? Is it not remarkable that the same idea can be thought of—as in (A)—as a *static* notion or—as in (B)—as a *dynamic* notion or—as in (C)—as a *geometric* notion? This remarkable feature is one reason why the idea of a function is an important one. Already the reader may expect that the study of functions will have a bearing on the study of dynamics (that is, motion), and on the study of curves in the plane. If the reader has also the feeling that the idea of a function can change a moving, or fluid, situation into a more easily scrutinized static situation, then much of what the ensuing chapters hold has been foreseen.

A surprising amount of mathematics consists in simply saying the same thing in many different ways, until it is finally said in a way that makes it simple. The problem in Example 1 of finding the least possible cost could be rephrased as either of the following problems:

(1) Find the least number that can possibly occur in the second column in (A).
(2) Find the lowest point ever hit on the vertical axis by $f(L)$ in (B).
(3) Find the second coordinate of the lowest point on the curve in (C).

Calculus will teach us how to do problem 3 above. In Chapter 3 we shall begin the study of a technique that often enables one to find with ease the lowest point on a curve.

For obvious reasons, Example 1 is called an *optimization problem*, where the optimum is achieved by *minimizing* a certain variable (the cost C). Let us now look at a second example, where the optimization problem that arises requires that a certain variable be *maximized*.

EXAMPLE 2. A farmer has a cow named Minerva. For her has been purchased 1200 feet of fencing to enclose *three* sides of a rectangular grazing area. The fourth side is bounded by a long barn and requires no fence. Find the largest possible grazing area that Minerva can have.

In this example the area varies with the design of the rectangle. Our task is to become familiar enough with *how* it varies in order to recognize the greatest possible area. We first pick at random a few possible designs and calculate the corresponding areas. There are lots of ways to use that 1200 feet of fencing.

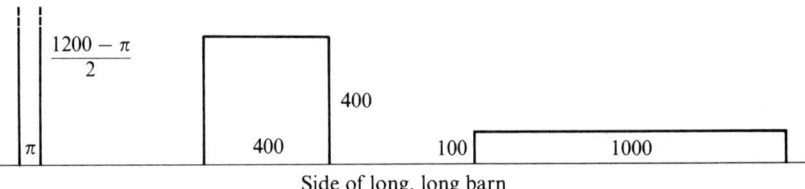

Side of long, long barn

EXERCISES

3.1. Suppose the side along the barn is 100 feet. Find the area enclosed. *Hint.* First figure out the lengths of the other sides.

3.2. Suppose the side along the barn is 400 feet. Find the area.

3.3. Suppose the side along the barn is 1000 feet. Find the area.

3.4. Suppose the side along the barn is π feet. Find the area. *Answer:* $600\pi - \frac{1}{2}\pi^2$ square feet.

§4. Words versus Algebra

Letting s stand for the length, in feet, of the side along the barn, and letting A stand for the area enclosed, in square feet, we have the following table:

s	$\xrightarrow{g}$	A
100		55,000
400		160,000
1000		100,000
π		$600\pi - \frac{1}{2}\pi^2$

From the exercises above, it is clear that the value of s completely determines the value of A. This means that A is a function of s. We want to become familiar with this function in order to recognize the largest possible area A

4. Words versus Algebra

that it can produce for Minerva. We begin by giving it a name. Let us denote this function by g. (If we have a function pop up, we are free to baptize it with any name we choose. However, it is conventional in most books to reserve the letters f, g, F, and G to designate functions.)

We now have $A = g(s)$. That is, $g(s)$ is the area A corresponding to the rectangle whose length along the side of the barn is s feet. That is,

$$g(s) = \text{area, in square feet, of } \boxed{}_{s} \qquad (4)$$

Equation (4) defines the function g in *words*. It is perfectly proper to define a function by writing out its rule in words. However, if the rule is really an *algebraic* rule in disguise, it behooves us to recognize it. What is the height of the rectangle in (4) whose base is s feet? It is $\frac{1}{2}(1200 - s)$. *Reason*: Having used s feet opposite the barn, we have $1200 - s$ feet left, of which half must go on each of the other sides. Thus, from (4) we can go on:

$$g(s) = \text{area of } \boxed{}_{s} \quad \frac{1200 - s}{2}$$

$$= \frac{s(1200 - s)}{2}$$

$$= 600s - \frac{s^2}{2}.$$

This shows that the function g, written out in words in equation (4), can be expressed as an equation in algebra:

$$g(s) = 600s - \frac{s^2}{2}. \qquad (5)$$

For obvious reasons, such a function is called an *algebraic* function. Almost all the functions we shall encounter in the first six chapters of this book will be algebraic functions, and it is important to learn to convert an equation in words to an equation in algebra, whenever it is possible to do so. There arise many functions like g, whose rules are expressed in words, but whose rules are really algebraic rules in disguise.

Exercises

4.1. Use the algebraic rule $g(s) = 600s - \frac{1}{2}s^2$ to calculate
  (a) $g(100)$.
  (b) $g(400)$.
  (c) $g(700)$.
  (d) $g(1000)$.
  (e) $g(\pi)$.
  (f) $g(x)$.
  (g) $g(x + \pi)$.
  (h) $g(x + h)$.
  (i) $g(2 + 3k)$.
  (j) $g(1/\pi)$.
  (k) $g(1/x)$.
  Answers: (c) 175,000. (g) $600(x + \pi) - \frac{1}{2}(x + \pi)^2$.

4.2. Read again the three ways (A), (B), and (C) of looking at a function.
   (a) Draw a few arrows, as in (B), picturing the function g as a correspondence going from a horizontal number line to a vertical number line:

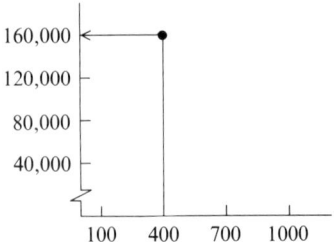

   (b) Plot a few points, as in (C), lying on the curve g:

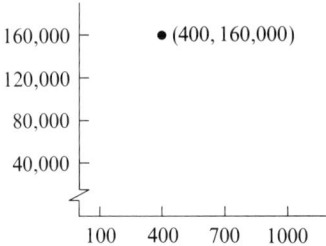

   (c) In Chapter 3 we shall learn an easy way to find the highest point on the curve g. Can you *guess* what the highest point might be?

§5. Domain and Range

Look again at equations (4) and (5) above. There is a subtle difference between them, despite the fact that both equations describe exactly the same rule of correspondence. The difference is this: In equation (4) it would make no sense (why?) to let the variable s have a value greater than 1200. Nor would it make any sense in (4) to let s take on a negative value. On the other hand, the algebraic rule $600s - \frac{1}{2}s^2$, given in equation (5), is well defined for *any* value whatever of the variable s. For instance, when s is -2, this algebraic rule gives

$$600(-2) - \frac{(-2)^2}{2} = -1202,$$

even though it is impossible to have a rectangle whose area is negative.

A way to avoid such confusion is to agree to specify, *at the outset*, as soon as a function is introduced, the collection of numbers on which the function is defined. This collection is called the *domain* of the function. The domain of our function *g*, as specified in words in equation (4), is then the collection of all permissible values of the variable *s*, which may be pictured like this:

(The *open* circles at the endpoints 0 and 1200 indicate that these values are *excluded* from the domain. We cannot get an honest-to-goodness rectangle if we permit *s* to equal either 0 or 1200.) Instead of drawing a picture of the domain, one could equally well specify the domain by writing the inequality

$$0 < s < 1200,$$

which says that the values of the variable *s* are restricted to lie between 0 and 1200.*

Once the domain of a function has been specified, one can then speak of the *range* of values assumed by the function. For the function given by $A = g(s)$, the domain consists of all permissible values of *s* and the range consists of all corresponding values of *A*. Since there are three ways of looking at a function, there are three ways of thinking about a function's domain and range:

(A) If the function is thought of as a pair of columns, then its domain is the collection of all numbers allowed to go in the first column and its range is the collection of all numbers in the second.

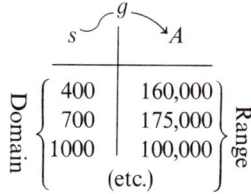

(B) If the function is thought of as a rule of correspondence, then its domain is the set of all numbers on which the rule acts, and its range is the set of all corresponding numbers.

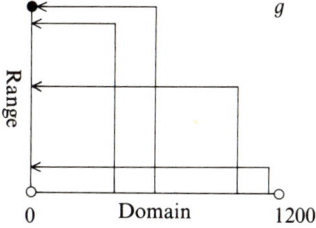

---

* Had we wished (we did not) to include, say, the point 0 and exclude 1200, we would have written $0 \leq s < 1200$ or drawn the picture with a *closed* circle at 0 and an open circle at 1200.

(C) If the function is thought of as a curve, its domain is the projection of the curve on the horizontal axis and its range is the projection of the curve on the vertical axis.

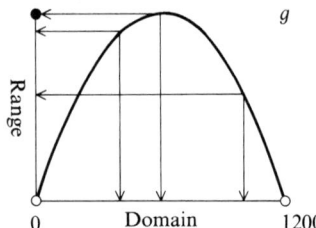

The domain must be specified before it makes any sense to speak of the range of a function. If the domain is altered, then the range will likely change as well. To find the range of a given function is a problem we shall not discuss until Chapter 3. By (C) above, we see that finding the range involves finding the highest and lowest points on a curve, a topic we shall meet in Chapter 3.

It is usually easy to specify the domain of a function, however. In the function of Example 1, given by

$$f(L) = 7L + \frac{48}{L},$$

it is natural to take the domain to be specified by the inequality

$$0 < L$$

(or $L > 0$, if you prefer), which says that the values of the variable L are restricted to be positive. This restriction is forced by equation (1), where the rule for f is written out in words.

If one does not like to write inequalities, then one should learn to draw pictures. The domain of the function f of Example 1 can be pictured as follows:

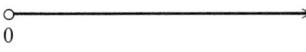

(The arrow indicates that the domain is not bounded on the right, but continues to include all positive numbers.)

Suppose a function is specified simply by giving an algebraic rule, such as $\sqrt{x+1}$. (The radical sign $\sqrt{}$ denotes the positive square root of what follows.) What shall we understand to be its domain? We shall agree to the following convention.

5. Domain and Range

Convention. *Unless otherwise specified, the domain of an algebraic rule shall be understood to be the collection of all numbers for which the rule makes sense.*

In applying this convention, one often has to remember two facts which ought to be familiar from arithmetic:

(1) It makes no sense to "divide by zero".
(2) It makes no sense to take the "square root" of a negative number.

Thus, the domain of the algebraic rule given by $\sqrt{x+1}$, unless otherwise specified, shall be understood to be the collection of all numbers for which $x + 1$ is not negative, that is, the collection of all numbers x for which

$$0 \leq x + 1,$$

which is the same as saying

$$-1 \leq x,$$

or drawing the picture

Domain of the rule $\sqrt{x+1}$
•────────────────→
−1

Since it makes no sense to divide by zero, the domain of the rule $(x^2 + x)/x$ is pictured as follows:

Domain of the rule $(x^2 + x)/x$
←────────○────────→
0

The rule given by $x + 1$, on the other hand, makes sense for *any number whatsoever*. By our convention, the domain of this rule (unless otherwise specified) shall be understood to be unrestricted:

Domain of the rule $x + 1$
←─────────────────→

We now make a point which the reader may think at first to be overly precise. The significance of this point will not be appreciated until later. The point is this: Although it is true that

$$\frac{x^2 + x}{x} = \frac{x(x+1)}{x}$$

$$= x + 1 \quad \text{if } x \neq 0,$$

the functions given by

$$F(x) = \frac{x^2 + x}{x}$$

and

$$G(x) = x + 1$$

are *not* the same. *Reason*: The functions F and G do not have the same domain. To say two functions are the same means they have the same graph, and, in particular, they must have the same domain.

EXERCISES

5.1. In Example 1 we found that the numbers 37 and 55 were in the range of f. Do you believe that every number between 37 and 55 is also in the range? Why might you think so?

5.2. In Example 2 we found that it was possible to enclose an area of 100,000 square feet and also possible to enclose an area of 160,000 square feet. From these facts, given the nature of the problem raised in Example 2, can you conclude that it is possible to enclose 130,000 square feet?

5.3. Apply the convention above to specify the domain of each of the following algebraic rules. (You may specify the domain either by an inequality or by a picture.)
(a) $\sqrt{x}$.
(b) $\sqrt{x-1}$.
(c) $1/x$.
(d) $1/(x-1)$.
(e) $7x + (48/x)$.
(f) $600x - \frac{1}{2}x^2$.
(g) $(h^2 + 2h)/h$.
(h) $h + 2$.
(i) $\sqrt{1+h^2}$.
(j) $L^2/(L^2 - 1)$.
(k) $(s-1)(s-2)$.
(l) $\sqrt{s-1}$.
Answers:
(e) $x \neq 0$:

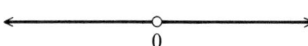

(g) $h \neq 0$:

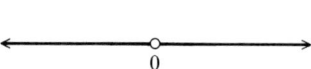

(h) h unrestricted:

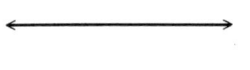

(l) $1 \leq s$:

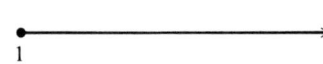

5.4. True or false? The function specified by the rule $(h^2 + 2h)/h$ is the same as the function specified by the rule $h + 2$.

§6. Optimization

In Example 1, the problem of finding the least cost was seen to be the same as another problem, that of finding the least number in the range of possible costs. To answer the question raised in Example 1, we need to find the least

6. Optimization

number in the range of f, where f is the function whose rule of correspondence and whose domain are specified succinctly by writing

$$f(L) = 7L + \frac{48}{L}, \quad 0 < L.$$

We shall find this number, once we have developed the appropriate tools of calculus.

In Example 2, the problem of finding the biggest possible area was seen to be the same as another problem, that of finding the largest number in the range of possible areas. To answer the question raised in Example 2, we need to find the largest number in the range of g, where g is the function whose rule of correspondence and whose domain are specified succinctly by writing

$$g(s) = 600s - \frac{1}{2}s^2, \quad 0 < s < 1200.$$

We shall find this number later, using calculus.

In our discussion of Examples 1 and 2, we have seen the first step in how to handle *optimization* problems. An optimization problem can always be spotted by the presence of a superlative. Whenever a problem requires that we find the *least*, or *most*, or *cheapest*, or *best*, or *closest*, etc., we know that we have an optimization problem on our hands. From our discussions in Examples 1 and 2, we may expect that any optimization problem will give rise to a function, and that the solution to the problem will involve finding the highest (or lowest) point on the curve determined by the function. Thus, by seeing the optimization problem in terms of variables, and by getting an algebraic rule relating one variable to another, the optimization problem is transferred to another problem, that of studying the curve determined by the rule, or function, relating the variables. This is the first step in solving optimization problems. This step takes a little while to master. Once it is mastered, however, the second step of finding the highest (or lowest) point on a curve can be accomplished with the study of only a little calculus.

Must every curve necessarily have a highest point and a lowest point? Certainly not. The curve f of Example 1 has no highest point. *Reason*: The range of costs is not bounded above. There exists no most expensive way to build that fence. The curve g of Example 2 has no lowest point. *Reason*: The grazing area is to be a rectangle and thus cannot have an area of zero, yet the area A ranges arbitrarily close to zero. There is no least possible grazing area for Minerva.

Exercises

6.1. Suppose, in Example 1, the pen was to enclose 30 square yards instead of 12, the costs of stone and wood remaining the same. Find an algebraic rule giving the cost C in terms of the length L of the front, and specify the domain of this rule.

6.2. Suppose, in Example 2, the farmer had 2000 feet of fencing instead of 1200, the other conditions of the problem remaining unchanged. Find an algebraic rule giving the area A in terms of the length s of the side along the barn, and specify the domain of this rule. *Answer*: $A = 1000s - \frac{1}{2}s^2$, $0 < s < 2000$.

§7. Purpose

What follows, gentle reader, is an unorthodox introduction to the notion of a *limit*. (If this is frightening, then be assured that an orthodox discussion is given in Section 9.) Calculus is, in a sense, the study of limits, yet this simple notion is also easily misunderstood, unless the student can make the proper distinction between two things which are easy to confuse. These two things we might call "purpose" and "action". The analogy we shall make, in hopes that it will make the idea of a limit easier to grasp, is this:

> *The "limit" of a function, at a point in or near its domain, is like the purpose of a human being, at a point in time.*

The reader may find that the word *limit* is almost exactly as easy (or as hard) to understand as the word *purpose*.

This analogy will be worth nothing at all unless the ordinary distinction between purpose and action is kept well in mind. These two notions, though often related, are quite different. Most of us can think of instances when our action did not reflect our purpose or of times when we wandered aimlessly to no purpose whatever. Sometimes, even with a purpose, one hesitates to act. Finally, there are the gratifying times when one has a purpose and acts accordingly.

A function, believe it or not, is just like a person in this respect, and one can learn a lot by inquiring into this aspect of the life of a function. At any point in the domain of a function we may compare its action (what it actually does at the point) with its purpose (what it seemed on the threshold of doing at the point). Often, just as in the lives of human beings, the action will agree with the purpose, giving a sense of "continuity". But there are several other possibilities that can occur. The action at some point may disagree with the purpose, or there may be no discernible purpose, or there may be purpose with no action, or there may be neither purpose nor action.

We study functions all the time in calculus, and we gradually learn that each function has a personality all its own. A function is something more than might be imagined from the description "a rule of correspondence", just as a human being is something more than "a featherless plantigrade biped mammal".

Let us try, while studying calculus, to feel ourselves into the world of functions, to see what they really are. Here is a fable. It is offered in fun. Take it seriously, but not too seriously.

7. Purpose

Lim: A Fable

> *The gods did not reveal all things to men at the start; but, as time goes on, by searching, they discover more and more.*
>
> Xenophanes
>
> *The lord whose is the oracle at Delphi neither reveals nor hides, but gives tokens.*
>
> Heraclitus

On the First Day all functions were created, and solemnly told the harsh facts of a functional existence:

> *Each function has been assigned his domain, to which he will be restricted eternally to live in accordance with the rule he has been given. During eternity, he must contemplate the purpose of his being, knowing that on the Last Day the gods may require him to state his purpose at some troublesome point. At that point the function must either state his purpose, or reply that no purpose exists. For at each point some functions have been given a purpose and some have not.* Remember the words of Xenophanes and Heraclitus.

Among the multitude of functions trembling on the First Day was g *of Example 2*. Charged with a Herculean task, g of Example 2 must take each s between 0 and 1200 and throw it to the corresponding A. The throw must be in accordance with the god-given rule

$$A = \text{area of a rectangle of sides } s \text{ and } \frac{1}{2}(1200 - s).$$

Up and down his domain, g carefully moves, throwing his s's until he knows by heart where each little s is supposed to go. He is glad the gods did not ask him to throw -2, or any negative number, or to throw any number exceeding 1200, because he would have no clue where the gods might want these numbers thrown. At last, clever g realizes that he has no purpose at the point -2, or at any negative number, or at any number exceeding 1200. Should the gods ask him, on the fearsome Last Day, of his purpose at the point -2, g would reply in his best courtly fashion:

$$\text{The purpose of } g, \text{ at the point } -2, \text{ does not exist!} \qquad (6)$$

Confidence begins to well up in g.

Yet soon g realizes that the gods have played a trick on him. "Ye gods!" exclaims g, "Why did ye not give me a closed domain?" Poor g is tantalized whenever he moves near the ends of his domain. When he moves to his left, toward 0, he is allowed to throw numbers that lie arbitrarily close to 0; nevertheless, he is not allowed to throw 0, since 0 is not in his domain. A similar frustration is felt when he moves to the right toward 1200.

Night and day, for what seemed like half of eternity, g continuously worried about the points 0 and 1200. Finally the gods had pity upon g and sent down to him a messenger, named Lim.

"Hail, long-suffering g, most favored of Minerva, hail!" shouted Lim.

"Who that?" responded g, so startled that he began dropping his s's.

"I'm the One Who Knows", replied Lim, and smiled smugly. "Remember the words of Xenophanes and Heraclitus."

"Get off my domain!" shouted *g*, thinking his intruder to be an oracular fanatic.

"Now, now, calm down", said Lim. "I have been sent to help you find your purpose, if you should need help at any point. Do you know your purpose at 1200?"

"Since I am restricted to my domain for all of eternity," said *g*, "*g*(1200) does not exist. I am not allowed to act at the point 1200."

"It is true that you are not allowed action at 1200," responded Lim, "but it is still possible that you may have purpose at that point, not to be fulfilled before the Last Day. Have you no clue what the gods want you to do at 1200 on the Last Day?"

Long-suffering *g* thought and thought and thought. He thought about his *s*'s near 1200, and about the *A*'s that corresponded to them:

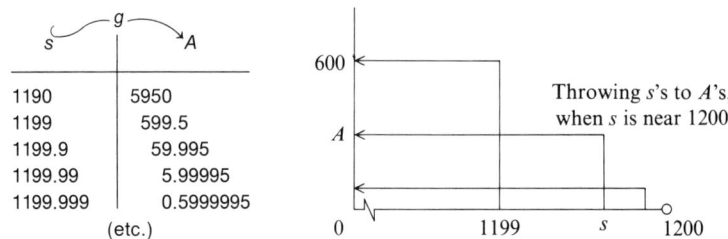

s	*A*
1190	5950
1199	599.5
1199.9	59.995
1199.99	5.99995
1199.999	0.5999995
(etc.)	

"As *s* gets closer to 1200, *A* gets closer to 0", exclaimed both Lim and *g* simultaneously. Then *g*, in deep tones, declaimed,

The purpose of g at 1200 *is to throw it to* 0. (7)

"Exactly," said Lim, "but why do you speak in such an old-fashioned way? The gods haven't talked like that for ages. Just use my name. Instead of your statement (7), just say,

Lim *g at* 1200 *is* 0,

and instead of (6), say

Lim *g at* −2 *does not exist.*

The gods will understand what you mean. They all know my name. I deliver their ambrosia on Thursdays."

EXERCISES

7.1. Is 1200 in the domain of *g*? *Answer*: No.

7.2. Does *g*(1200) exist? *Answer*: No, *g*(1200) is undefined, because 1200 is not in the domain of *g*. There is no action of the function *g* at the point 1200.

7.3. Does Lim *g* at 1200 exist? *Answer*: Yes. Lim *g* at 1200 is 0, because as $s \to 1200$, $g(s) \to 0$. (The arrow is an abbreviation for *approaches*, or *gets closer and closer to* or *tends to*.)

7.4. What is Lim g at -2? *Answer*: Lim g does not exist at -2, because $g(s)$ does not exist when s is close to -2, giving g no clue as to a purpose at -2.

7.5. What is Lim g at 1202?

7.6. What is $g(1202)$? *Answer*: Since the function g is not allowed to act at the point 1202, $g(1202)$ does not exist.

7.7. What is $g(0)$?

7.8. What is Lim g at 0?

7.9. What is Lim g at 1200.1? *Answer*: It does not exist.

7.10. What is Lim g at -0.1?

§8. Continuity: Purpose versus Action

"Aha!" said g, "I understand now everything about a function's purpose."

"That is doubtful," replied Lim, "for you are still likely to confuse *purpose* with *action*. What, for example, is your purpose at the point 600?"

"Lim g at 600 is 180,000," responded g without hesitation, "because $g(600)$ is 180,000."

"Aha!" said Lim, "A right answer, *but for a wrong reason*. Just as I expected. When the gods inquire about your purpose at 600, they have in mind something more subtle than you imagine. To reply that $g(600)$ is 180,000 is to state your *action* at the point 600. But action need not necessarily agree with purpose. (At the point 1200 you have no action, yet you do have purpose.)"

"To find your purpose at 600, the first thing you must do is to *forget entirely about your action* at 600. You may as well pretend that 600 has been removed from your domain. Then you proceed just as before. What does A approach as s approaches 600?"

Long-suffering g thought and thought and thought. What, indeed, would be his purpose at 600 if 600 were removed from his domain?

The point 600, being in the interior of the domain, can be approached by values of s either slightly smaller or slightly larger than 600:

s	$\xrightarrow{g}$ A	s	$\xrightarrow{g}$ A
500	175,000	700	175,000
550	178,750	650	178,750
590	179,955	610	179,955
598	179,998	602	179,998
599	179,999.5	601	179,999.5
599.9	179,999.995	600.1	179,999.995
(etc.)		(etc.)	
Letting s approach 600 from the left $(s \to 600^-)$		Letting s approach 600 from the right $(s \to 600^+)$	

Whether s tends to 600 from the left, or the "minus side" ($s \to 600^-$), or whether s tends to 600 from the "plus side" ($s \to 600^+$), the corresponding values of A tend to 180,000. Since $A \to 180,000$ as $s \to 600$ (from either side),

$$\text{Lim } g \text{ at } 600 \text{ is } 180,000. \tag{8}$$

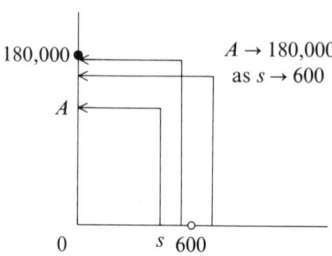

"Now let me get this straight", said g. "To find my purpose at 600, I first pretend that 600 has been removed from my domain, and then see what happens to A as s tends to 600. This is the way I figure out that statement (8) is true. Isn't there an easier way to do it?"

"Yes," said Lim, "if you are not afraid to use your common sense. Just look at the rule you were given, $g(s) = 600s - \frac{1}{2}s^2$, and note that it is described in terms of some simple algebraic operations. Look at what happens to each of them in turn, as $s \to 600$. Common sense should tell you that, as $s \to 600$, it must follow that $s^2 \to (600)^2$ and $600s \to 600(600)$. Therefore, as $s \to 600$,

$$A = g(s) = 600s - \frac{1}{2}s^2 \to 600(600) - \frac{1}{2}(600)^2.$$

Thus, $A \to 180,000$."

"I really feel great at 600," said g, "whereas at 1200 I become so frustrated."

"That is because, at 600, your action agrees with your purpose:

$$g(600) = 180,000 \quad \text{(by applying the rule } g \text{ to 600)}$$
$$= \text{Lim } g \text{ at } 600 \quad \text{[by (8)]}.$$

Like any creature, you experience the wholesome feeling of *continuity* at any point where action and purpose exist and agree. Whenever there is not agreement between action and purpose, or whenever one or both are missing, the anxieties of discontinuity emerge. At 1200, friend g, you behave discontinuously. You have a purpose:

$$\text{Lim } g \text{ at } 1200 \text{ is } 0,$$

but you do not act accordingly:

$$g(1200) \text{ does not exist.}$$

Everyone is frustrated by discontinuity."

"Let me leave you with this idea, to ponder as you will. To say that a function is continuous at a certain point means that, at the point, the function has both purpose and action, and they agree."

9. Limits

Definition. A function G is said to be **continuous** at a point x provided that the following three conditions are satisfied:

(1) G(x) exists.
(2) Lim G at x exists.
(3) G(x) = Lim G at x.

"That is all the help I can give you," said Lim, "for I must now depart. I have to collect 37 chariotloads of ambrosia before Thursday."

"Wait!" shouted g. "Can't you help my friend f of Example 1? She lives on the domain 0 < L. What a curve!"

"Remember the words of Xenophanes and Heraclitus!" said Lim, and departed without another word.

EXERCISES

8.1. What is g(400)?

8.2. What is Lim g at 400?

8.3. Is g continuous at the point 400?

8.4. Is g continuous at the point 0?

8.5. Is g continuous at the point −2?

8.6. Is g continuous at every point in its domain? *Answer*: Yes.

8.7. Consider the function f of Example 1. Is f continuous
   (a) at 0?
   (b) at −2?
   (c) at 2?
   Give reasons justifying your answers.

§9. Limits

In everyday language the word *limit* has virtually the same meaning as *bound*. In calculus, however, it has a rather different meaning. The *limit* of a function, at a certain point, is (roughly speaking) what the function, at that point, is on the threshold of doing.* If c is the point in question, then the limit of f at c is symbolized by

$$\text{Limit}_{x \to c} f(x), \qquad (9)$$

or by

$$\text{Lim } f \text{ at } c,$$

and is found by investigating the action of f at points near c, *while completely ignoring the value of f at c.* Before any further explanation is given, it should be emphasized that $f(c)$, the value of f at c, may well be entirely unrelated to the limit of f at c. [If it happens that they are the same, that is,

---

* The word *limit* is kin to the Latin word *limen*, which means "threshold".

that $f(c) = \text{Limit}_{x \to c} f(x)$, then the function f is said to be **continuous** at the point c.]

How does one find the limit of a function at a point? The symbolism (9) is designed to suggest the method of doing this. We simply ask for the limiting value of $f(x)$, as we imagine the values of the variable x taken closer and closer to (but not equal to) the number c. The arrow in "$x \to c$" is supposed to suggest x approaching c ever more closely.

Some examples should serve to clarify things. The reader is asked simply to use common sense in thinking about what happens as the value of a variable gets closer and closer to a fixed number c.

EXAMPLE 3. Let $F(x) = (x^2 + x)/x$, with domain specified by the inequality $x \neq 0$. Find Lim F at 4.

Here we are asked to find

$$\text{Limit}_{x \to 4} \frac{x^2 + x}{x}, \tag{10}$$

and it is obvious how this is to be done, simply by reading the formula (10) in words: We are asked to find the limiting value of the expression $(x^2 + x)/x$ as x tends to 4. What happens to this expression as $x \to 4$? Common sense tells us that $x^2 \to 16$, so that the expression $(x^2 + x)/x$ approaches $(16 + 4)/4$, which is equal to 5. Therefore,

$$\text{Limit}_{x \to 4} \frac{x^2 + x}{x} = 5,$$

which answers the question raised in Example 3, and also shows, incidentally, that F is continuous at 4. (Why?) □

EXAMPLE 4. Let $G(x) = x + 1$, with unrestricted domain. Find Lim G at 0.

This is even easier than the preceding example. As $x \to 0$, common sense says that $(x + 1) \to 1$. Therefore,

$$\text{Limit}_{x \to 0} (x + 1) = 1. \qquad □$$

EXAMPLE 5. Let $F(x) = (x^2 + x)/x$, with domain specified by $x \neq 0$. Find Lim F at 0.

Here we are asked to find

$$\text{Limit}_{x \to 0} \frac{x^2 + x}{x},$$

and it is *not* obvious, at first, how this is to be done. As $x \to 0$, both the numerator $x^2 + x$ and the denominator x also approach 0. What is to be done?

What is to be done is to realize that the question raised in Example 5 is exactly the same as the question raised in Example 4, whose answer, we

9. Limits

have seen, is 1. Why are these two questions exactly the same? Because Lim F at 0, remember, is to be found by letting x tend to 0, but never allowing x to equal 0. If $x \neq 0$, though, then

$$F(x) = \frac{x(x+1)}{x} = x+1,$$

so that the limit of F at 0 is the same as the limit of $x + 1$:

$$\operatorname*{Limit}_{x \to 0} \frac{x^2 + x}{x} = \operatorname*{Limit}_{x \to 0} (x+1) = 1,$$

by Example 4. (What has just been illustrated in this example is not hard, but it is subtle. Reread this example, and also the last remark in Section 5, to make sure you understand it.) ☐

A picture is the best way to illustrate why Examples 4 and 5 must have the same answer:

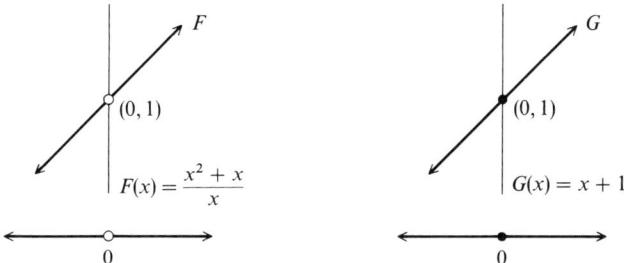

The curves F and G are identical, except when x is 0. Since the limit of a function at the point 0 is independent of the action at 0, F and G have the same limit at 0.

Here is another example, using h instead of x as the variable.

EXAMPLE 6. Find $\operatorname{Limit}_{h \to 0}((h^2 + 2h)/h)$.

This is like Example 5, where it is not immediately obvious whether the limit exists at 0. Both the numerator $h^2 + 2h$ and the denominator h tend to 0 as $h \to 0$. However, *if h is not equal to 0*, then we may divide by h, so

$$\frac{h^2 + 2h}{h} = \frac{h(h+2)}{h} = h + 2.$$

This shows that the algebraic rule given by $(h^2 + 2h)/h$ is exactly the same as the algebraic rule $h + 2$, *provided h is not* 0. Since the limit at 0 is independent of the action at 0, these two rules have the same limit at 0:

$$\operatorname*{Limit}_{h \to 0} \frac{h^2 + 2h}{h} = \operatorname*{Limit}_{h \to 0} (h+2) = 2. \qquad \square$$

When investigating the limit of a function at a point, one may encounter any of the following situations:

(I) The limit exists and agrees with the action of the function at the point.
(II) The limit exists, but the function does not act accordingly.
(III) The limit does not exist.

If case (I) occurs, the function is said to be continuous at the point. This is illustrated in Examples 3 and 4. Case (II) is illustrated in Examples 5 and 6. Case (III) will be illustrated in Examples 7 and 8.

EXAMPLE 7. Find $\text{Limit}_{x \to 0}(7x + (48/x))$.

This limit does not exist. As $x \to 0$, the first term, $7x$, is "well-behaved", tending to 0, but the second term, $48/x$, does not tend to a limit, since it becomes large-positive as x tends to 0 from the right, and it becomes large-negative as x approaches 0 from the left:

x	$7x + 48/x$		x	$7x + 48/x$
1	55		-1	-55
0.1	480.7		-0.1	-480.7
0.01	4800.07		-0.01	-4800.07
	(etc.)			(etc.)

Letting x approach 0 from the right $(x \to 0^+)$

Letting x approach 0 from the left $(x \to 0^-)$

□

EXAMPLE 8. The Post Office has discovered that the cost of sending a letter by mail varies in terms of the weight w of the letter. Accordingly, the number of stamps to be affixed to a letter is a function of w. One stamp is required if the weight w is 2 ounces or less; two stamps if $2 < w \leq 4$; three stamps if $4 < w \leq 6$; etc. Let us call this function F, so that

$F(w) = $ the number of stamps on a letter of weight w.

Find Lim F at 4.

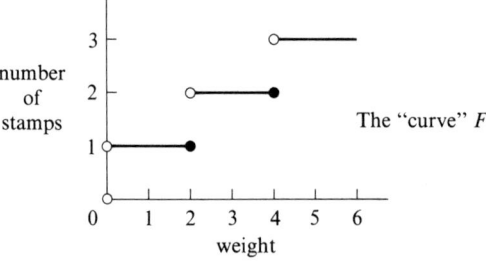

The "curve" F

9. Limits

Lim F at 4 does not exist. As w tends to 4 from the left, the number of stamps $F(w)$ tends to 2; whereas, when w tends to 4 from the right, $F(w)$ tends to 3:

w	$F(w)$
3.9	2
3.99	2
3.999	2
(etc.)	

$[F(w) \to 2, \text{ as } w \to 4^-]$

w	$F(w)$
4.1	3
4.01	3
4.001	3
(etc.)	

$[F(w) \to 3, \text{ as } w \to 4^+]$

The limit does not exist at 4, because we get different "answers" when we approach 4 from different sides. □

Exercises

9.1. Find the indicated limits:
(a) $\text{Limit}_{x \to 0}((x^2 - 4x)/x)$.
(b) $\text{Limit}_{x \to 0}(x - 4)$.
(c) $\text{Limit}_{x \to 1}((x^2 - 1)/(x - 1))$.
(d) $\text{Limit}_{h \to 0}(3h^2/h)$.
(e) $\text{Limit}_{h \to 0}(3h/h)$.
(f) $\text{Limit}_{h \to 0}(3/h)$.
(g) $\text{Limit}_{x \to -1}((x^2 - 1)/(x + 1))$.
(h) $\text{Limit}_{t \to 3}(5/(6 + t))$.

9.2. Consider each of the following algebraic rules, and tell whether it is continuous at the indicated point c:
(a) $(x^2 - 4x)/x$; $c = 0$. *Answer*: Not continuous.
(b) $x - 4$; $c = 0$.
(c) $(x^2 - 1)/(x - 1)$; $c = 1$.
(d) $5/(6 + t)$; $c = 3$. *Answer*: Continuous.

9.3. Consider the "Post Office function" defined in Example 8.
(a) The function F is defined by a rule stated in words. Do you think it is likely that this rule is an algebraic rule in disguise? *Hint.* Do you think an algebraic rule could have a graph like the "curve" F?
(b) Does Lim F at 2 exist?
(c) Does $F(2)$ exist?
(d) Is F continuous at 2?
(e) Is F continuous at 3? *Answer*: Yes.
(f) A politician asserts that "the scale of charges imposed by the Post Office upon its customers exhibits unnatural and unjustifiable discontinuities at 2-ounce intervals." Explain, in more detail, what the politician means.

9.4. (*A philosophical question to be pondered for a while before being answered*) Is discontinuity unnatural? That is, must the rules that come from laws of nature necessarily be continuous? (Man-made rules, like the Post Office function, are often discontinuous, at least at some points.)

§10. Summary

Variables, functions, and *limits* were ideas that came of age in the seventeenth century. Fermat (pronounced fer-MAH) was probably the first to see the real importance of limits. *Continuity* is an old philosophical term that drew new interest from Leibniz (pronounced LĪP-nits), who was the first to use the word *function*.

These notions were not particularly well defined by their inventors, who were content to describe things in intuitive terms. The word *function* at first referred only to an *algebraic* rule, which is automatically continuous at each point in its domain.

Problem Set for Chapter 1

1. Consider Example 1 once more. We chose to look at it in terms of the variables C and L. The cost variable C cannot be avoided, since the problem involves finding the minimum of this variable. However, instead of choosing L, the length of the front, as our second variable, we might just as well have chosen W, the depth of the pen.

   (a) Write an algebraic rule expressing C in terms of W.
   (b) What is the domain of the rule in (a)?
   (c) Plot a few points on the graph of the equation in (a) that expresses C in terms of W.
   (d) Write an equation that relates W and L. *Hint.* What is their product?
   (e) Go back to equation (2) and, in it, replace L by $12/W$ and simplify. Do you get the same equation as you got in part (a) above? Why did it work out that way?

2. In Example 1, change the word *least* to *greatest*. With this modification, respond to the question raised.

3. Consider Example 2 once more. We chose to look at it in terms of the variables A and s. There is no getting around the variable A, since it must be maximized in order to answer the question raised. However, instead of s, we might just as well have chosen the other dimension w of the rectangle to be our other variable.

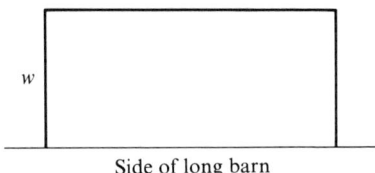

Side of long barn

(a) Write an algebraic rule expressing A in terms of w.
(b) What is the domain of the function whose rule is given in part (a)? (Be careful.)
(c) Plot a few points on the graph of the equation in (a).
(d) Write an equation relating w and s. *Hint*. What is the sum of $2w$ and s?
(e) In the equation $A = 600s - \frac{1}{2}s^2$, replace s by $1200 - 2w$, and simplify. You should get the same answer as in part (a). Why?

4. In Example 2, change the word *largest* to *smallest*. With this modification, respond to the question raised, bearing in mind that no honest-to-goodness rectangle has an area of 0.

5. Some curves determine functions and some do not. Does a circle ever determine a function?

6. Do all straight lines determine functions? If not, give an example of one that doesn't.

7. Some algebraic equations determine functions and some do not. Consider the algebraic equation $x^2 + y^2 = 1$.
(a) Is $(0, 1)$ on the graph of this equation?
(b) Is $(0, -1)$ on the graph of this equation?
(c) Does the algebraic equation $x^2 + y^2 = 1$ determine a function?

8. Does the algebraic equation $y = \sqrt{1 - x^2}$ determine a function? If so, what is its domain?

9. One way to specify a function is to draw the curve it determines. For each of the curves below, specify the domain and the range. (Specify either by drawing pictures or by writing an inequality, whichever is easier.)

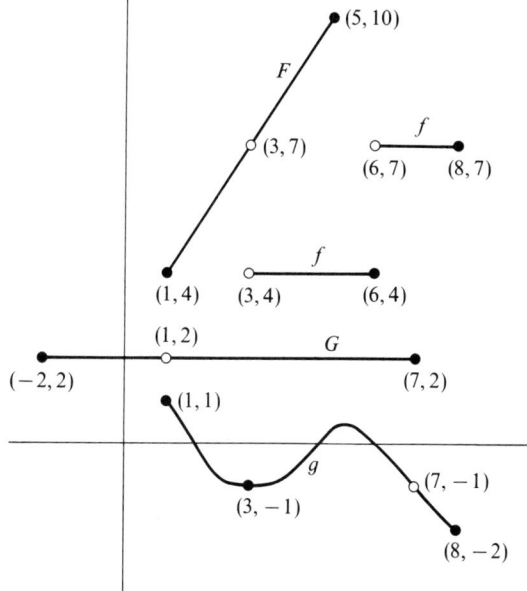

10. Referring to the functions F, f, G, and g pictured in the preceding problem, find
    (a) Lim F at 3.
    (b) Lim f at 6.
    (c) Lim G at 1.
    (d) Lim g at 3.
    (e) Lim g at 7.
    (f) $F(3)$.
    (g) $f(6)$.
    (h) $G(1)$.
    (i) $g(3)$.
    (j) $g(7)$.

11. Still referring to the functions F, f, G, and g of problem 9, answer the following questions.
    (a) Is F continuous at 3?
    (b) Is f continuous at 6?
    (c) Is G continuous at 1?
    (d) Is g continuous at 3?
    (e) Is g continuous at 7?

12. The domain of the "Post Office function" of Example 8 is specified by the inequality $0 < w$. What is its range?

13. The functions of Examples 5 and 6 (in Section 9) have the same domain. It has a hole in it, at the point 0:

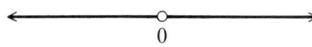

    (a) What is the range of the function of Example 5?
    (b) What is the range of the function of Example 6?

14. Suppose the numbers 1 and 3 are known to be in the range of a certain function. Must the range then necessarily contain all numbers between 1 and 3? *Hint.* Look at your answer to problem 12.

15. Suppose the numbers 1 and 3 are known to be in the range of a certain function, and suppose the function is continuous at every point in its domain. Must the range necessarily contain all numbers between 1 and 3? *Hint.* Look at your answer to problem 13(b). What if, in addition, the domain has no "holes" in it?

16. In the corner of a large courtyard a rectangular enclosure is to be built. To pay for the material, $240 has been allocated. This is to be used to pay for both the stone fence, which costs $6 per meter, and the wood fence, which costs $2 per meter. The area A of the enclosure will vary with the way the enclosure is built.

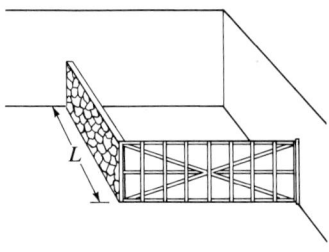

    (a) Let L be the length of the stone fence. How much money will be left to spend for wood?

(b) Let L be the length of the stone fence. How long will the wood fence be? *Hint.* The answer to part (a) tells you how much money is left for wood.

(c) Let L be the length of the stone fence. Find an algebraic rule giving the area A in terms of L, and specify the domain of this rule.

17. There is often more than one way to choose your variables. In problem 16,
    (a) Let x be the amount of money spent on stone. How much is left to spend on wood, and how long, therefore, is the wooden fence?
    (b) Let x be the amount of money spent on stone. Find an algebraic rule giving A in terms of x, and specify the domain of this rule.

18. A metal container, in the form of a rectangular solid, is to be constructed. The base is to be square, there is to be no top, and the volume of the container (the product of its three dimensions) is to be 12 cubic meters. Suppose the material for the sides costs $2 per square meter, and the material for the base costs $3 per square meter.

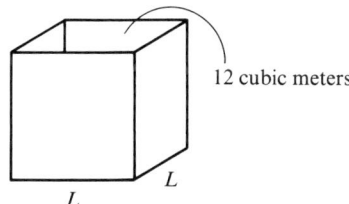

(a) Let C be the cost of the material for the container, and let L be the length of a side of the square base (in meters). Find the cost C if L is 2. *Hint.* First note that the height of the container must be 3 (why?) if L is 2.
(b) Find the cost C if L is π. *Hint.* First note that the height of the container must be $12/\pi^2$ if L is π.
(c) Find an algebraic rule giving C in terms of L, and specify its domain.

19. (*This problem is like the preceding one, except that we have a specified amount of material, instead of a specified volume.*) Suppose that we have 120 square feet of material, out of which is to be constructed a square base and four sides of a rectangular container. (The container is to have no top.) Let L be the length of a side of the base.
    (a) If H is the height of the container, then it is true that $120 = L^2 + 4LH$. (Why?) Solve this equation for H, to get H in terms of L.
    (b) The volume V of the container is the product of its three dimensions, so $V = L \cdot L \cdot H$. Use part (a) to get V in terms of L alone.
    (c) In part (b) we have V as a function of L. What is the domain of this function?

2 The Spirit of Greece

About 2500 years ago, a Greek named Pythagoras walked by Homer's fabled wine-dark sea, until he came to know what would inform his whole life. Arithmetic, with grand contempt for the slippery pebbles' uncertain support, boldly vaulted from the earth—and geometry was drawn out of the stars. Mathematics sprang from this marriage.

In childhood, mathematics was nurtured by that early spirit rising from the shores of the Mediterranean, exhorting Greeks to walk like giants, to wrest secrets from the gods. The history of mathematics ever since has been bound up with the workings of that spirit. What was it like, back then? Whose was the voice that howled above the seashore, as Pythagoras lengthened his stride? What moved the train—Eudoxus to Archimedes—that followed Pythagoras down the shore?

Let us study the workings of that train, to come to know its spirit. We ourselves move in it. We are still transported by a caravan of Greeks—who felt mathematics spring from head to shoulder—upon whose shoulders we too stand.

The purpose of the present chapter is to remind ourselves of history. Some readers may have the urge instead to jump flat-footed into an attack upon the problem that arose in Chapter 1: the problem of how the highest and lowest points on a curve can be easily found. Those readers may jump to Chapter 3, but they are warned that flat-footed jumps are awkward without a good foundation from which to leap. Studying history helps to build foundations.

§1. The Philosophy of Pythagoras

Real mathematics begins with Pythagoras (ca. 569–500 B.C.), although small steps were taken earlier by the Sumerians, Babylonians, and Egyptians. Some would say that Thales, who taught Pythagoras, deserves as much

1. The Philosophy of Pythagoras

credit; but Thales, hailed widely as the "father of philosophy", surely enjoys honor enough already. Why the spark of mathematics should suddenly glow so brightly (and why the flame would die with the coming of the Romans) is still a mystery.

Geometry became increasingly the dominant theme in Greek mathematics, and Pythagoras set the style. Yet Pythagoras was at first more attracted to arithmetic. He and his followers, the *Pythagoreans*, founded a small society that virtually worshipped numbers. One short sentence is all it takes to sum up the philosophy of Pythagoras:

All is number!

Pythagoras, it is said, invented the word *philosophy*, which literally means "love of wisdom". He sought wisdom by studying numbers. *Number*, to the Pythagoreans, referred to the ideas they pictured by the sequence

$$\bullet \quad \bullet\bullet \quad \bullet\bullet\bullet \quad \bullet\bullet\bullet\bullet \quad \bullet\bullet\bullet\bullet\bullet \quad \text{(etc.)}$$

Today we call these numbers *positive integers*. Exactly what the Pythagoreans meant by asserting that all is number is not entirely clear. At the least they meant that numbers are connected with many things that, at first, seem totally unrelated to numbers. For example, the musical tones produced by plucked strings seem at first to have nothing at all to do with numbers. Yet it was Pythagoras himself, so legend has it, who discovered what we now call *thirds, fifths, octaves*, etc., because of the numbers that are naturally associated with their relative pitches. Elementary facts about music are such common knowledge today that we surely underrate their significance. In the sixth century B.C., their discovery must have been astonishing. Imagine! Numbers have something to do with music!

"Perhaps numbers have something to do with everything!" thought Pythagoras. "Perhaps everything is number...." At least this offered a viable alternative to Thales' philosophy. Thales thought everything was water.

The Pythagoreans bequeathed much to Western culture. Who has not heard of the Pythagorean theorem?

The Pythagorean Theorem. *In a right triangle, the square built on the hypotenuse has the same area as the combined area of the squares on the other two sides.*

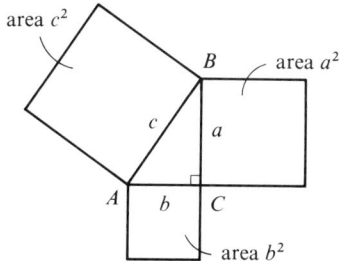

PROOF. From the vertex of the right angle, drop a perpendicular to the hypotenuse, hitting the hypotenuse at Q.

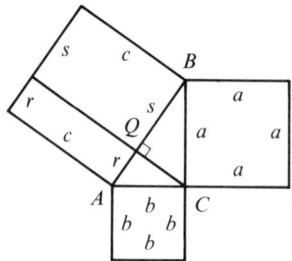

The hypotenuse is then split in two, as indicated, so that

$$c = s + r. \tag{1}$$

Since $\triangle AQC$ is similar to $\triangle ACB$ (why are they similar?), we have

$$\frac{r}{b} = \frac{b}{c}. \tag{2}$$

Since $\triangle BQC$ is similar to $\triangle BCA$, it follows that

$$\frac{s}{a} = \frac{a}{c}. \tag{3}$$

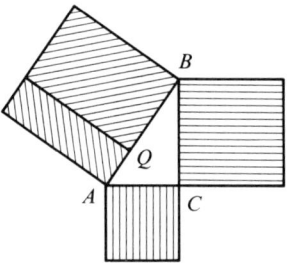

From equation (2) it follows that $b^2 = rc$ (showing that the two figures with vertical markings are equal in area). From equation (3) it follows that $a^2 = sc$ (showing that the two figures with horizontal markings are equal in area). From the equations $a^2 = sc$ and $b^2 = rc$, together with equation (1), we have

$$a^2 + b^2 = sc + rc = (s + r)c = c \cdot c = c^2.$$

Therefore, $a^2 + b^2 = c^2$, Q.E.D.

The proof just given shows exactly how the square built on the hypotenuse can be split into two areas that are equal, respectively, to the squares built on the sides. Pythagoras probably gave a more elementary proof, perhaps like the one that is outlined in problem 3 at the end of this chapter.

1. The Philosophy of Pythagoras

The Pythagorean theorem is applicable to a surprising variety of situations, as indicated in the exercises that follow.

Exercises

1.1. In the plane, indicate the position of the points P and Q, whose coordinates are given below, and use the Pythagorean theorem to find the distance from P to Q.
(a) $P = (3, 37)$, $Q = (4, 40)$.
(b) $P = (1, 55)$, $Q = (6, 50)$.
(c) $P = (4, 40)$, $Q = (2, 38)$.
(d) $P = (\pi, \pi^2)$, $Q = (\pi^3, \pi^4)$.
Answers:
(a) Dist P to Q is $\sqrt{10}$, by the Pythagorean theorem:

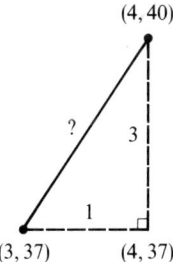

(d) Dist (π, π^2) to (π^3, π^4) is $\sqrt{(\pi^3 - \pi)^2 + (\pi^4 - \pi^2)^2}$.

1.2. Use the Pythagorean theorem to find a formula for the distance from (x_1, y_1) to (x_2, y_2). *Answer*: Dist (x_1, y_1) to (x_2, y_2) is $\sqrt{(x_2 - x_1)^2 + (y_2 - y_1)^2}$. (This is called the **distance formula** and should be memorized.)

1.3. (a) Use the distance formula to find the distance from $(0, 0)$ to $(3, 4)$.
(b) What is wrong with the following "calculation"?

$$\text{Dist } (0, 0) \text{ to } (3, 4) = \sqrt{3^2 + 4^2}$$
$$= 3 + 4 = 7.$$

1.4. Use the distance formula to find the distance from $(0, 0)$ to (x, y). Be sure you know why your answer is *not* equal to $x + y$. *Answer*:

$$\text{Dist } (0, 0) \text{ to } (x, y) = \sqrt{(x - 0)^2 + (y - 0)^2}$$
$$= \sqrt{x^2 + y^2}.$$

1.5. Consider this sentence:

The distance from $(0, 0)$ to (x, y) is 5.

Rewrite this sentence as a sentence (that is, an equation) using only algebraic symbols. *Hint*. In algebra, the word *is* may be translated "equals".

1.6. Consider this sentence:

The point (x, y) lies on the circle of radius 5 with center at $(0, 0)$.

Rewrite this sentence as an equation in algebra. *Hint.* This is precisely the problem posed in exercise 1.5 (why?). The answer is the same as the answer to 1.5.

1.7. Consider this sentence:
$$x^2 + y^2 = 25.$$
Rewrite this, using the distance formula, as a sentence in words. *Answer*: The point (x, y) lies on the circle of radius 5 with center at $(0, 0)$.

1.8. Consider each of the following equations, and rewrite it as a sentence in words.
  (a) $x^2 + y^2 = 49$.
  (b) $x^2 + y^2 = 1$.
  (c) $x^2 + y^2 = 4$.

1.9. Consider this sentence:
$$\text{The distance from } (\pi, 3) \text{ to } (x, y) \text{ is } 7.$$
Rewrite this as an equation. *Answer*: $(x - \pi)^2 + (y - 3)^2 = 49$.

1.10. Write an equation which says that (x, y) lies on the circle of radius 7 with center at $(\pi, 3)$.

1.11. Write an equation which says that (x, y) lies on the circle
  (a) of radius 3 with center at $(2, 5)$.
  (b) of radius $\sqrt{2}$ with center at $(1, 0)$.
  (c) of radius 3 with center at $(-2, 5)$.
  (d) of radius π with center at $(-\pi, 3\pi)$.
  (e) of radius r with center at (a, b).
  Answer: (c) $(x + 2)^2 + (y - 5)^2 = 9$.

1.12. Consider each of the following equations, and rewrite it as a sentence in words.
  (a) $(x + 2)^2 + (y - 5)^2 = 9$.
  (b) $(x - 2)^2 + (y + 5)^2 = 9$.
  (c) $x^2 + (y - 2)^2 = 3$.
  (d) $(x + \pi)^2 + (y - 3\pi)^2 = 2$.
  Answer: (d) The point (x, y) lies on the circle of radius $\sqrt{2}$ with center at $(-\pi, 3\pi)$.

1.13. A lighthouse is located 3 miles away from a straight shoreline. An electric plant is 13 miles downshore (as indicated in the figure, where P is the nearest shore point to the lighthouse). Suppose that undersea cable costs $7,000 per mile, whereas underground cable costs $2,000 per mile.

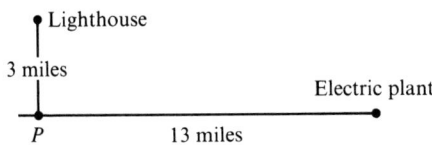

  (a) What is the cost (in thousands) of running cable undersea to P, then underground to the power plant?

(b) What is the cost (in thousands) of running cable undersea from the lighthouse to a point Q 4 miles downshore, then underground to the plant?

(c) Find an algebraic rule giving the cost (in thousands) of running the cable undersea from the lighthouse to a point x miles downshore, then underground to the plant. *Answer*: $C = 7\sqrt{9 + x^2} + 2(13 - x)$.
(d) What is the domain of the function expressed in words in part (c)?

§2. Geometry versus Arithmetic

Pythagoras perhaps had little notion that the Pythagorean theorem could be relevant in such a variety of contexts as we have just seen in the exercises in Section 1. In particular, the idea that an *algebraic* equation (like $x^2 + y^2 = 25$) could be identified with a *geometric* curve (the circle of radius 5 with center at the origin) is an idea whose value the Greeks never fully realized. The importance of this interplay between algebra and geometry was first seen by two seventeenth-century Frenchmen, Pierre de Fermat and René Descartes. It was they who developed *analytic geometry*, the name given to the study of this interplay, whose goal is the attainment of a synthesis of algebra and geometry.

Ever since A.D. 1637, when Descartes wrote *La Géométrie*, it has been common knowledge that curves can have equations and that equations determine curves. Why did the Greeks fail to utilize this means of approaching problems in geometry? The answer is simple. The Greeks knew that their curves had equations. However, they developed little abbreviative symbolism, and therefore had to write the "equations" out in words. For the Greeks, sometimes only a wondrous wealth of limiting clauses could describe adequately the mathematician's latest and prettiest discovery:

> Let a cone be cut by a plane through the axis, and let it also be cut by another plane cutting the base of the cone in a straight line perpendicular to the base of the axial triangle, and further let the diameter of the section be parallel to one side of the axial triangle; then if any straight line be drawn from the section of the cone parallel to the common section of the cutting plane and the base of the cone as far as the diameter of the section, its square will be equal to the rectangle bounded by the intercept made by it on the diameter in the direction of the vertex of the section and a certain other straight line. . . .
>
> Apollonius, *Conics*, ca. 200 B.C.

See?

By contrast, our modern system, which uses Descartes's coordinates and his abbreviative notation, is almost magically efficient. In modern terms, the long statement of Apollonius simply says,

> Given a parabola, a Cartesian coordinate system can be introduced in which the parabola has an equation of the form $y = cx^2$, where c is a certain constant. ...

It is hard to overestimate the value of appropriate symbolism. The Greeks never had it, and they developed only a little algebra. Their powers were concentrated upon geometry.

Why did the Greeks prefer to couch their mathematics in *geometry*? Why not let *number* play the key role in mathematics, particularly since Pythagoras would base everything upon number? The reason has to do with the discovery by the Pythagoreans of *irrational* quantities, a discovery that might be interpreted as disproving their own philosophy!

> It is told that those who first brought out the irrationals from concealment into the open perished in shipwreck, to a man. For the unutterable and the formless must needs be concealed. And those who uncovered and touched this image of life were instantly destroyed and shall remain forever exposed to the play of the eternal waves.
>
> Proclus

Pythagoreans want to explain everything by numbers. Trouble starts when one tries to explain the simplest elements of geometry by numbers. How does one account for points on a line in terms of numbers? This appears easy at first, but the appearance is deceptive. On a line segment a *unit length* is first chosen, and then to each ratio of integers is associated a point, in a natural way that is now familiar to every schoolchild. The ratio $\frac{3}{4}$, for example, names the point obtained by dividing the unit length into 4 equal parts and then taking 3 of them. At first, it appears that every point on the line can be named in this way, by using ratios of integers, or *rational numbers*.

The Pythagoreans, however, discovered to their distress that there was a certain point P that could be accounted for by *no rational number whatever*!

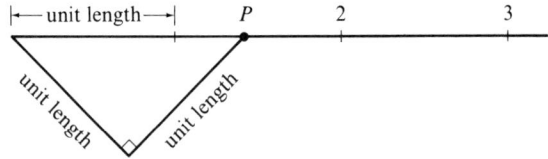

Consider the point P situated on the line as indicated above. The number associated with P would measure the length of the hypotenuse of a right

2. Geometry versus Arithmetic

triangle whose legs each have a length of one unit. Therefore, by the Pythagorean theorem, *the square of the number associated with P must be equal to* 2. The shock was felt when somehow, out of the Pythagorean school, around 500 B.C., came the following remarkable statement and proof.

Theorem. *There is no rational number whose square is 2.*

PROOF. (The proof uses the indirect, or *reductio ad absurdum* method, which consists in showing that an absurd conclusion results from supposing the theorem false.) Suppose that the theorem stated above is false, i.e., suppose there *is* a rational number whose square is 2. Then, by canceling out any common factors in the numerator and denominator, we should have a rational number a/b *in lowest terms* whose square is 2. We should then have integers a and b satisfying:

$$a \text{ and } b \text{ have no common divisor;} \qquad (4)$$

$$\frac{a^2}{b^2} = 2. \qquad (5)$$

From equation (5) it follows that

$$a \text{ is even.} \qquad (6)$$

Reason: If a were an odd number, then a^2 would be odd; yet equation (5) tells us that $a^2 = 2b^2$, showing that a^2 is even, being twice another integer.

Since we know that a is even, we know that a must be equal to twice some other integer. Calling this other integer k, we then have $a = 2k$, or $a^2 = 4k^2$, so that equation (5) becomes

$$\frac{4k^2}{b^2} = 2, \qquad (7)$$

where k and b are integers. From equation (7) it follows that

$$b \text{ is even.} \qquad (8)$$

Reason: If b were an odd number, then b^2 would be odd, yet equation (7) tells us (when it is solved for b) that $b^2 = 2k^2$, showing that b^2 is even.

The absurd conclusion is evident when statements (4), (6), and (8) are compared. This shows that an absurd conclusion is a consequence of assuming the theorem false. Therefore, the theorem must be true. Q.E.D.

The Greeks evidently saw no way to explain the point P by a number, since they could conceive of no "number" other than a rational number. Today, most students have no qualms over associating the point P with the number defined by a *never-ending* decimal expansion beginning

$$1.414\ldots.$$

The Greeks preferred to shy away from infinite processes whenever they could. Here, they took refuge in the "safe" framework of geometry. The point P offers, of course, no difficulty at all to geometry. It is just as simple an object as any other geometric point. Although the difficulty with irrational numbers surfaces in geometry as a problem with *incommensurable* line segments, the brilliant Eudoxus (ca. 408–355 B.C.) showed how things could be handled. It seemed that geometry could handle something that arithmetic could not, and the Greeks came to regard geometry more highly than arithmetic. For over two thousand years mathematics was couched in the framework of geometry.

What might have happened if, instead, the Greeks had emphasized measurement by numbers and had studied the way that numerical quantities relate to one another in a given setting? Then the development of the calculus and of modern science might have been accelerated radically. For this reason it may not be an exaggeration to say that the irrationality of $\sqrt{2}$ had a profound effect upon the history of mankind.

EXERCISES

2.1. Is 1.41 a rational number? *Answer*: Yes, since it is equal to 141/100, a ratio of integers.

2.2. Is 2 a rational number? *Hint*. $2 = 2/1$.

2.3. Is it true that $\sqrt{2} = 1.414$?

2.4. Is $\sqrt{4}$ irrational?

2.5. Give a *reductio ad absurdum* argument showing that $5\sqrt{2}$ is irrational. *Answer*: If it were rational, it would be equal to some ratio m/n of integers, which leads to a contradiction, as follows: $5\sqrt{2} = m/n$ implies that $\sqrt{2} = m/5n = $ a rational number, contradicting the theorem just proved.

2.6. Prove that $3\sqrt{2}$ is irrational.

2.7. Give a *reductio ad absurdum* argument showing that $\tfrac{3}{4}\sqrt{2}$ is irrational.

2.8. Prove that $(a/b)\sqrt{2}$ is irrational, if a and b are nonzero integers.

2.9. Give a *reductio ad absurdum* argument showing that $3 + \sqrt{2}$ is irrational. *Answer*: If $3 + \sqrt{2}$ were rational, we would have $3 + \sqrt{2} = m/n$, which, when solved for $\sqrt{2}$, yields $\sqrt{2} = (m - 3n)/n = $ a rational number, which is a contradiction.

2.10. Is the sum of a rational number and an irrational number always irrational?

2.11. Is the sum of two rational numbers always rational?

2.12. Is the sum of two irrational numbers always irrational? *Hint*. Consider the sum of $3 + \sqrt{2}$ and $3 - \sqrt{2}$.

2.13. Is the product of two rationals always rational?

2.14. Is the product of two irrationals always irrational?

2.15. Find an irrational number lying between 0 and 1/1000. *Hint.* Use your answer to exercise 2.8 and choose *b* very large.

2.16. Are there *infinitely many* irrationals between 0 and 1/1000?

2.17. Are there infinitely many irrationals between any two rational numbers?

2.18. (*For more ambitious students*) The long statement of Apollonius given in this section is, believe it or not, one of the prettiest theorems of geometry, once it is understood. Read pp. 203–204 of *The World of Mathematics*, edited by James R. Newman, Simon and Schuster, 1956, where Apollonius' proof may be found.

§3. Plato, Aristotle, and Mathematics

Western philosophy has been described as a series of footnotes to the writings of Plato. Yet, as we know from the warning on his gate ("Let no one ignorant of geometry enter here!"), Plato's philosophy was influenced by his conception of mathematics. The Pythagorean spirit lives in Plato.

Mathematics, in Plato's time (430–349 B.C.), was enjoying vigorous activity, the leading figure being Plato's colleague Eudoxus, whose work we shall meet in Chapter 6. Plato could not have failed to be impressed with the remarkable achievements of Eudoxus and with the fact that Eudoxus might have created (or is *discovered* a better word?) something that would last forever.

The *eternal* was of paramount interest to Plato, transitory things being of less value. What is important is the power to prevail against the ravages of time. To Plato, mathematics seemed to possess this power. The theorems of the Pythagoreans will live, even if the Greek language should die. Plato became enamoured of mathematics.

The theorems of mathematics are eternally true, thought Plato, and significant theorems will retain their value not only for the next few thousand years, but literally *forever*. "Geometry will draw the soul toward truth," said Plato, "and create the spirit of philosophy." All knowledge should perhaps aspire to the state attained by mathematics. Here, beyond the realm of immediate practicality, lies the true spirit of pure thought.

Plato went even further. As an illustration, consider the question whether the Pythagorean theorem was true *before Pythagoras came upon it*. Plato would reply strongly in the affirmative and would assert that the theorem had always been true. It had been built into the universe, and Pythagoras was just the first one to see it clearly.

Plato believed that the Pythagorean theorem existed, in some sense, long before Pythagoras. The connection between the ideas involved was there always. It was waiting to be discovered, as is often said nowadays, just as America was waiting to be discovered by Columbus. Plato began to think that all enduring knowledge must be like this. Knowledge consists of ideas, or eternal forms, and their great web of connections.

To put a significant piece of this knowledge into down-to-earth terms, so that all can understand, is a noble undertaking. Socrates, Plato's teacher, undertook to explain the idea of *justice*, an idea that is still imperfectly understood. Plato tackled the virtually impossible task of explaining the *good*, the *true*, and the *beautiful*, and to see the interrelationship between these ideas. In the more concrete realm of mathematics, Euclid (ca. 365–275 B.C.) tried to uncover the interrelationship between all the ideas of geometry that were known up to his time.

However naive Plato's outlook may appear, it cannot be denied that such efforts as these have inspired to this day many more seemingly impossible undertakings.

The aspect of Plato's philosophy just described is sometimes pictured as follows. The ideas, or eternal forms, already exist, floating in the "Platonic heaven", just beyond our grasp. Perhaps, as the Pythagoreans believed, we ourselves existed in a former life when we might have known these ideas before, but we are born with an imperfect knowledge of them. In order to remember ourselves, we must study philosophy: Only a lover of wisdom can climb high enough to swing around heaven, and slide back down to earth with a new perspective. A Platonist today might hold the view that "liberal arts" consists of ideas brought down to earth by swingers. □

The excitement and enthusiasm that Plato found in the study of mathematics may be contrasted with the tone taken by his great student Aristotle (384–322 B.C.). It has been suggested, perhaps unkindly, that Aristotle was never enthusiastic about anything. But it is certainly true that Aristotle saw in mathematics nothing to inspire such a flight of imagination as was taken by Plato. Aristotle was more of a scholar, or critic, than a speculator. The capacity to systematize knowledge, to bring order through reason, was of the highest importance. The value of mathematics, to Aristotle, lies in its exemplification of this capacity to a degree unmatched in any other discipline. Aristotle seems really to have been more interested in logic than in mathematics.

Aristotle's views on logic had great influence. They tend to be reflected in the style of Euclid, whose *Elements*—the greatest textbook of modern times—appeared about 300 B.C. Euclid seemed to show that the towering edifice of geometry was simply the consequence of logic unerringly applied to "self-evident" propositions, or *axioms*. The value of Euclid's work lies not so much in the announcement of previously unknown theorems (many of the theorems in the *Elements* were known before Euclid was born), but rather in the masterful logical organization of a great body of knowledge by the *axiomatic method*. Aristotle endorsed this method, which seems to have been introduced several centuries earlier by Thales, who taught Pythagoras.

The axiomatic method consists in stating clearly one's initial assumptions (axioms) and deducing all else by means of logic. The method results in a

3. Plato, Aristotle, and Mathematics

writing style that is demanding, austere, and—to some—supremely beautiful:

> Euclid alone has looked on Beauty bare.
> Let all who prate of Beauty hold their peace,
> And lay them prone upon the earth and cease
> To ponder on themselves, the while they stare
> At nothing, intricately drawn nowhere
> In shapes of shifting lineage; let geese
> Gabble and hiss, but heroes seek release
> From dusty bondage into luminous air.
> O blinding hour, O holy, terrible day,
> When first the shaft into his vision shone
> Of light anatomized! Euclid alone
> Has looked on Beauty bare. Fortunate they
> Who, though once only and then but far away,
> Have heard her massive sandal set on stone.
>
> Edna St. Vincent Millay*

Western civilization has consumed over a thousand editions of Euclid's *Elements*. It is no surprise that traces of the axiomatic method can be detected in many nonmathematical writings:

> We hold these truths to be self-evident . . .
> . . . a new nation . . . dedicated to the proposition that . . .

Thomas Jefferson and Abraham Lincoln were among Euclid's admirers. Lincoln considered his reading of Euclid an indispensable part of his education. The following passage is from a biographical sketch written for the 1860 presidential campaign.

> He studied and nearly mastered the six books of Euclid since he was a member of Congress.
> He began a course of rigid mental discipline with the intent to improve his faculties, especially his powers of logic and language. Hence his fondness for Euclid, which he carried with him on the circuit till he could demonstrate with ease all the propositions in the six books; often studying far into the night, with a candle near his pillow, while his fellow-lawyers, half a dozen in a room, filled the air with interminable snoring.[†]

What is it about Euclid that attracts? Is it not the cold, unexcited certainty with which tower upon tower of seemingly irrefutable arguments are built? No work could be more dispassionate than Euclid's *Elements*. Yet this same quality has probably repelled as often as it has attracted. For many people, Euclid is too severe, too lacking in enthusiasm, "faultless to a fault". Whatever enthusiasm Euclid felt for mathematics he restrained in writing the *Elements*. Aristotle would have wanted it that way. Scholarship must stand up under the cold, steady eye.

---

* Sonnet XLV, from *Collected Poems*, Harper and Row. Copyright 1923, 1951 by Edna St. Vincent Millay and Norma Millay Ellis.

[†] This passage is cited in E. T. Bell's *Men of Mathematics*, p. xvi.

EXERCISES

3.1. Latin translations of Euclid have made famous the abbreviation Q.E.D. If you don't know already, use a dictionary to find out what this abbreviation stands for, in both Latin and English.

3.2. Here are some comments on Euclid made by an eminent modern mathematician. Are his criticisms justified?

> *Euclid's work ought to have been any educationist's nightmare. The work presumes to begin from a beginning; that is, it presupposes a certain level of readiness, but makes no other prerequisites. Yet it never offers any "motivations", it has no illuminating "asides", it does not attempt to make anything "intuitive", and it avoids applications to a fault. It is so "humorless" in its mathematical purism that, although it is a book about "Elements", it nevertheless does not unbend long enough in its singlemindedness to make the remark, however incidentally, that if a rectangle has a base of 3 inches and a height of 4 inches then it has an area of 12 square inches. Euclid's work never mentions the name of a person; it never makes a statement about, or even an (intended) allusion to, genetic developments of mathematics ... In short, it is almost impossible to refute an assertion that the Elements is the work of an unsufferable pedant and martinet.**

3.3. In the next section, we shall make use of the following proposition, which is typical of propositions in the *Elements*. Can you give a proof of this proposition?

> *If two parallel lines are cut by a transversal, then the alternate interior angles are equal.*

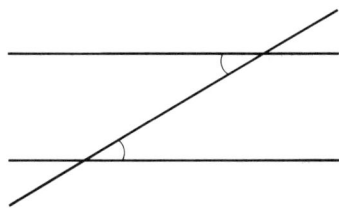

(Prove that the indicated angles are equal. Look up a proof in a geometry book if you have trouble.)

§4. Measuring the Earth

The root *geo-* means "earth", and *geometry* literally means "earth-measurement". Eratosthenes (ca. 276–195 B.C.) did just that, with the aid of Euclid's proposition about alternate interior angles (see exercise 3.3). In the third century B.C., Eratosthenes convinced himself that the earth's circumference

---

* Salomon Bochner, *The Role of Mathematics in the Rise of Science*, p. 35. (Copyright © 1966 by Princeton University Press and reprinted by permission of the Press.)

is about 50 times the distance from Alexandria to Aswan. (Aswan was known as "Syene" then. Its distance from Alexandria is about 500 miles.)

We indicate part of Eratosthenes' reasoning, leaving the rest to the reader. Eratosthenes thought that Alexandria was due north of Aswan. (It is not quite due north. Locate the two cities on a globe.) In Aswan there was a deep well that had an unusual feature. The sun shone straight down the well, casting no shadow at all, once every year: at the summer solstice. The sun, at noon on June 22, is directly overhead in Aswan, so that the sunlight beaming down a well is headed for the center of the earth. At the same time, in Alexandria, Eratosthenes observed the shadow cast by an upright stick, and measured an angle of slightly more than 7 degrees, or about $\frac{1}{50}$ a complete revolution. (One-fiftieth a complete revolution is, in degrees, equal to $(\frac{1}{50})(360)$ degrees, or 7.2°.)

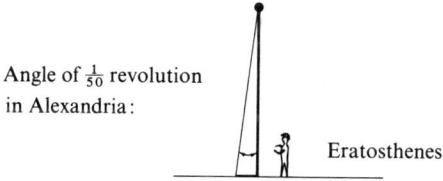

Angle of $\frac{1}{50}$ revolution in Alexandria:

Eratosthenes

EXERCISES

4.1. Explain how, from the facts above, one might infer that the circumference of the earth is about 25,000 miles.

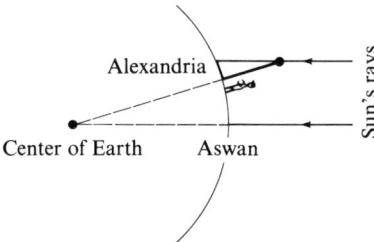

4.2. Look up the circumference of the earth in an almanac or encyclopedia. Was Eratosthenes' calculation close to being accurate?

4.3. In A.D. 1492 Columbus had his own idea of the earth's size. Did he think it larger or smaller than Eratosthenes believed it to be? What would have happened if Columbus had believed Eratosthenes' calculation to be correct?

§5. Archimedes versus the Romans

Had Eratosthenes lived in another time, his cleverness might have earned him a greater reputation. He was known, however, by the nickname "Beta", for it was his lot to live in the shadow of a deeper mathematician, Apollonius (ca. 260–200 B.C.), and to be virtually eclipsed by the incomparable Archimedes (287–212 B.C.).

Residing within Archimedes of Syracuse was a mysterious, driving force that compelled him to contemplate mathematics with his whole being. Archimedes' contemporaries spoke in wonder of his "raging Siren", his "familiar demon", his "muse", or his "spirit".

Archimedes discovered a significant part of the calculus. The notion of a *limit* was well understood by Archimedes, though he did not call it by name. His understanding of the essential idea is implicit in the papers he wrote. What is even more remarkable, Archimedes had a better grasp of this notion than the seventeenth-century mathematicians who invented the term.

In addition to his mathematics, which includes papers two thousand years ahead of his times, Archimedes developed the theory of floating bodies into the science now called *hydrostatics*. He was also an inventor of ingenious and useful devices such as a water pump, elaborate compound pulleys utilizing the law of the lever to remarkable advantage, and a mechanical contraption that is said to have described accurately the motions of the heavenly bodies. But he was, above all, a pure mathematician.

Archimedes' "violence from within" never abandoned him, even in great old age. When Rome attacked Syracuse in 214 B.C., Archimedes, despite his seventy years of age, went into action. To repel the Roman legions of Marcellus, he invented and deployed all manner of weapons. As Plutarch recorded, Archimedes' catapults and other devices scared the pluperfect hell out of the Romans:

> In fine, when such terror had seized upon the Romans that, if they did but see a little rope or a piece of wood from the wall, instantly crying out, that there it was again, Archimedes was about to let fly some engine at them, they turned their backs and fled; Marcellus desisted from conflicts and assaults, putting all his hope in a long siege.
>
> Yet Archimedes possessed so high a spirit, so profound a soul, and such treasures of scientific knowledge, that though these inventions had now obtained him the renown of more than human sagacity, he yet would not deign to leave behind him any commentary or writing on such subjects; but, repudiating as sordid and ignoble the whole trade of engineering, and every sort of art that lends itself to mere use and profit, he placed his whole affection and ambition in those purer speculations where there can be no reference to the vulgar needs of life; studies, the superiority of which to all others is unquestioned, and in which the only doubt can be whether the beauty and grandeur of the subjects examined, or the precision and cogency of the methods and means of proof, most deserve our admiration. It is not possible to find in all geometry more difficult and intricate questions, or more simple and lucid explanations. Some ascribe this to his natural genius; while others think that incredible effort and toil produced these, to all appearances, easy and unlaboured results. No amount of investigation of yours would succeed in attaining the proof, and yet, once seen, you immediately believe you would have discovered it; by so smooth and so rapid a path he leads you to the conclusion required.

5. Archimedes versus the Romans

> And thus it ceases to be incredible that (as is commonly told of him) the charm of his familiar and domestic Siren made him forget his food and neglect his person, to that degree that when he was occasionally carried by absolute violence to bathe or have his body anointed, he used to trace geometrical figures in the ashes of the fire, and diagrams in the oil on his body, being in a state of entire preoccupation, and, in the truest sense, divine possession with his love and delight in science. His discoveries were numerous and admirable; but he is said to have requested his friends and relations that, when he was dead, they would place over his tomb a sphere contained in a cylinder, inscribing it with the ratio which the containing solid bears to the contained.
>
> Plutarch, translated by John Dryden

Archimedes, at the age of seventy-five, was killed by sword. He had ignored the orders of Marcellus' soldier and had continued to study the lines and curves drawn in the sand. This was in 212 B.C. Apollonius and Eratosthenes died not long after, and the glory of Greece was soon to pass.

The long period of Roman domination now began. The Roman culture was inclined more toward engineering than toward mathematics. Romans valued only the parts of mathematics that could be applied in everyday life. Cicero, while he restored Archimedes' gravesite in Sicily and praised his name, nevertheless wrote with pride of how the Romans assimilated into Roman culture only the "practical" parts of mathematics.

> With the Greeks geometry was regarded with the utmost respect, and consequently none were held in greater honor than mathematicians, but we Romans have restricted this art to the practical purposes of measuring and reckoning.
>
> Cicero, *Tusculan Disputations*

One is surely free, like Cicero, to embrace utilitarianism if one wishes; but who is wise enough to say today what might be useful tomorrow? Consider, for instance, the simple proposition of Euclid regarding alternate interior angles. Many, upon seeing this proposition, would immediately discard it as useless. Yet Eratosthenes used it to measure the earth.

As a more striking example, consider the calculus, which was to grow out of seventeenth-century mathematics. It is still prized highly today for its utility, even by some who value relevance and applicability of knowledge more than knowledge itself. Calculus is indispensable to the modern engineer. Yet Archimedes, who scorned the utilitarian as no one else, had unlocked some of the secrets of the calculus in the normal course of his studies.

Today the other secrets of the calculus seem to us not far from Archimedes. But Roman engineers added to mathematics little of value. For nearly two thousand years the puzzling riddles lay right where Archimedes fell, in the reddening sand, amongst pebbles, lines, and curves.

At last there came colossal shrieks and shouts of surprise and delight in the seventeenth century, when suddenly Archimedes' Siren raged anew. The great man's spirit had prevailed against the Romans, and even today it is heard, howling in jubilation, *Eureka*!

Problem Set for Chapter 2

1. The Pythagoreans, by arranging pebbles in clever ways, could draw remarkable conclusions. To illustrate:
   (a) What is the sum of the first two odd numbers?
   Answer: • plus •• is ••• , the square of two.
   (b) What is the sum of the first three odd numbers?
   Answer: • plus •• plus ••• is ::: , the square of three.
   (c) What is the sum of the first four odd numbers?
   (d) What is $1 + 3 + 5 + 7 + 9 + 11$, the sum of the first six odd numbers?
   (e) What is the sum of the first n odd numbers?
   (f) What is $1 + 3 + 5 + \cdots + 99$? (Here, the three dots stand for the sum of the odd numbers between 5 and 99.)

2. The Pythagoreans knew that an easy way to count the number of pebbles arranged in the shape of a triangle is to view the triangle as half of a rectangle:
   (a) What is the sum of the first two numbers?
   Answer: • plus : is .:, which is half of ::,
   so the sum of the first two numbers is $\frac{1}{2}(2)(3)$.
   (b) What is the sum of the first three numbers? Answer: It is $\frac{1}{2}(3)(4)$, or half the dots in a 3 by 4 rectangle.
   (c) What is $1 + 2 + 3 + 4 + 5 + 6 + 7$?
   (d) What is $1 + 2 + 3 + \cdots + 100$? (Here, the three dots stand for the integers between 3 and 100.)
   (e) What is the sum of the first n positive integers?

3. The following principle is self-evident: *If the same amount is taken away from two figures having equal area, then the two modified figures have equal area.* It is thought that Pythagoras might have employed this principle, as follows:

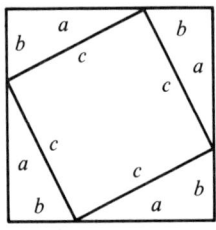 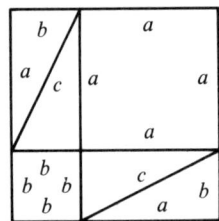

   The two large squares have equal area. Take away the four right triangles, use the principle above, and it becomes clear that $c^2 = a^2 + b^2$. *Eureka*!

4. Undersea cable costs $11,000 per mile, whereas underground cable costs $7,000 per mile. An island and a power plant are located as indicated, and cable is to be run between them.

Problem Set for Chapter 2

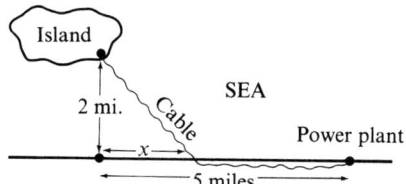

Let C be the cost (in thousands) of the cable, and let x be as indicated. Find an algebraic rule expressing C in terms of x, and specify the domain of this rule. (In Chapter 4 we shall discuss how to find the cheapest way of laying the cable.)

5. Find the center and the radius of the circle corresponding to each of the following algebraic equations:
   (a) $x^2 + (y - 2)^2 = 7$.
   (b) $(x + 3)^2 + (y - \sqrt{2})^2 = 10$.
   (c) $x^2 + y^2 = 10$.
   (d) $5x^2 + 5y^2 = 10$.

6. Write an algebraic equation corresponding to each of the following circles: the circle with
   (a) radius 3, center at $(3, -4)$.
   (b) radius $\sqrt{5}$, center at $(0, 3)$.
   (c) radius 9, center at $(\pi, \sqrt{7})$.

7. Give a *reductio ad absurdum* argument to prove that
   (a) $\sqrt{8}$ is irrational.
   (b) $\sqrt{18}$ is irrational.
   Hint. (a) First note that $\sqrt{8} = \sqrt{4 \cdot 2} = \sqrt{4} \cdot \sqrt{2} = 2\sqrt{2}$. Then proceed, using the fact that $\sqrt{2}$ is known to be irrational.

8. (In this question, you are not asked to give rigorous justification for your answers, only to test your intuition about the notion of area. It may be helpful to think about what happens to a figure in a photograph, when the photograph is enlarged.)
   (a) Suppose, in a plane with figures drawn, each length is tripled (i.e., blown up by a factor of 3). For instance, a square of size 2 by 2 becomes a square of size 6 by 6. By what factor is the *area* of each figure blown up? *Answer:* Areas of squares, and therefore all (why?) areas, are blown up by a factor of 9.
   (b) Suppose, as in part (a), that each length is blown up by a factor of r. By what factor is each area magnified?
   (c) The number π is **defined** as follows:

   π is the area of a circle of radius 1 unit.

   Use your answer to part (b), plus a little imagination, to find the area of a circle of radius r units.
   (d) How could you convince someone that π is less than 4? *Hint.* Consider a square circumscribed about the unit circle. What is the area of the square?
   (e) How could you convince someone that π exceeds 3? (A certain state legislature once considered seriously passing a law declaring that π was equal to 3 in that state. What are the objections to such a law?)
   (f) Archimedes knew, beyond any doubt, that π lies somewhere between $3\frac{10}{71}$ and $3\frac{1}{7}$. Can you imagine how he might have been able to determine this?

9. The formula for the area A of a circle [see problem 8(c)], gives A as a function of r. What is the domain of this function? What is its range? Plot a few points so that you can make a rough sketch of the curve determined by this function.

10. The Pythagorean theorem deals with squares constructed on the three sides of a right triangle. What about *semicircles* instead? Prove that the area of the semicircle with hypotenuse as diameter is equal to the sum of the areas of the semicircles on the other two sides. *Hint.* Use the equation $A = \pi r^2$ as an aid in finding the areas of the semicircles. Then use the Pythagorean theorem.

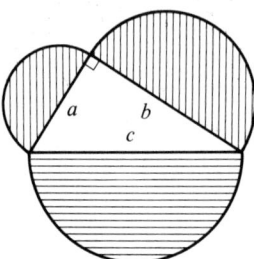

11. (*This is a famous result of Hippocrates of Chios, a member of the Pythagorean school.*) In the figure below, the hypotenuse of the right triangle is also the diameter of the circle in which the triangle is inscribed.

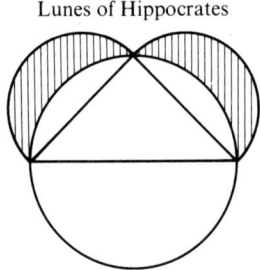

Lunes of Hippocrates

Prove that the combined area of the two "lunes" (with vertical markings) is equal to the area of the right triangle. *Hint.* From problem 10 we know that the area marked vertically in the lower figure is equal to the area marked horizontally. Take away the cross-hatched area from both figures, and use the principle given at the beginning of problem 3. *Eureka!*

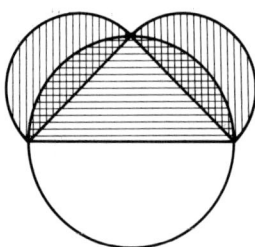

12. Consider the figure below, where the two triangles have a vertex in common and the lengths of their bases are equal. Prove that the triangles have the same area.

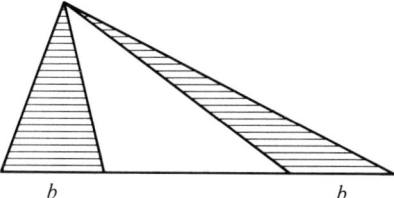

13. Here is a problem that intrigued the Greeks. Given a figure, construct a *triangle* whose area is the same as that of the figure. (For example, given the lunes of problem 11, Hippocrates found a triangle of the same size.) Do this for a *regular octagon*. (*Regular* means all sides have the same length and all angles made by adjacent sides are equal.) *Hint.* Stare at the figure below, and use the result of the preceding problem. (Both figures can be thought of as being made up of eight triangles.)

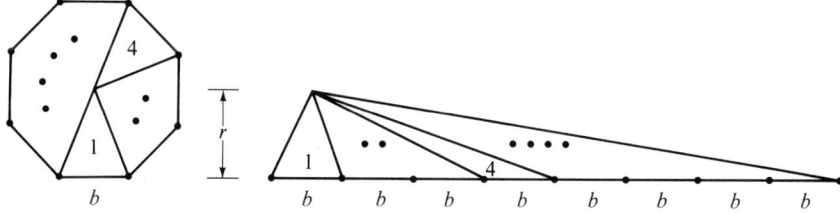

14. By a *regular polygon of n sides* is meant a figure in the plane bounded by n equal sides with n equal angles. (Problem 13 dealt with a regular polygon of 8 sides.) Let r denote the perpendicular distance from the center of a regular polygon to a side. Show that the area of a regular polygon is equal to the area of a triangle whose height is r and whose base is equal in length to the perimeter of the polygon. *Hint.* Stare at the figure below, and use the same reasoning as you did in the preceding problem.

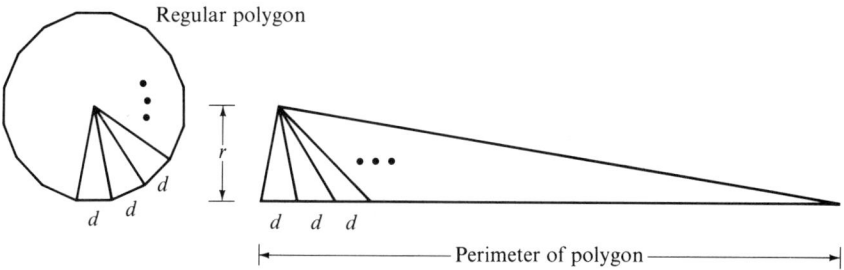

15. (*Here the reader is asked simply to make a guess after considering the evidence.*) Keep in mind that the equality of areas of the regular polygon and the corresponding triangle pictured in problem 14 holds, no matter how many sides the polygon has. This equality of areas holds for a polygon of a billion sides, for instance. Keeping this in mind, stare at the two figures below. One is a circle of radius r, and the other is a triangle of height r whose base is equal in length to the circumference of the circle.

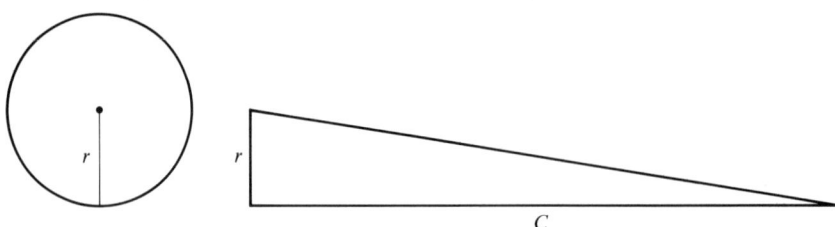

Now make a guess as to which of the following is true:
(a) The area of the circle exceeds the area of the triangle.
(b) The area of the triangle exceeds that of the circle.
(c) The area of the triangle equals the area of the circle.

16. (*For more ambitious students*) The amount and the type of reasoning which constitute an irrefutable argument in mathematics have never been fixed. Some things that the seventeenth century took as *obvious* (i.e., requiring no proof) the twentieth century and also the ancient Greeks accepted only after a careful demonstration from basic principles had been given. If you believe that the statement in part (c) of problem 15 is "obviously" true, then you are in the good company of some of the keenest minds of the seventeenth century. They would reason that equality between areas of polygons and triangles carries over "in the limit", a circle being regarded as the limit of polygons that approach it more and more closely.

    On the other hand, you may feel that the statement (c) requires a clear proof, because you have only made an educated guess that it is true. If so, then you are at home with Archimedes and with most twentieth-century mathematicians who would think so too. Archimedes proved 15(c) by showing that 15(a) leads to a contradiction, as does 15(b). Can you?

17. The statement in (c) of problem 15 is true. Using it, and letting C stand for the circumference of a circle of radius r, prove that
    (a) $\pi r^2 = \frac{1}{2} Cr$.
    (b) $C = 2\pi r$.

18. The algebraic rule $C = 2\pi r$ defines the circumference of a circle as a function of its radius. What is the domain of this function? What is its range? Plot a few points on the curve determined by this function. Is the "curve" really a straight line?

19. The number π is irrational, but this fact was not proved until the nineteenth century, and the proof is a little sophisticated. However, numbers like $\sqrt{3}$, $\sqrt{5}$, $\sqrt{6}$, $\sqrt{7}$, $\sqrt{8}$, $\sqrt{10}$, etc., have been known to be irrational since antiquity. Pick a couple of these and try your hand at proving them irrational.

20. Do a little outside reading about the Greeks, and particularly about Archimedes. For example, read pp. 19–34 of E. T. Bell's *Men of Mathematics*, Simon and Schuster, New York, 1937.
    (a) What did Archimedes mean when he said, "Give me a place to stand on, and I will move the earth!"
    (b) "Eureka! Eureka!" shouted the streaking sage of Syracuse. Why?
    (c) What does E. T. Bell mean when he says that modern mathematics was born with Archimedes and died with him for over two thousand years?

21. The words below are derived from Greek. Look them up in a good dictionary and find out their literal meaning:
    (a) *arithmetic.*
    (b) *geometry. Answer:* "Earth-measurement".
    (c) *mathematics.*
    (d) *philosophy. Answer:* "Love of wisdom".
    (e) *enthusiasm.*

22. *Rome itself stands for the impress of organization and unity upon diverse fermenting elements. Roman Law embodies the secret of Roman greatness in its Stoic respect for intimate rights of human nature within an iron framework of empire. Europe is always flying apart because of the diverse explosive character of its inheritance, and coming together because it can never shake off that impress of unity it has received from Rome. The history of Europe is the history of Rome curbing the Hebrew and the Greek, with their various impulses of religion, and of science, and of art, and of quest for material comfort, and of lust of domination, which are all at daggers drawn with each other. The vision of Rome is the vision of the unity of civilization.*

                                                                                          A. N. Whitehead

    What did the Romans do for mathematics?

3 Sherlock Holmes Meets Pierre de Fermat

Given a curve, such as the one below, how can one locate its lowest point? This problem arose naturally in Chapter 1, along with the analogous problem of finding the highest point on a curve. Both problems can be solved by the same method, to which we now turn.

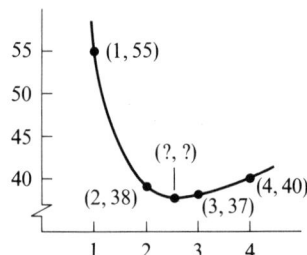

§1. Rising and Falling Lines

We must agree first how to use the words *rising* and *falling*, for there is danger of misunderstanding. Is the curve above *rising* or *falling* as it passes through the point (1, 55)? The answer depends upon whether one thinks of the curve as being traced out from left to right or in the reverse direction. So that we all speak the same language, let us agree to think of any function's curve as being traced out *from left to right* (or *from west to east*, if you prefer). The curve above is then falling as it passes through (1, 55), and rising as it passes through (4, 40).

Before going further, we had better mention the simplest curves of all: straight lines. Below are pictured *falling, horizontal,* and *rising* lines.

1. Rising and Falling Lines

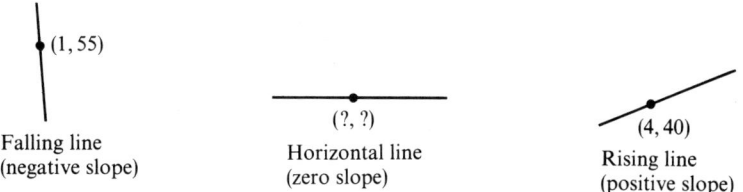

Falling line
(negative slope)

Horizontal line
(zero slope)

Rising line
(positive slope)

These three lines, if superimposed upon the curve pictured on page 50, may give the reader a hint as to the method we shall develop. At each point P on a curve, we shall seek *the line through P that most closely approximates the curve near P*. This line will be called the **tangent line** to the curve at P. The discussion of tangent lines begins in Section 3, and most of this chapter is devoted to their study.

What does the study of tangent lines to curves have to do with the problem stated in the first sentence of this chapter? Look again at the curve above. It is pretty clear, is it not, that the lowest point occurs where the tangent line is *horizontal*, that is, where the tangent line slopes neither up nor down.

We must give a precise meaning to the word *slope*.

Definition. The **slope** of the line joining (x_1, y_1) and (x_2, y_2) is given by

$$\frac{y_2 - y_1}{x_2 - x_1}, \quad \text{provided } x_1 \neq x_2.$$

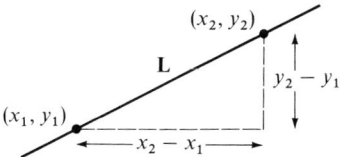

For example, the slope of the line joining $(4, 6)$ and $(5, -3)$ is given by

$$\frac{-3 - 6}{5 - 4} = -9.$$

The slope of a line is a number that measures how fast the line *rises* (or, when the slope is negative, how fast the line *falls*.) If L is the line joining (x_1, y_1) and (x_2, y_2), then

$$\frac{y_2 - y_1}{x_2 - x_1} = \text{Slope } L,$$

so that we have

$$y_2 - y_1 = (\text{Slope } L)(x_2 - x_1). \tag{1}$$

The relation expressed in equation (1) will be useful later in finding an equation of the line L.

EXERCISES

1.1. Find the slope of the line joining (1, 2) and (3, 7), and draw a picture of this line. *Partial answer*: Slope is 5/2.

1.2. Find the slope and draw a picture of the line joining (1, 2) and (3, −2). Is this line *rising* or *falling*?

1.3. Find the slope and draw a picture of the line joining (1, 2) and (5, 2). Is this line *rising* or *falling*?

1.4. Find the slope and draw a picture of the line joining (1, 2) and (1, 5). *Partial answer*: Slope is undefined.

1.5. Using geometry, show that the slope of a line is independent of which pair of points is chosen to calculate the slope. That is, in the figure below, show that the slope from P to Q is equal to the slope from R to S. *Hint.* The slope is simply a ratio of two sides of a triangle. Prove that the triangles are similar.

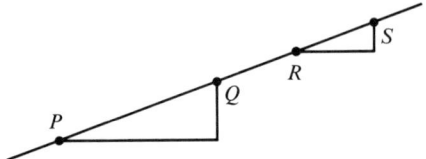

1.6. (a) Find the slope of the line **L** joining (1, 2) and (3, 5).
(b) Find the slope of the line joining (1, 2) and (300, 450).
(c) Using your answers to (a) and (b), decide whether the point (300, 450) lies *above*, *on*, or *below* the line **L** joining (1, 2) and (3, 5).
(d) Is the point (301, 452) on this line **L**? How do you know?

§2. Linear Functions

It is easy to see, as illustrated in exercises 1.1–1.3, that a line is

rising if its slope is *positive*,
falling if its slope is *negative*,
horizontal if its slope is *zero*.

Some curves (and we shall understand a line to be an especially simple kind of curve) determine functions, and some do not. Any *nonvertical* line does determine a function. (Why?) Such a function is called a **linear** function.

The slope of a line tells us something about the linear function it determines. It tells us how much the function "stretches". What does this mean? Look at the figure below, where a line of positive slope is pictured, and consider the function determined by this line. If the domain is the interval from x_1 to x_2 and if the corresponding range extends from y_1 to y_2, then by what factor is the domain stretched as it is sent into the range?

2. Linear Functions 53

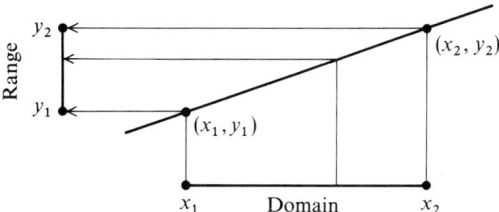

From the figure, the length of the range is $y_2 - y_1$, and the length of the domain is $x_2 - x_1$. Equation (1) above thus says:

Length of range = (slope of line)(length of domain).

The slope of the line thus gives the factor by which a linear function stretches lengths. A line of slope 3, for instance, determines a linear function that sends any interval into an interval three times as long. A line of slope 3, considered as a function, has a "stretching factor" of 3.

The preceding discussion applies to lines of *positive* slope. Suppose the slope of a line is negative, say -3. Then the linear function determined by the line still has a stretching factor of 3, but intervals in the domain are "flipped upside down" before they land in the range.

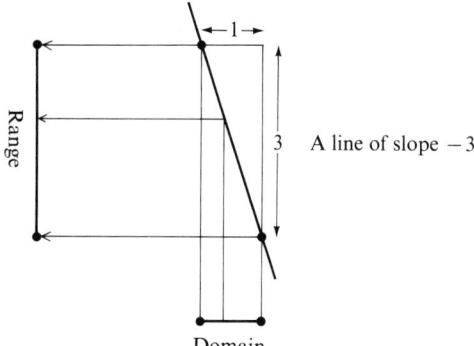

A line of slope -3

The notion of slope makes it easy to write or to recognize equations of nonvertical lines. The following exercises illustrate this.

EXERCISES

2.1. Translate into words the algebraic equation $y - 2 = \frac{3}{2}(x - 1)$. *Answer*: By equation (1), this says "(x, y) lies on the line of slope 3/2 passing through $(1, 2)$." [This is an equation, then, of the line described in exercise 1.6(a).]

2.2. Translate into words the algebraic equation $y - 4 = 3(x - 2)$, and sketch the line determined by this equation.

2.3. Translate into words the algebraic equation $y + 4 = -2(x - 3)$, and sketch the line determined. *Hint*. First rewrite the equation as $y - (-4) = -2(x - 3)$, then use equation (1).

2.4. Translate into words each of the following equations, and sketch the line determined.
 (a) $y = 2x + 4$. *Hint*. Rewrite as $y - 4 = 2(x - 0)$.
 (b) $3x + 4y = 6$. *Hint*. First solve for y. Then proceed as you did in part (a).
 (c) $y = \pi x + \sqrt{2}$.
 (d) $y = 5$. *Hint (if needed)*. Rewrite as $y - 5 = 0(x - 0)$, and use equation (1).

2.5. The **slope-intercept** form of the equation of a line is
$$y = bx + c,$$
where b and c are constants.
 (a) Rewrite this equation as $y - c = b(x - 0)$. Find the slope of the line determined by this equation, and find both coordinates of the point where the line meets the y-axis.
 (b) Describe the curve determined by the function given by $f(x) = 3x + 5$. *Answer*: The graph of f is a line passing through $(0, 5)$ with slope 3, since the algebraic rule $3x + 5$ is in slope-intercept form.
 (c) Describe the curve determined by each of the following rules:
  (i) $-2x - 5$.
  (ii) $x - 1$.
  (iii) $5 - x$.

2.6. Find an algebraic equation for the line of slope 3 passing through $(0, \pi)$. *Answer*: By equation (1), a point (x, y) lies on this line if and only if $y - \pi = 3(x - 0)$, or (simplifying) $y = 3x + \pi$.

2.7. Find an algebraic equation for the line of slope 3 passing through $(\pi, 0)$.

2.8. (a) Find the slope of the line joining $(4, 6)$ and $(3, 8)$.
 (b) Using your answer to (a), find an equation of the line joining $(4, 6)$ and $(3, 8)$.

2.9. Find an equation of the line joining
 (a) $(0, 0)$ and $(1, -2)$.
 (b) $(3, 4)$ and $(4, 7)$. *Answer*: $y = 3x - 5$.
 (c) $(3, 4)$ and $(7, 4)$.
 (d) $(3, 4)$ and $(3, 7)$. *Hint*. This is made simple, not hard, by the fact that the slope is undefined. Use common sense.

§3. The Principle of Elimination

In the preceding section we have made an essentially complete investigation of the simplest kind of function. We have learned that any function given by a rule of the form
$$bx + c$$
is a linear function. Its graph is a line of slope b passing through the point $(0, c)$ on the vertical axis.

The next simplest kind of function is a *quadratic* function, arising when a linear expression $bx + c$ is modified by a term involving a square: A

3. The Principle of Elimination

quadratic function is given by an algebraic rule of the form

$$ax^2 + bx + c, \quad \text{where } a \neq 0.$$

The behavior of quadratic functions is not hard to study. To investigate that behavior, *and to learn at the same time how to find tangent lines to curves,* let us consider the simplest quadratic function of all. This is, of course, the **squaring function** given by

$$f(x) = x^2.$$

Plotting a lot of points on the graph of the squaring function shows that it looks something like this:

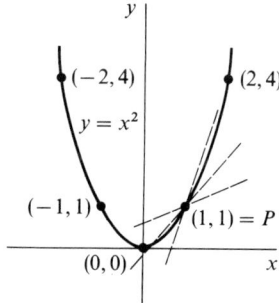

There are many lines through $P = (1, 1)$. How can we pick out the line tangent to the curve at P?

We are ready to move toward attacking the problem stated in the first sentence of this chapter. We have already hinted that the solution of that problem involves the study of tangent lines to curves. Our task now is to figure out exactly what a tangent line is. So far, we have only made the (rather vague) statement that the tangent line to a curve at a point P is the line through P that most closely resembles the curve near P. With this meager thread to hold on to, *how can one determine the slope of the tangent line to the squaring function at the point $P = (1, 1)$?*

This is a challenging question, even for the keen mind of a master sleuth. Let us therefore enlist the aid of the great detective:

Sherlock Holmes's Principle. *When you have eliminated the impossible, whatever remains, however improbable, must be the truth.*

The answer to a question is among what remains after wrong answers have been set aside. By this principle of elimination, the tangent line is the line left when all "nontangent" lines have been discarded. Will this be helpful to us? We shall see. Let us first look at some exercises to test whether this principle of elimination is well understood.

EXERCISES

Apply Sherlock Holmes's principle to each of the following situations.

3.1. Winnie the Pooh's honey is gone. Everyone but Tigger has a valid alibi that proves his innocence. *Answer*: By Holmes's principle, Tigger stole the honey, *provided it was stolen*.

3.2. A survey shows that Peter Pan is a citizen of no country questioned in the survey, and England is the only country not questioned. *Answer*: By Holmes's principle, Peter Pan is a citizen of England, *provided that Peter Pan is a citizen of some country*.

3.3. The county seat of Yoknapatawpha County, Mississippi, is none other than the city of Jefferson.

3.4. $1984 + h$ is not the title of a famous book, if h is not equal to 0. *Answer*: No famous book has a numerical title, except possibly *1984*.

3.5. If $h \neq 0$, then the area of a circle of radius 1 is not $\pi + h$. *Answer*: The area of a circle of radius 1 is none other than π.

3.6. If $x \neq 5$, then x is not the solution of a certain problem in arithmetic. *Answer*: The solution of the problem, by Holmes's principle, is 5, *provided the problem has a solution*.

3.7. If $h \neq 0$, then $h + 2$ is not the answer to a certain problem in arithmetic.

3.8. If $h \neq 0$, then $(h^2 + 2h)/h$ is not the answer to a certain problem. *Hint.* $(h^2 + 2h)/h = h + 2$ if $h \neq 0$.

3.9. If $h \neq 0$, then $(h^2 + 4h)/h$ is not the answer to a certain problem. *Answer*: By Holmes's principle, the answer must be 4, *provided the problem has an answer (and provided the answer is a number)*.

3.10. If $h \neq 0$, then $(h^2 + 9h)/h$ is not the answer to a certain problem.

§4. The Slope of a Tangent Line

We are prepared to begin our detective work. To employ the principle of Sherlock Holmes, we must attain skill at finding wrong answers, in order to eliminate them. Let us recall the question:

$$\text{What is the slope of the tangent line to the curve } y = x^2 \text{ at the point } P = (1, 1)? \qquad (2)$$

How can we get a wrong answer to this question? Look once again at the graph of the squaring function near P. A line through P that cuts the curve twice will *not* be tangent at P, it would seem. The tangent line at P will touch the curve only at P.

4. The Slope of a Tangent Line

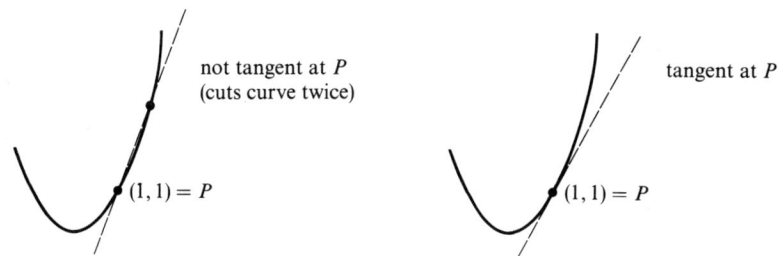

Now we have a clue. To obtain a *wrong* answer to question (2), we need only find the slope of a line joining P to another point on the graph of the squaring function. This graph consists of each point in the plane whose second coordinate is the square of its first coordinate. Another point on the curve, then, is $(1 + h, (1 + h)^2)$ if h is not equal to zero. (If h is 0, this "other" point would coincide with P.) The slope of the nontangent line joining $(1, 1)$ and $(1 + h, (1 + h)^2)$ is given by

$$\frac{(1 + h)^2 - 1}{1 + h - 1} = \frac{1 + 2h + h^2 - 1}{h} = \frac{2h + h^2}{h}.$$

We now know a host of wrong answers to question (2), for if $h \neq 0$, then $(2h + h^2)/h$ is the slope of a line that is *not* tangent at P. Note that this expression simplifies to $2 + h$.

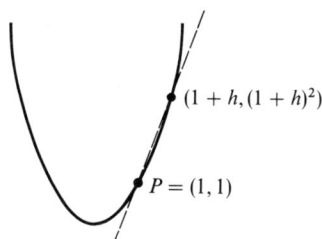

This line, not tangent at P if $h \neq 0$, has a slope equal to $2 + h$

What is the answer to question (2), now that we know that $2 + h$ is not the answer, if $h \neq 0$. The only number not eliminated is 2. By Sherlock Holmes's principle, the answer to question (2) must be 2, provided the question has an answer. That is,

$$\text{the slope of the tangent line to } y = x^2 \text{ at } (1, 1) \text{ is } 2, \tag{3}$$

provided the curve $y = x^2$ has a tangent line at $(1, 1)$. Elementary, dear Watson! □

Holmes's method illustrates the curious fact that it is possible to get the right answer by first considering how to get wrong answers. Let us try another question.

$$\text{What is the slope of the tangent line to the curve } y = x^2 \text{ at the point } (-2, 4)? \qquad (4)$$

Let us consider how to get wrong answers to question (4). A wrong answer is the slope of the line joining $(-2, 4)$ and $(-2 + h, (-2 + h)^2)$ if $h \neq 0$. The slope of this nontangent (or *secant*, as a line cutting a curve twice is often called) is given by

$$\frac{(-2 + h)^2 - 4}{-2 + h + 2} = \frac{4 - 4h + h^2 - 4}{h}$$

$$= \frac{-4h + h^2}{h}$$

$$= -4 + h \quad \text{if } h \neq 0.$$

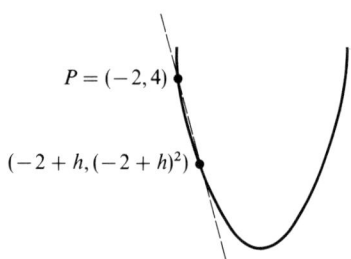

This secant has a slope equal to $-4 + h$

What is the answer to question (4), now that we know $-4 + h$ is not the answer if $h \neq 0$? The only number not eliminated is -4. By Holmes's principle, the answer must be -4, provided there is an answer. That is,

$$\text{the slope of the tangent line to } y = x^2 \text{ at } (-2, 4) \text{ is } -4, \qquad (5)$$

provided the curve $y = x^2$ has a tangent line at $(-2, 4)$.

EXERCISES

Apply the principle of elimination to each of the following.

4.1. What is the slope of the tangent line to the curve $y = x^2$ at $(0, 0)$? *Answer:* It is 0, provided there is a tangent line.

4.2. What is the slope of the tangent line to the curve $y = x^2$ at the point $(2, 4)$?

4.3. What is the slope of the tangent line to $y = x^2$ at (π, π^2)? *Answer:* 2π, if there is a tangent line.

4.4. What is the slope of the tangent line to the curve $y = x^2 + 3$ at the point $(1, 4)$?

4.5. What is the slope of the tangent line to the curve $y = x^2 + 3x$ at the point $(1, 4)$? *Hint.* A wrong answer to this question is given by

$$\frac{(1+h)^2 + 3(1+h) - 4}{h} = 5 + h \quad \text{if } h \neq 0.$$

4.6. What is the slope of the tangent line to the curve $y = x^2 + 3x + 2$ at the point $(1, 6)$? *Answer:* 5, if there is one.

4.7. What is the slope of the tangent line to the curve $y = x^2 + 3x + 2$ at the point $(\pi, \pi^2 + 3\pi + 2)$?

§5. Fermat's Method and the Derivative

As clever as Holmes's method is, it has serious drawbacks, as illustrated in problem 17 at the end of this chapter. One worrisome thing about this method is that things are left hanging a bit at the end. How do we know whether a curve has a tangent line at a certain point? What is needed is a clear definition.

Pierre de Fermat used the notion of *limit* to invent a workable definition of the *slope of a tangent line to a curve*. It is only a slight modification of the method we have just employed, but by it the drawbacks to Holmes's method are removed.

Fermat described the following method of finding the slope of the tangent line to a curve f at a given point $P = (c, f(c))$ on the curve. First find the slope of the line joining $(c, f(c))$ and $(c + h, f(c + h))$, where $h \neq 0$. Although this slope, which is given by

$$\frac{f(c + h) - f(c)}{h}, \tag{6}$$

is likely *not* the desired slope of the tangent line, it clearly approximates the desired slope as h is taken nearer to zero. It is natural, then, to *define* the slope of the tangent line at $(c, f(c))$ to be the number (if there is one) that expression (6) is trying to become as h approaches zero.

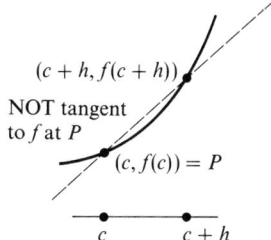

This nontangent line has slope given by expression (6)

Definition. The **slope of the tangent line** to the curve f at the point $(c, f(c))$ is defined to be
$$\underset{h \to 0}{\text{Limit}} \frac{f(c+h) - f(c)}{h}.$$

Fermat's idea is simple yet subtle. The "right answer" is the limiting value of wrong answers that approximate it ever so closely. Here are several examples to illustrate Fermat's method.

EXAMPLE 1. Find the slope of the tangent line to the curve $y = x^2$ at the point $(1, 1)$.

Here the function is given by $f(x) = x^2$, and the point P is $(1, f(1))$. According to Fermat's method, the slope of the tangent line at $(1, f(1))$ is given by

$$\underset{h \to 0}{\text{Limit}} \frac{f(1+h) - f(1)}{h} = \underset{h \to 0}{\text{Limit}} \frac{(1+h)^2 - 1}{h}$$

$$= \underset{h \to 0}{\text{Limit}} \frac{h^2 + 2h}{h}$$

$$= \underset{h \to 0}{\text{Limit}} (h + 2)$$

$$= 2.$$

Note that nothing is left hanging at the end. Since the limit exists, there is a tangent line, and its slope is equal to that limit. By Fermat's definition, the existence of a tangent line is tantamount to the existence of the limit of expression (6). □

EXAMPLE 2. Find the slope of the tangent line to the curve $y = x^2$ at the point (π, π^2).

Here we have the squaring function again, given by $f(x) = x^2$, and the point P is $(\pi, f(\pi))$. By Fermat's method, the slope of the tangent line is

$$\underset{h \to 0}{\text{Limit}} \frac{f(\pi+h) - f(\pi)}{h} = \underset{h \to 0}{\text{Limit}} \frac{(\pi+h)^2 - \pi^2}{h}$$

$$= \underset{h \to 0}{\text{Limit}} \frac{\pi^2 + 2\pi h + h^2 - \pi^2}{h}$$

$$= \underset{h \to 0}{\text{Limit}} \frac{2\pi h + h^2}{h}$$

$$= \underset{h \to 0}{\text{Limit}} (2\pi + h)$$

$$= 2\pi.$$
□

5. Fermat's Method and the Derivative

EXAMPLE 3. Find the slope of the tangent line to the curve $y = x^2$ at the point (x, x^2).

This is so similar to Example 2 that the reader can probably guess the answer. The answer is $2x$, for the same reason that the answer to the preceding example is 2π. This is seen by a calculation identical to that of Example 2, with x replacing π:

$$\underset{h \to 0}{\text{Limit}} \frac{f(x+h) - f(x)}{h} = \underset{h \to 0}{\text{Limit}} \frac{(x+h)^2 - x^2}{h}$$

$$\vdots$$

$$= \underset{h \to 0}{\text{Limit}} (2x + h) = 2x.$$

(The reader is asked to fill in the missing steps in this calculation.) □

The work of Examples 1–3 may be summarized in a table:

x	y	Slope of tangent line at (x, y)
1	1	2
π	π^2	2π
x	x^2	$2x$
-1	1	?

If we recall the definition of a function in terms of a pair of columns, then we see that the *first* and *third* columns above determine a *new* function. This new function, derived from the original function f, will be denoted by f' and called the **derivative** of f. From the third line of the table above, we see that the rule determining f' is simply the "doubling" rule, sending x to $2x$. That is, we see that

$$\text{if } f(x) = x^2, \quad \text{then } f'(x) = 2x. \tag{7}$$

Or, in words, *the derivative of the squaring function is the doubling function.*

EXAMPLE 4. Find the slope of the tangent line to the curve $y = x^2$ at the point $(-1, 1)$.

Now there is no need to go back to Fermat's method, because *the derivative gives us the general slope-predicting rule*. All that is asked here is that the question mark in the preceding table be filled in appropriately, and that is now easy. The answer is $f'(-1)$, which is equal to -2, since f' is the doubling function. □

EXAMPLE 5. Find the slope of the tangent line to the curve $y = x^2$ at the point $(4, 16)$.

The answer is $f'(4)$, which is equal to 8. □

The function f sends x into y. What does the function f' do? It is convenient to let y' stand for the long phrase "Slope of the tangent line at (x, y)". Then the function f' stands x into y'. Thus, equation (7) says exactly the same thing as

$$\text{if } y = x^2, \quad \text{then } y' = 2x.$$

EXERCISES

5.1. Fill in the missing steps in Example 3.

5.2. Find the slope of the tangent line to the curve $y = x^2$ at the point $(3, 9)$
   (a) by using Fermat's method, going through all the steps to find the limit of $((3 + h)^2 - 9)/h$ as h approaches 0.
   (b) by using the shortcut method of Example 5, knowing that the derivative of the squaring function is the doubling function.

5.3. What does statement (5) of Section 4 say in terms of y'? *Answer*: It says, "Given $y = x^2$, then y' is -4 when x is -2."

5.4. What does statement (3) of Section 4 say about y'?

5.5. What does the answer to exercise 4.5 say about y'? *Answer*: It says, "Given $y = x^2 + 3x$, then y' is 5 when x is 1 (assuming there is a tangent line)."

5.6. What does the answer to exercise 4.6 say about y'?

§6. The Interplay between a Function and Its Derivative

The derivative f' is useful for many reasons. One reason (we shall see others later) is that f' gives information about the behavior of the original function f. To illustrate this, let us continue to study the squaring function f, whose derivative, we have seen, is the doubling function.

First, note that f' is just as "good" a function as f. The equation $y' = f'(x)$ determines a curve too! In this case the rule for f' is the *linear* expression

$$2x,$$

which we should recognize immediately to be pictured as a line of slope 2, passing through the origin $(0, 0)$ in the x-y' plane.

x	y	y'
1	1	2
x	x^2	$2x$
π	π^2	2π
s	s^2	$2s$

6. The Interplay between a Function and Its Derivative

To see the interplay between f and f', it is convenient to picture the curve f' on a separate coordinate system (the x-y' plane) and to compare it with the curve f in the x-y plane.

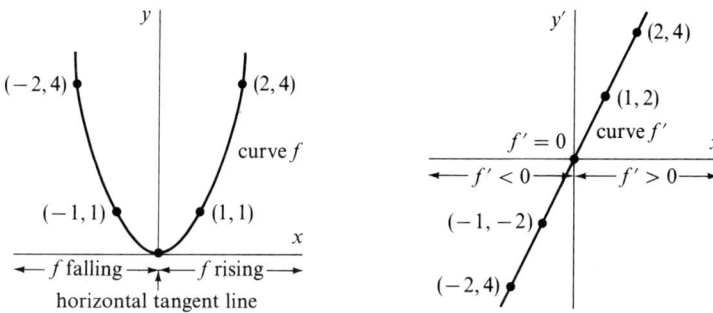

At a point where the curve f is *falling*, the tangent line must have a *negative* slope. Hence, if f is falling at the point $(c, f(c))$, then $f'(c)$ must be negative. Similarly, when f is *rising*, then f' must be *positive*. And when the curve f has a *horizontal tangent line*, then f' must be zero.

Exercises

6.1. Find both coordinates of a point on the curve $y = x^2$ where the slope of the tangent line is 3. *Answer*: We are required to fill in the question marks correctly in the following table:

x	y	y'
?	?	3
x	x^2	$2x$

When $y' = 3$, we have $2x = 3$; so $x = \frac{3}{2}$. When $x = \frac{3}{2}$, $y = (\frac{3}{2})^2 = \frac{9}{4}$. Therefore, the slope of the tangent line is 3 at the point $(\frac{3}{2}, \frac{9}{4})$.

6.2. Find both coordinates of a point on the curve $y = x^2$ where the slope of the tangent line is
   (a) -2.
   (b) 0.
   (c) 10.
   (d) 5.

6.3. Suppose the slope of the tangent line to a curve is -1 at a certain point. Is the curve *rising* or *falling* as it passes through that point?

6.4. Find an equation of the tangent line to the curve $y = x^2$ at the point $(3, 9)$. *Answer*: When x is 3, y' is 6; so the slope of the tangent line is 6. An equation of the line of slope 6 through $(3, 9)$ is $y - 9 = 6(x - 3)$.

6.5. Find an equation of the tangent line to the curve $y = x^2$ at the point
  (a) $(1, 1)$.
  (b) $(-1, 1)$.
  (c) (π, π^2).
  Answer: (c) $y - \pi^2 = 2\pi(x - \pi)$.

§7. Solving Optimization Problems with Derivatives

Compare the equation $y = x^2$ with the equation $A = s^2$. Both equations determine *the same function*. Why? Because both equations define exactly the same rule, the *squaring rule*. The curve in the x-y plane of the equation $y = x^2$ is identical with the curve in the s-A plane of the equation $A = s^2$. Since the derivative of the squaring function is the doubling function, it is clear that

$$\text{if } A = s^2, \quad \text{then } A' = 2s. \tag{8}$$

By the same token, we know, for example, that

$$\text{if } y = L^2, \quad \text{then } y' = 2L.$$

Changing only the *names* of the variables doesn't alter the function, or its derivative, at all.

Let us find another quadratic function to play with. In Example 2 of Chapter 1 we encountered the personable function g given by the quadratic rule

$$-\frac{s^2}{2} + 600s.$$

What is the rule for g', the derivative of g?

s	A	A'
400	160,000	
700	175,000	
s	$-\frac{1}{2}s^2 + 600s$	?

Can you *guess* the rule for g', before we work it out below? *There is nothing wrong with guessing.* Consider the facts. Statement (8) tells us that from the expression s^2 in the second column we derive the expression $2s$ in the third column. On the basis of this, what would you guess to be derived from the expression $-\frac{1}{2}s^2$? As for the expression $600s$, that is easy. This is

7. Solving Optimization Problems with Derivatives

just a linear expression of slope 600, leading one to expect that from the expression 600s in the second column we would derive the expression 600 in the third. From these facts, what would you guess:

$$\text{If } g(s) = -\frac{s^2}{2} + 600s, \text{ then } g'(s) = ?$$

To verify your guess, go back to the definition of the derived function. By definition, $g'(s)$ is the slope of the tangent line to the curve g at the point $(s, g(s))$. Using Fermat's method to calculate that slope, we have

$$g'(s) = \underset{h \to 0}{\text{Limit}} \; \frac{g(s+h) - g(s)}{h}$$

$$= \underset{h \to 0}{\text{Limit}} \; \frac{-\frac{1}{2}(s+h)^2 + 600(s+h) - (-\frac{1}{2}s^2 + 600s)}{h}$$

$$= \underset{h \to 0}{\text{Limit}} \; \frac{-\frac{1}{2}s^2 - sh - \frac{1}{2}h^2 + 600s + 600h + \frac{1}{2}s^2 - 600s}{h}$$

$$= \underset{h \to 0}{\text{Limit}} \; \frac{-sh - \frac{1}{2}h^2 + 600h}{h}$$

$$= \underset{h \to 0}{\text{Limit}} \left(-s - \frac{h}{2} + 600 \right)$$

$$= -s + 600.$$

Thus we see that $g'(s) = -s + 600$. In other words,

$$\text{if } A = -\frac{s^2}{2} + 600s, \text{ then } A' = -s + 600.$$

s	A	A'
?	?	0
s	$-\frac{1}{2}s^2 + 600s$	$-s + 600$

We now have enough information to determine the highest point (?, ?) on the curve g, for this point must occur where the slope of the tangent is zero. This is easy, for $A' = 0$ when

$$-s + 600 = 0,$$

$$s = 600.$$

Thus, at the point $(600, 180{,}000)$, the curve g has a horizontal tangent line. How do we know this is the *highest* point? Look at the derivative. The curve $A' = -s + 600$ is a linear curve of slope -1, and A' is 0 when s is

600. The derivative looks like this:

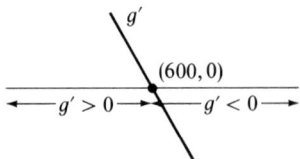

Therefore the curve g must be rising to the left of 600 and falling to the right of 600. This means that, at $s = 600$, the maximal A is attained.

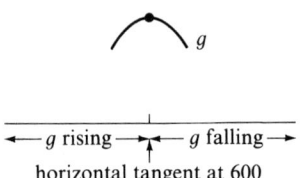

The optimization problem arising in Example 2, Chapter 1, is now solved. The maximal area is 180,000 square feet, attained when the length s along the barn is 600 feet:

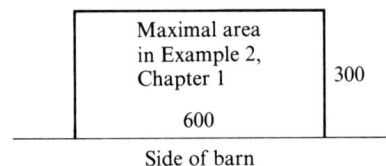

Let us attack a problem similar to the one just disposed of: to show how easy an optimization problem can become when calculus is applied.

EXAMPLE 6. A farmer has 300 meters of fencing to enclose three sides of a rectangular area. The fourth side is bounded by a long barn and requires no fence. What is the largest area she can enclose?

We want to maximize the area A, which varies in terms of the length s along the barn. Letting G denote the function that arises, we have

$$A = G(s) = \text{area (in square meters) of } \boxed{}$$

$$= 150s - \frac{s^2}{2},$$

where the domain is specified by the inequality $0 < s < 300$.

To solve this optimization problem, we must find the highest point (?, ?) on the curve G, which gives the area A as a function of s. Toward this end,

7. Solving Optimization Problems with Derivatives

we take the derivative:

$$\text{If } A = 150s - \frac{s^2}{2}, \text{ then } A' = 150 - s \quad (\text{why?}).$$

Thus A' is 0 when s is 150, and we have found the point on the curve G where the tangent line is horizontal:

s	A	A'
150	11,250	0
s	$150s - \frac{1}{2}s^2$	$150 - s$

In all likelihood, the point (150, 11,250) is the highest point on the curve G. To *prove* that it is, look at the sign of the derivative on either side of 150. The derivative G' is given by the linear rule $150 - s$, and thus looks like this:

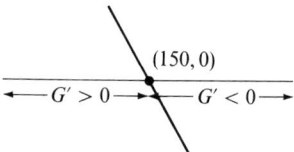

Therefore, the curve G must be rising to the left of 150 and falling to the right. This shows that, at $s = 150$, G attains its maximum.

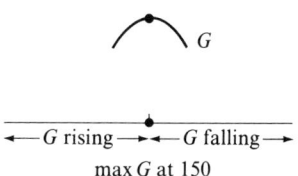

The maximal area is 11,250 square meters. □

EXERCISES

7.1. The derivative of the squaring function is the doubling function. The slope of the line $bx + c$ is b. Use these facts and try your hand at *guessing* answers to the following:
   (a) If $y = x^2 + 3x$, what is y'?
   (b) If $y = s^2 - 57s$, what is y'?
   (c) If $A = 4s^2 + 60s$, what is A'?
   (d) If $y = 5x^2 + 13x - 7$, what is y'?
   (e) If $y = ax^2 + bx + c$, what is y'?

7.2. In each of (a) through (e) of exercise 7.1, use Fermat's method to verify the correctness of your guess. *Answer*: (a) Given $y = f(x) = x^2 + 3x$, by Fermat's method we have

$$y' = f'(x) = \lim_{h \to 0} \frac{f(x+h) - f(x)}{h}$$

$$= \lim_{h \to 0} \frac{(x+h)^2 + 3(x+h) - (x^2 + 3x)}{h}$$

$$= \lim_{h \to 0} \frac{2xh + h^2 + 3h}{h}$$

$$= \lim_{h \to 0} (2x + h + 3)$$

$$= 2x + 3.$$

7.3. In each of (a) through (e) of exercise 7.1, find both coordinates of the point on the quadratic where the tangent line is horizontal. *Answer*: (a) At the point $(-3/2, -9/4)$, y' is zero.

7.4. Find both coordinates of the highest point on the curve $A = 1200w - 2w^2$, with domain $0 < w < 600$. (This is the function which arose in problem 3 at the end of Chapter 1.)

7.5. A farmer has 4000 feet of fencing to enclose *three* sides of a rectangular area (the fourth side being bounded by a long fence already standing). Find the largest area that can be enclosed, and specify the dimensions that should be used to attain maximal area.

7.6. A farmer has 4000 feet of fencing to enclose *four* sides of a rectangular area. What dimensions should be used to maximize the area enclosed?

7.7. By working through the following steps in turn, *find a pair of positive numbers whose sum is 10 and whose product is as large as possible*.
   (a) We want to maximize their product. Let P denote their product. What is P if the first number is 2? (First find the second number, using the fact that the sum of the two numbers must be 10.)
   (b) What is P if the first number is π?
   (c) What is P if the first number is x?
   (d) Your answer to (c) yields a quadratic rule giving P as a function of x. What is P'?
   (e) What is the domain of the function you found in part (c)? (Remember that *both* numbers must be positive.)
   (f) Find both coordinates of the highest point on the graph of the quadratic function of part (c).
   (g) Answer the question of problem 7.7 with a complete sentence.

7.8. Express the number 10 as the sum of two positive numbers in such a way that the sum of the *square* of the first and *three times* the second is as *small* as possible. *Hint*. This is similar to exercise 7.7.

7.9. Work through the following steps in turn, in order to answer the question at the end.
   (a) In the x-y plane, draw the line $y = 3x + 2$. Also indicate the position of the point $(4, 0)$.

8. Definition of the Derivative

(b) Find the *square* of the distance between the point (4, 0) and the point on the line $y = 3x + 2$ whose first coordinate is π. (First find the second coordinate, then find the square of the distance by the Pythagorean theorem.)

(c) Find the square of the distance between the point (4, 0) and the point on the line $y = 3x + 2$ whose first coordinate is x. *Answer*: $10x^2 + 4x + 20$.

(d) The rule written down in the answer to part (c) is a quadratic function. Find the value of x that yields the minimum of this function.

(e) Find both coordinates of the point on the line $y = 3x + 2$ that is closest to the point (4, 0). *Answer*: $(-1/5, 7/5)$.

7.10. Find both coordinates of the point on the line $y = 5 - 2x$ that is closest to the point (0, 0).

§8. Definition of the Derivative

Calculus relies greatly upon derivatives. We therefore seek rules enabling us to write down quickly the derivative of any function we might meet. We have already found such a rule for writing down the derivative of any quadratic function:

$$\text{If } y = ax^2 + bx + c, \text{ then } y' = 2ax + b.$$

(Another way of expressing the same thing is, "If $f(x) = ax^2 + bx + c$, then $f'(x) = 2ax + b$.") By virtue of this simple rule, there is no need to go through all the details of Fermat's method in order to find the derivative of a quadratic. In the next chapter, however, we shall meet more complicated algebraic functions, such as are given by the rules $1/x$ (the *reciprocal function*), x^3 (the *cubing function*), $\sqrt{x}$ (the *square root function*), etc. To find their derivatives, we must be clear about the definition of the derivative.

If f is any function, the rule defining its derivative f' is given below. The derivative is defined so that, at a point x, the derivative f' gives the slope of the tangent line to the curve f at the point $(x, f(x))$. Since Fermat's method gives this slope, we have the following definition.

Definition. Given a function f, and a point x in its domain, the **derivative** f' is defined by the rule

$$f'(x) = \lim_{h \to 0} \frac{f(x+h) - f(x)}{h}.$$

Note that the definition of the derivative incorporates all three basic notions: *variable, function, limit*.

To calculate f' directly from this definition is sometimes tedious, requiring several lines of computation. However, as in Section 7, it is possible to guess and to verify shortcut rules of finding derivatives. This will be the business of Chapter 4. To understand that chapter, it is necessary to understand the preceding definition and to recognize a derivative when it is staring you in the face. That is the point of the following exercises.

EXERCISES

8.1. Consider each of the expressions below, and show that you recognize it as a derivative.

(a) $\mathop{\text{Limit}}\limits_{h \to 0} \dfrac{f(\pi + h) - f(\pi)}{h}$.

(b) $\mathop{\text{Limit}}\limits_{h \to 0} (1/h)(f(x + h) - f(x))$.

(c) $\mathop{\text{Limit}}\limits_{h \to 0} (1/h)(g(1 + h) - g(1))$.

(d) $\mathop{\text{Limit}}\limits_{h \to 0} (1/h)(F(s + h) - F(s))$.

Partial answer: The expression (a) is equal to $f'(\pi)$, and (d) is equal to $F'(s)$.

8.2. Let $f(x) = x^3$, $F(x) = \sqrt{x}$, $g(x) = 1/x$. Which of the following is equal to $f'(x)$? to $F'(x)$? to $g'(x)$?

$\mathop{\text{Limit}}\limits_{h \to 0} \dfrac{\sqrt{x + h} - \sqrt{x}}{h}$, $\mathop{\text{Limit}}\limits_{h \to 0} \dfrac{1}{h}\left(\dfrac{1}{x + h} - \dfrac{1}{x}\right)$, $\mathop{\text{Limit}}\limits_{h \to 0} \dfrac{(x + h)^3 - x^3}{h}$.

8.3. (*For more ambitious students*) Evaluate each of the limits in exercise 8.2. (Answers may be found in Chapter 4.)

8.4. (*For more ambitious students*) Consider the function f given by $f(L) = 7L + (48/L)$.
  (a) What is $f(L + h)$?
  (b) Simplify the expression $f(L + h) - f(L)$, as much as possible, by combining fractions with the use of a common denominator.
  (c) Divide your answer to (b) by h, where $h \neq 0$. *Answer*: $7 - (48/L(L + h))$.
  (d) Find $f'(L)$, by taking the limit of your answer to (c), as h tends to 0.
  (e) Solve Example 1 of Chapter 1.

§9. Classifying Quadratics: the Quadratic Formula

The reader has probably heard of the *quadratic formula*, which is the answer to question (9) below. This formula was known long before calculus was developed, but our study in this chapter of the calculus of quadratics may cast a new light upon it. We have seen that the quadratic function given by $f(x) = ax^2 + bx + c$ has a horizontal tangent line at the point

$$\left(\dfrac{-b}{2a}, \dfrac{-b^2 + 4ac}{4a}\right).$$

9. Classifying Quadratics: the Quadratic Formula

This followed, as in exercise 7.3(e), from setting the derivative $f'(x)$ equal to zero and solving for x. The graph of the quadratic f might look like this:

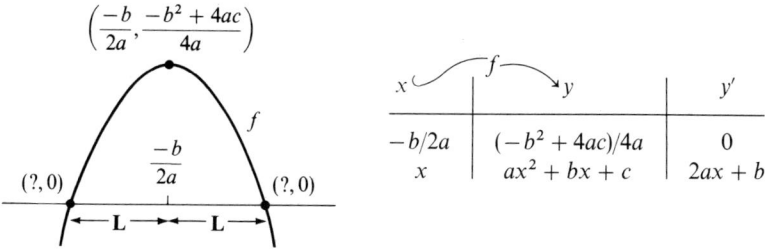

For any quadratic f we have thus answered the question, *when is $f'(x)$ equal to zero?* (*Answer*: At the "critical point" $-b/2a$.) We now ask a different question.

x	y
?	0
?	0
x	$ax^2 + bx + c$

When is $f(x)$ equal to zero? (9)

The clue to answering question (9) lies in the apparent symmetry of the curve above. We are inclined to guess that the question has *two answers*, each lying the same distance **L** from the critical point $-b/2a$. There ought to be, then, some number **L** such that

$$f(x) = 0 \quad \text{when } x = \begin{cases} \dfrac{-b}{2a} + \mathbf{L}, \\ \dfrac{-b}{2a} - \mathbf{L}. \end{cases}$$

All that remains is to find this number **L**. Since $f((-b/2a) + \mathbf{L}) = 0$, we have

$$a\left(\frac{-b}{2a} + \mathbf{L}\right)^2 + b\left(\frac{-b}{2a} + \mathbf{L}\right) + c = 0. \tag{10}$$

In (10), when the first term is squared out, a cancellation results (the reader is asked to perform the calculations), and eventually we get

$$4a^2\mathbf{L}^2 = b^2 - 4ac. \tag{11}$$

Equation (11) bears some scrutiny. We are trying to find **L**, with a, b, and c being given. Note that the left-hand side of (11) is a square, since

$4a^2L^2 = (2aL)^2$, and therefore cannot be negative. If $b^2 - 4ac$ should be negative, then there is no number **L** satisfying (11). On the other hand, if $b^2 - 4ac$ is nonnegative, we can take its square root to solve for **L**. From (11) there are two possible paths:

Case I. If $b^2 - 4ac < 0$, then there is no number **L** satisfying equation (11), and hence there is no number **L** satisfying (10).

Case II. If $b^2 - 4ac \geq 0$, then by taking square roots we get

$$2a\mathbf{L} = \sqrt{b^2 - 4ac},$$

$$\mathbf{L} = \frac{\sqrt{b^2 - 4ac}}{2a}.$$

We now know a formula for **L** and we know that the answer to question (9) is given by $x = (-b/2a) \pm \mathbf{L}$. Putting these facts together yields the *quadratic formula* in the theorem below.

Theorem on Quadratics. *The equation*

$$ax^2 + bx + c = 0 \qquad (a \neq 0)$$

has solutions given by

$$x = \frac{-b \pm \sqrt{b^2 - 4ac}}{2a} \qquad \text{(quadratic formula)}$$

provided that the discriminant $b^2 - 4ac$ *is not negative. The equation has no solution if the discriminant is negative.*

For example, consider the equation

$$-16x^2 - 50x + 200 = 0.$$

The discriminant here is $(-50)^2 - 4(-16)(200) = 15{,}300$, whose square root is approximately 123.7. The quadratic formula says the solutions are

$$x = \frac{-(-50) \pm \sqrt{15{,}300}}{-32} \approx \begin{cases} -5.43. \\ 2.30. \end{cases} \qquad \square$$

As another example, consider the equation

$$-16x^2 + 50x - 200 = 0.$$

The discriminant here is $(50)^2 - 4(-16)(-200) = -10{,}300$, which is negative, showing that the equation has no solutions. $\qquad \square$

If the discriminant is equal to zero then the "two" solutions meld into one. In this case, **L** is zero, making the critical point $-b/2a$ into a "double root" of the quadratic equation. This happens in the equation

$$x^2 - 4x + 4 = 0,$$

9. Classifying Quadratics: the Quadratic Formula

where the discriminant is $(-4)^2 - 4(1)(4) = 0$, and the quadratic formula yields

$$x = \frac{-(-4) \pm 0}{2} = 2$$

as the only solution. □

There is nothing mysterious going on here. A little reflection shows that any quadratic function falls into one of the following six classifications.

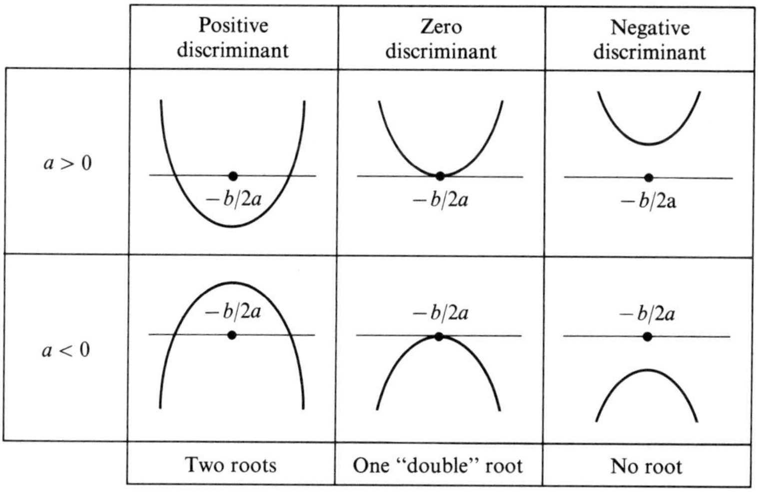

	Positive discriminant	Zero discriminant	Negative discriminant
$a > 0$	⌣ with two x-intercepts, vertex at $-b/2a$	⌣ tangent to x-axis at $-b/2a$	⌣ above x-axis, $-b/2a$
$a < 0$	⌢ with two x-intercepts, vertex at $-b/2a$	⌢ tangent to x-axis at $-b/2a$	⌢ below x-axis, $-b/2a$
	Two roots	One "double" root	No root

EXERCISES

9.1. Solve the equation $x^2 - 4x + 4 = 0$ by factoring.

9.2. Solve the equation $x^2 - x - 6 = 0$
 (a) by the quadratic formula.
 (b) by factoring $x^2 - x - 6$ into the product of $x + 2$ and $x - 3$.

9.3. Solve the equation $x^2 - x - 4 = 0$.

9.4. Factor the quadratic $x^2 - x - 4$ into the product of two linear expressions. *Answer*: $x^2 - x - 4 = (x - \frac{1}{2}(1 + \sqrt{17}))(x - \frac{1}{2}(1 - \sqrt{17}))$.

9.5. Solve the equation $x^2 - 6x + 13 = 0$.

9.6. Show all the steps of an algebraic derivation of equation (11), beginning with equation (10).

9.7. For each of the six categories of quadratics pictured above, give an example.

9.8. Consider once again Example 1 of Chapter 1. Show that it is impossible to build the fence described there for a cost C of $35. *Hint*. If $C = 35$, then $7L + (48/L) = 35$, so upon multiplying through by L we get

$$7L^2 + 48 = 35L,$$
$$7L^2 - 35L + 48 = 0.$$

Use the theorem on quadratics to show that, no matter what the length L of the front fence is, this equation cannot be satisfied.

9.9. What is the least positive value of C for which the equation $7L^2 - CL + 48 = 0$ has a solution?

§10 Three Frenchmen

The influence of France was increasingly felt throughout Europe in the seventeenth century. This influence was particularly strong in mathematics. France nurtured no fewer than three mathematical minds of the first rank, in addition to many lesser lights.

Blaise Pascal (1623–1662), who at the age of eighteen invented the first calculating machine, might have been unsurpassed as a mathematician, had his other great talents not drawn him elsewhere. Even so, he helped give birth to projective geometry and to the theory of probability, and he came very close to discovering the fundamental theorem of calculus (to be discussed in Chapter 6). In fact, Leibniz hit upon the fundamental theorem while reading a mathematics paper by Pascal.

Little need be said here of René Descartes (1596–1650), for half the world already knows his name. We have noted earlier that he developed analytic geometry and made it widely known through his writings. Without analytic geometry the step up to the calculus would be formidable indeed. Isaac Newton was to say, "If I have seen further than Descartes, it is by standing on the shoulders of giants." One of those giants was, of course, Descartes himself.

Another giant was Pierre de Fermat (1601–1665). Fermat occupies a special place in the hearts of those who love mathematics. His appeal is that of the *amateur* who can outdo the professionals. Fermat developed analytic geometry in 1629, but did not publicize the fact, and Descartes got all the credit with a paper published in 1637. In correspondence with Pascal, Fermat was an equal partner in creating the theory of probability. He corrected mistakes that Descartes and Pascal made, in fields where they were acknowledged as masters, and was rarely himself in error. Fermat's real love was the theory of numbers, which was revolutionized by his accomplishments.

Unlike Descartes and Pascal, Fermat was restrained in expressing himself, and his work is known mainly through his letters and through the notes he was accustomed to make in the margins of books. The "true inventor of differential calculus" was this quiet man of Toulouse.

Problem Set for Chapter 3

1. Use Fermat's method (not a shortcut rule) to show that the derivative of $x^2 - 6x + 13$ is given by $2x - 6$.

Problem Set for Chapter 3

2. Consider the function defined by $f(x) = x^2 - 6x + 13$.
   (a) Fill in the question marks in the following table.

x	y	y'
0	?	?
?	?	-4
5	?	?
π	?	?
?	?	0

   (b) Use the first line of the table to find the slope of the tangent line to the curve f at the point $(0, 13)$.
   (c) Is the curve f rising or falling as it passes through the point $(0, 13)$?
   (d) Write an equation of the tangent line to the curve f at the point $(0, 13)$.
   (e) For what values of x is $f'(x)$ positive?
   (f) For what values of x is the curve f rising?
   (g) For what values of x is the curve f falling?
   (h) Find both coordinates of the point where the tangent line to the curve f is horizontal.
   (i) Sketch the curve f in the x-y plane, making sure your sketch is in accordance with your answers to the preceding three questions.
   (j) If the domain of f is taken to be all values of x satisfying the inequality $0 \le x \le 5$, what is the range?
   (k) What is the range of f if the domain is given by the inequality $0 < x < 5$?
   (l) What is the range if the domain is $0 \le x < 2$?
   (m) What is the range if the domain is unrestricted?

3. Consider the function defined by $f(x) = 2x^2 - 6x$.
   (a) What is the slope of the tangent line to the curve f at the point $(0, 0)$?
   (b) Is the curve f rising or falling at $(0, 0)$?
   (c) For what values of x is the curve f falling?
   (d) Sketch the curve f.
   (e) What is the range of f if the domain is $0 \le x \le 4$?
   (f) What is the range of f if the domain is $0 < x < 1$?
   (g) What is the range of f if the domain is unrestricted?

4. Consider the function defined by $f(x) = 8 - x^2 + 3x$.
   (a) What is the range if the domain is $0 \le x < 2$?
   (b) What is the range if the domain is $-2 < x < 0$?

5. Consider the function defined by $f(x) = 8 - 3x$.
   (a) What is the range if the domain is $0 \le x \le 3$?
   (b) What is the range if the domain is $-2 < x < 5$?

6. Find both coordinates of the point on the line $y = 2x - 3$ that is closest to $(0, 0)$.

7. Find both coordinates of the point on the line $y = 6 - x$ that is closest to $(-2, -4)$.

8. Express the number 20 as the sum of two numbers in such a way that the sum of the second and the square of the first is as small as possible.

9. A *Norman window* is in the shape of a rectangle surmounted by a semicircle. If the perimeter of the window is 16 feet, find the dimensions which allow the most light to pass through the window.

10. Although the point $(2, -1)$ is not on the quadratic $y = x^2 - 2x + 3$, there are two tangent lines to this quadratic that pass through this point. Find an equation of either one of these lines.

11. A wire 500 centimeters long is cut in two. The first part is bent into the circumference of a circle, and the second is bent into the perimeter of a square. How should the wire be cut in order that the combined areas of the circle and the square be as small as possible?

12. Use Fermat's method to show that if $f(x) = 7x - 9$, then $f'(x) = 7$.

13. Use Fermat's method to show that if $f(x) = 7$, then $f'(x) = 0$.

14. Consider the quadratic function given by $f(x) = x^2 + 2x + 7$. When $x = 0$, then $f(x) = 7$. What is $f'(x)$ when $x = 0$? Does your answer contradict the result of problem 13? Explain.

15. Reread problem 16 in the problem set at the end of Chapter 1. Find the maximal area that can be enclosed.

16. Carry out the following steps in order to accomplish the last step.
    (a) The points $(1, 1)$ and $(1 + h, (1 + h)^3)$ lie on the curve $y = x^3$. Find the slope of the line joining these points (assuming, of course, that $h \neq 0$).
    (b) Simplify your answer to part (a) by using the fact that $(1 + h)^3 = 1 + 3h + 3h^2 + h^3$.

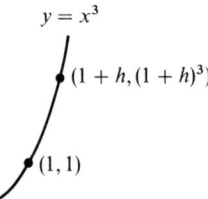

    (c) Take the limit, as $h \to 0$, of your answer to part (b), and thus show that the slope of the tangent line to the curve $y = x^3$ at $(1, 1)$ is 3.

17. (*This problem is supposed to show why the "Sherlock Holmes method" of finding tangent lines will not always work. Actually, Descartes proposed a closely related method, but it had to be discarded in favor of Fermat's approach.*) The Sherlock Holmes method of finding tangents rests upon the belief that a line joining two points on a curve cannot be tangent to the curve. (This happens to be true for quadratic curves.)
    (a) Using the result of problem 16(c), write an equation of the tangent line at $(1, 1)$ to the curve $y = x^3$.
    (b) Does the point $(-2, -8)$ lie on the line of part (a)?
    (c) Does the point $(-2, -8)$ lie on the curve $y = x^3$?

Problem Set for Chapter 3

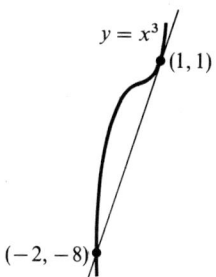

(d) If the Sherlock Holmes method were applied to the curve $y = x^3$ at the point $(1, 1)$, would it work?

18. Consider each of the following functions, and match it with its derivative. [The derivative of the function pictured in (a), for example, is pictured in (d).]

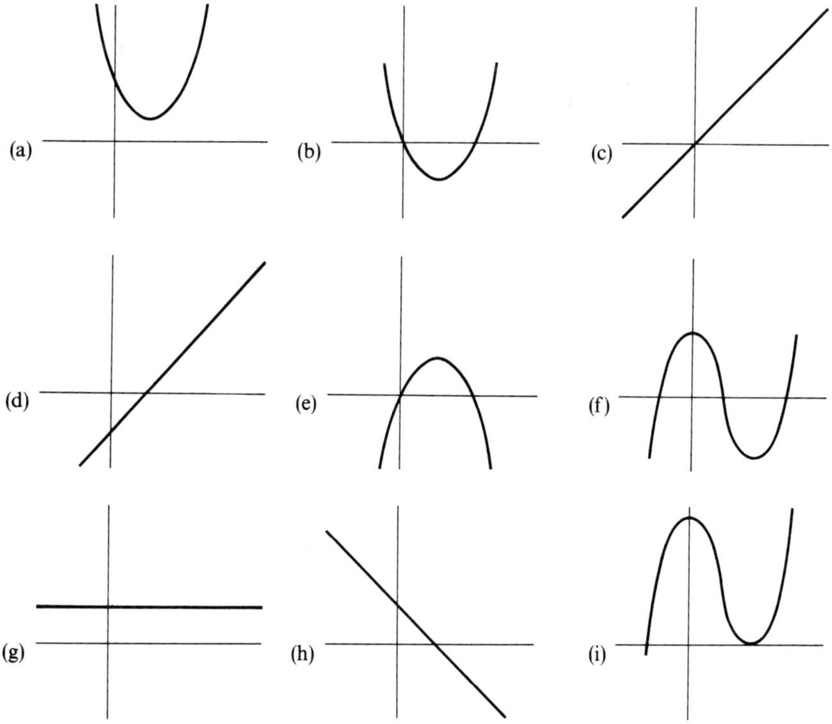

(a) (b) (c)

(d) (e) (f)

(g) (h) (i)

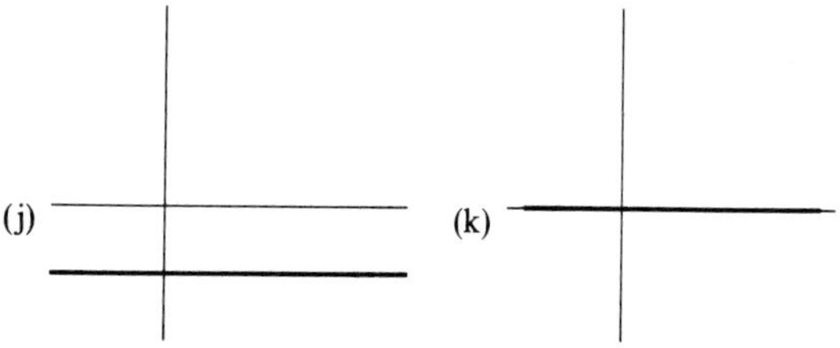

(j) (k) [The curve in (k) coincides with the horizontal axis.]

4 Optimistic Steps

What is calculus? It is the study of the interplay between a function and its derivative. There are quite a few aspects to this interplay, some of which may be surprising. In this chapter we shall learn more about the use of derivatives in solving optimization problems. To do this efficiently, the major part of the chapter is concentrated upon the development of shortcut rules for finding derivatives.

§1. The Derivative of the Reciprocal Function

If f is a function, then $f'(x)$ is defined as the limit of the *difference quotient*

$$\frac{f(x+h) - f(x)}{h} \tag{1}$$

as h tends to zero. (*Do you see why expression (1) is a quotient of differences?*) In order to find this limit, it is often necessary to use a little algebra to write the difference quotient in a simple way.

Before proceeding, let us review very briefly the algebra of simplifying fractions by combining them with the use of a common denominator. For instance,

$$\frac{1}{5} - \frac{1}{4} = \frac{4}{5 \cdot 4} - \frac{5}{5 \cdot 4} = \frac{4-5}{5 \cdot 4} = \frac{-1}{20},$$

$$\frac{1}{\pi+2} - \frac{1}{\pi} = \frac{\pi}{(\pi+2)\pi} - \frac{\pi+2}{(\pi+2)\pi} = \frac{\pi - (\pi+2)}{(\pi+2)\pi} = \frac{-2}{(\pi+2)\pi},$$

1. The Derivative of the Reciprocal Function

and, by the same token,

$$\frac{1}{x+h} - \frac{1}{x} = \frac{x}{(x+h)x} - \frac{x+h}{(x+h)x} = \frac{x-(x+h)}{(x+h)x} = \frac{-h}{(x+h)x}. \qquad (2)$$

Whenever the rule for f involves division, the difference quotient (1) is often filled with fractions that need to be combined by using a common denominator.

EXAMPLE 1. Find $f'(x)$, if $f(x) = 1/x$.

Here, the algebraic rule $1/x$ involves division (and, of course, is undefined when x is 0). Let us first use a common denominator to simplify the expression

$$f(x+h) - f(x)$$

before dividing by h and taking the limit. Since $f(x) = 1/x$ and $f(x+h) = 1/(x+h)$, equation (2) shows that

$$f(x+h) - f(x) = \frac{-h}{(x+h)x}.$$

Dividing by nonzero h yields

$$\frac{f(x+h) - f(x)}{h} = \frac{-1}{(x+h)x}.$$

Using this simplified expression for the difference quotient makes it easy to take the limit:

$$f'(x) = \lim_{h \to 0} \frac{f(x+h) - f(x)}{h}$$

$$= \lim_{h \to 0} \frac{-1}{(x+h)x}$$

$$= \frac{-1}{(x+0)x}$$

$$= \frac{-1}{x^2}.$$

The derivative of $1/x$ is therefore $-1/x^2$. The expression $1/x$ is called the *reciprocal* of x. We now know the derivative of the reciprocal function.

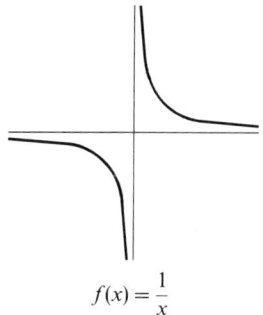

$$f(x) = \frac{1}{x}$$

The domain of f does not include 0. Note that f' is always negative (because the curve f is always falling).

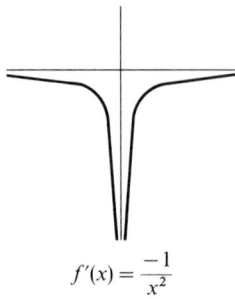

$$f'(x) = \frac{-1}{x^2}$$

EXERCISES

1.1. If $f(t) = 1/t$, what is $f'(t)$? *Answer:* $f'(t) = -1/t^2$, *because f is simply the reciprocal function. Remember that changing only the name of the variable does not alter the function.*

1.2. If $f(L) = 1/L$, what is $f'(L)$?

1.3. If $A = 1/s$, what is A'?

1.4. In each of the following, first make a guess as to the expression giving the derivative. Then verify your guess by Fermat's method.
   (a) $y = 48/L$.
   (b) $C = 7L + (48/L)$.
   (c) $y = 5/(x - 7)$.
   (d) $y = 5/(7 - x)$.

1.5. Consider the function given by $f(x) = 1/(\pi x + 7)$. Carry out the following steps to find $f'(x)$.
   (a) Simplify the expression $f(x + h) - f(x)$ by using a common denominator.
   (b) Divide your answer to (a) by nonzero h.
   (c) Find the limit, as h tends to zero, of your answer to (b). *Answer:* $f'(x) = -\pi/(\pi x + 7)^2$.

1.6. Suppose $f(x) = 1/g(x)$, where g is some given function. [For instance, exercise 1.5 dealt with the case where $g(x) = \pi x + 7$.] Since f is expressed in terms of g, the difference quotient of f can be expressed in terms of g as well. Show that the difference quotient of f can be expressed by

$$\frac{f(x+h) - f(x)}{h} = \frac{-1}{g(x+h)g(x)} \cdot \frac{g(x+h) - g(x)}{h}.$$

Hint. Begin by writing

$$f(x+h) - f(x) = \frac{1}{g(x+h)} - \frac{1}{g(x)},$$

and combine the fractions on the right. Then divide by h.

§2. General Rules for Reciprocals and for Constant Multiples

Now that we know the derivative of $1/x$, it is natural to ask about the derivatives of similar expressions involving reciprocals. For example, what is the derivative of $1/x^2$? Or of $1/(x^2 + 2x)$? Or, more generally, what is the derivative of $1/g(x)$, where g is some given function?

To answer this question, first recall that $g'(x)$ is the limit, as h approaches zero, of the difference quotient

$$\frac{g(x+h) - g(x)}{h}. \tag{3}$$

In order that the quotient (3) tend to a limit, the numerator of the quotient must tend to 0 as h tends to 0. *Reason*: The denominator h tends to 0; if the numerator did not, then the quotient would "blow up" as h approached 0, and consequently the limit would not exist.

When the numerator in (3) tends to 0, we have

$$\text{Limit}_{h \to 0} g(x+h) = g(x). \tag{4}$$

This fact will be useful in just a moment.

We can now answer the question raised at the beginning of this section: *What is the derivative of* $1/g(x)$? The derivative of a function is the limit of its difference quotient. The difference quotient of $1/g(x)$ is simplified in exercise 1.6, showing that *the derivative of* $1/g(x)$ *is equal to*

$$\text{Limit}_{h \to 0} \frac{-1}{g(x+h)g(x)} \cdot \frac{g(x+h) - g(x)}{h} = \frac{-1}{g(x)g(x)} g'(x),$$

where (4) has been used (*how*?) in evaluating this limit. We have just proved the following rule to be valid.

General Rule for Reciprocals. *The derivative of* $1/g(x)$ *is*

$$\frac{-1}{(g(x))^2} g'(x)$$

if the function g has a derivative.

If we suppress writing the variable, this rule can be expressed in a very compact way. It says

$$\left(\frac{1}{g}\right)' = \frac{-1}{g^2} g'. \tag{5}$$

For example, *what is the derivative of* $1/(4 - 3x)$? By (5),

$$\left(\frac{1}{4-3x}\right)' = \frac{-1}{(4-3x)^2}(4-3x)' = \frac{3}{(4-3x)^2},$$

since $(4 - 3x)' = -3$. □

For another example, *what is the derivative of* $1/(\pi x + 7)$? By (5),

$$\left(\frac{1}{\pi x + 7}\right)' = \frac{-1}{(\pi x + 7)^2}(\pi x + 7)' = \frac{-\pi}{(\pi x + 7)^2},$$

which agrees with the answer to exercise 1.5. □

A rule that is much easier to prove involves multiplication by constants. We have essentially guessed this rule already, in Chapter 3, when we guessed that the derivative of $\frac{1}{2}s^2$ ought to be one-half the derivative of s^2:

$$\left(\frac{s^2}{2}\right)' = \frac{1}{2}(s^2)'$$

$$= \frac{1}{2}(2s) = s.$$

It takes little imagination to guess that there ought to be a general rule involving multiplication by constants, like $\frac{1}{2}$.

Rule for Constant Multiples. *The derivative of* $c \cdot g(x)$ *is* $c \cdot g'(x)$ *if* c *is a constant and the function* g *has a derivative.*

For example, *what is the derivative of* $100/(\pi x + 7)$? By the rule for constant multiplies,

$$\left(\frac{100}{\pi x + 7}\right)' = 100\left(\frac{1}{\pi x + 7}\right)' = \frac{-100\pi}{(\pi x + 7)^2},$$

where the second equality comes from the general reciprocal rule. □

The reader is asked to prove the rule for constant multiples, in a problem at the end of this chapter.

EXERCISES

2.1. Find y' if $y = 1/(7 - x)$.
   Answer: $y' = (1/(7 - x))' = 1/(7 - x)^2$.

2.2. Use the general rule for reciprocals to find the derivatives of the following functions:
   (a) $f(x) = 1/x^2$.
   (b) $g(x) = 1/(x^2 + 2x)$.
   (c) $F(x) = 1/(6 - 3x)$.
   (d) $G(x) = 1/(2x^2 - 3x + 4)$.
   Answer: (b) $g'(x) = (-2x - 2)/(x^2 + 2x)^2$.

2.3. Apply the general rule for reciprocals to find the derivative of $1/x$. Does your answer agree with the answer obtained in Section 1, where Fermat's method was used?

2.4. Find y' if $y = 1/t^2$. Answer: $y' = -2/t^3$.

2.5. Find C' if $C = 1/(L^2 + 4)$.

2.6. Use the general rule for reciprocals, together with the rule for constant multiples, to find the derivatives of the following functions, expressed as algebraic rules:
(a) $5/x^2$.
(b) $14/(x^2 + 2x)$.
(c) $\pi/(6 - 3x)$.
(d) $100/(2x^2 - 3x + 4)$.
(e) $5/(4 - 3s)$.
(f) $48/L$.
(g) $-1/t^2$.
(h) $6/(\pi - \theta)$.

§3. The Sum Rule and the Second Derivative

There is an easy rule involving the sum of two functions. We have essentially guessed this rule already, in Chapter 3, when we guessed that the derivative of $ax^2 + bx + c$ ought to be equal to the sum of the derivatives of ax^2 and of $bx + c$:

$$(ax^2 + bx + c)' = (ax^2)' + (bx + c)'$$
$$= 2ax + b.$$

One would surely suspect that this is a special case of a general rule.

Rule for Sums. *The derivative of $f(x) + g(x)$ is equal to $f'(x) + g'(x)$ if the functions f and g have derivatives.*

This rule is true, but its proof is left to the reader as a problem at the end of this chapter (problem 37).

One often has to use several rules at once.

EXAMPLE 2. Find the derivative of the function given by the algebraic expression $6x^2 + (17/(x^2 + 3x))$.

Here we have a sum, and the sum rule says $(f + g)' = f' + g'$. Therefore,

$$\left(6x^2 + \frac{17}{x^2 + 3x}\right)' = (6x^2)' + \left(\frac{17}{x^2 + 3x}\right)' \quad \text{(by sum rule)}$$

$$= 6(x^2)' + 17\left(\frac{1}{x^2 + 3x}\right)' \quad \text{(by rule for constant multiples)}$$

$$= 6(2x) + 17\frac{-1}{(x^2 + 3x)^2}(2x + 3),$$

by the rule for quadratics, together with the reciprocal rule. The answer may be simplified, if desired, to $12x - ((34x + 51)/(x^2 + 3x)^2)$. □

The calculation in Example 2 required several steps. The reader will find that, with practice, it is easy to combine these steps into one:

The derivative of $5x^2 + (6/(x^2 - 2))$ is $5(2x) + 6(-1/(x^2 - 2)^2)(2x)$.
The derivative of $7L + (48/L)$ is $7 + 48(-1/L^2)(1)$.
The derivative of $(10/x) - (45/x^2) - 5x^2 + x - \pi$ is $10(-1/x^2)(1) - 45(-1/x^4)(2x) - 5(2x) + 1$. □

Having taken one derivative, we have nothing preventing us from taking a second derivative. The **second derivative** (the derivative of f') is denoted by f''. We now have $y = f(x)$, $y' = f'(x)$, and $y'' = f''(x)$.

EXAMPLE 3. Find the first and second derivatives of the function given by

$$f(x) = 2x^2 + \frac{3}{x} - 4.$$

Here, the first derivative is given by

$$f'(x) = (2x^2 + \frac{3}{x} - 4)'$$

$$= 4x - \frac{3}{x^2},$$

and the second derivative is given by

$$f''(x) = \left(4x - \frac{3}{x^2}\right)'$$

$$= 4 - 3\frac{-1}{x^4}(2x)$$

$$= 4 + \frac{6}{x^3}.$$ □

A function is given by a pair of columns, and its derivative adds a third column to consider. The second derivative gives us still another column to play with. For example, in the function f above, if we let x equal 1, we get

x	y	y'	y''
1	1	1	10

The first two columns tell us that the curve f goes through the point $P = (1, 1)$. The third column tells us that the tangent at P has a positive slope, so the curve f is rising as it goes through P. What does the fourth column tell us? As we shall see in the next section, the positive second derivative tells us that the curve f, as it passes through P, looks rather like a smile.

EXERCISES

3.1. For the function f of Example 3, fill in the question marks appropriately.

x	y	y'	y''
2	?	?	?
−1	?	?	?
$\frac{1}{2}$	?	?	?

3.2. Find the *second* derivative of each of the following functions.
 (a) $f(x) = 2x^2 + 3x - 5$.
 (b) $g(x) = -3x^2 + 4x - \sqrt{2}$. Answer: $g''(x) = -6$.
 (c) $f(L) = 7L + (48/L)$.
 (d) $g(s) = 600s - \frac{1}{2}s^2$.
 (e) $G(t) = t^2 - 6t + (5/(t-3))$. Answer: $G''(t) = 2 + ((10t - 30)/(t^2 - 6t + 9)^2)$.
 (f) $F(x) = (10/x) - 5x^2 + x - \pi$.

§4. The Second Derivative and Concavity

The second derivative f'' gives the same sort of information about f' as the first derivative f' gives about f. Indirectly, then, the second derivative says something about the behavior of the original function f. Let us try to find out exactly what f'' tells about f.

First we must agree on some terminology to describe how a curve is "curving". There are several terms in use for this (the phrases *concave upwards* and *concave downwards* are common descriptions), but they do not seem to be immediately suggestive of what they are intended to describe. To remedy this, let us depart from common terminology and make up our own way of describing how a curve curves.

We have already agreed, on the first page of Chapter 3, to think of a curve as being traced out from left to right. If we thought of the curve as describing a road on a road map, then the pencil point tracing out the curve moves in a generally eastward direction. Pretend that you, on your motorcycle, have been shrunk to the size of that pencil point tracing out the curve. As you journey eastward, there is a simple way you can describe how the road is curving. All that need be said is whether you are leaning to your *left* or to your *right*, in order to keep your motorcycle on the road.

The functions f and g below curve in opposite directions. Note that their second derivatives f'' and g'' have opposite signs.

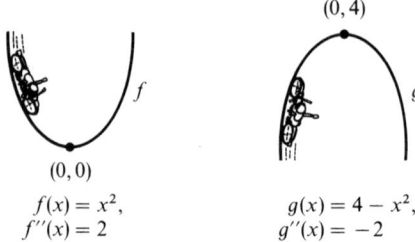

$f(x) = x^2$,
$f''(x) = 2$

$g(x) = 4 - x^2$,
$g''(x) = -2$

In the case of the curve $y = x^2$, you must always lean to your *left* to stay on the curve; you lean always to your *right* to stay on the curve $y = 4 - x^2$. The first curve has a *positive* second derivative; the second curve has a *negative* second derivative. These are simple examples, in that the second derivative is constant in both, whereas we shall see that generally the second derivative will change sign when the "road" starts to curve the other way. Nevertheless, these examples give us a clue to the truth: *When f'' is positive, the curve f is bending to the left; when f'' is negative, the curve f is bending to the right.*

Why should it be this way? Focus attention on a particular point P lying on a curve f. Then $P = (c, f(c))$ for some number c. A little reflection shows that

$$\text{if } f''(c) < 0, \quad \begin{array}{l}\text{then the curve } f \text{ is bending to} \\ \text{the right as it passes through} \\ \text{the point } P = (c, f(c)).\end{array} \qquad (6)$$

To see this, all one needs to recall is that when the derivative of a function is negative, then the values of the function are *decreasing* (because its curve is falling). Keeping this in mind, and remembering that f'' is the derivative of f', one can see the plausibility of statement (6), as follows. Suppose $f''(c) < 0$. Then the values of f' are decreasing, i.e., the slopes of the tangent lines to the curve f are decreasing, as the curve f passes through $(c, f(c))$. But decreasing slopes of the tangent lines near this point imply that the curve is bending to the right. Near P the curve f must look like one of the following if $f''(c)$ is negative:

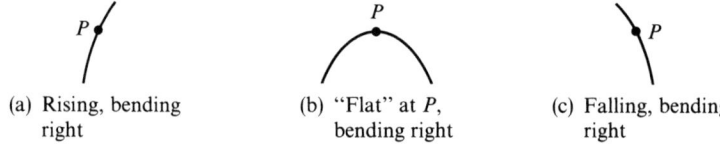

(a) Rising, bending right

(b) "Flat" at P, bending right

(c) Falling, bending right

4. The Second Derivative and Concavity

Similar reflection shows that, near P, the curve f must look like one of the following if $f''(c)$ is positive:

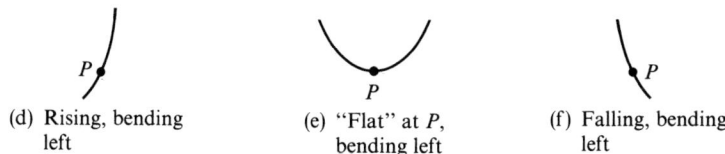

(d) Rising, bending left

(e) "Flat" at P, bending left

(f) Falling, bending left

The upshot of the preceding discussion is this. While the sign of the *first* derivative tells whether the curve is rising or falling, the sign of the *second* derivative tells whether it is bending to the left or to the right. The second derivative f'' tells which way the curve f is curving. □

When the second derivative is negative, as in the figures (a), (b), and (c) above, we have described the curve as "bending to the right". It is described as *concave down* in most books, and figures (d), (e), and (f) are described as *concave up*. The definition of concavity in these terms is given in a problem at the end of the chapter. There is little point in learning these terms, however, if you are interested in studying calculus only for a semester or so. In fact, it might be better to describe (d), (e), and (f) as "smiles", and call (a), (b), and (c) "frowns", and just remember that *a positive second derivative always draws a smile*.

Knowing both derivatives of a function at a point gives us a fairly good idea of what the curve looks like nearby. The word *local* (as opposed to *global*) is used in mathematics to describe this kind of information; it tells us what the road looks like only in a small neighborhood of a point as we roar through on our motorcycle. What adventures may lie elsewhere on the road remain to be seen.

EXAMPLE 4. Describe the local behavior of the curve $y = x^2 + (8/x)$ as the curve passes through

(a) $(1, 9)$.
(b) $(2, 8)$.
(c) $(-2, 0)$.

It is intended that we sketch the curve locally near each of these points, so as to indicate whether the curve is rising or falling, "smiling" or "frowning", as it passes through. From the first and second derivatives

$$y' = 2x - \frac{8}{x^2} \quad \text{and} \quad y'' = 2 + \frac{16}{x^3},$$

we can fill in the following table.

x	y	y'	y''
1	9	−6	18
2	8	2	4
−2	0	−6	0

It is really only the *sign* of y' and y'' that we need. From the first line of the table we see that y' is negative and y'' is positive. The curve is then falling and smiling as it goes through $(1,9)$. It must resemble the curve sketched in figure (f) above, with $P = (1,9)$.

From the second line of the table, with both derivatives positive, we see that the curve must resemble the one of figure (d), with $P = (2,8)$.

The third line of the table, with y' negative, shows the curve falling as it passes through $(-2,0)$, but how do we interpret the fact that y'' is zero? We must look at the sign of y'' for x just less than -2 and for x just greater than -2. Doing this reveals that the sign of y'' switches from positive to negative. This means that, at the point $(-2,0)$, the curve stops bending left and starts bending right (or, if you prefer, the concavity switches from *up* to *down*). *Such a point, where the curve stops bending one way and starts bending the other way, is called a* **point of inflection**.

All the information gleaned above is in this picture:

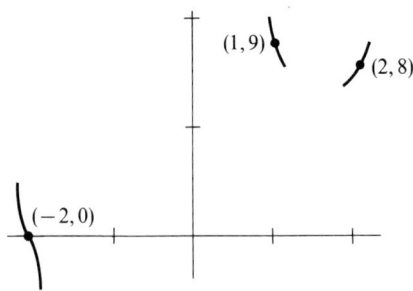

The curve $y = x^2 + (8/x)$, pictured locally in neighborhoods of three points. The point $(-2,0)$ is a point of inflection. □

At a point of inflection the second derivative must be zero (*why?*). However, the second derivative can be zero at points other than inflection points. A straight line has no inflection points, but the second derivative of a linear function is always zero.

Exercises

4.1. Consider the function given by $f(x) = x^2 - 3x + 2$. Describe the local behavior of the curve f as it passes through
  (a) $(0,2)$.
  (b) $(2,0)$.
  (c) $(1,0)$.

5. The Rule for Squares

4.2. Describe the local behavior of the curve $y = 10x - (4/x)$ near
   (a) $(1, 6)$.
   (b) $(2, 18)$.
   (c) $(-\frac{1}{2}, 3)$.
   Partial answer: The curve looks like figure (a), page 86, with $P = (1, 6)$.

4.3. Describe the local behavior of the curve $C = 7L + (48/L)$ near
   (a) $(2, 38)$.
   (b) $(2.5, 36.70)$.
   (c) $(3, 37)$.

4.4. Describe the local behavior of the curve $y = x^2 + (8/x)$ near
   (a) $(-1, -7)$.
   (b) $(\frac{4}{3}, 7\frac{7}{9})$.
   (*This is the curve of Example 4.*)

4.5. From the meager sketch in Example 4 above, we see that the curve appears to have a local minimum between $x = 1$ and $x = 2$. Find the x-coordinate of this local minimum. Is this a global minimum as well, i.e., is this the lowest point on the entire curve?

4.6. Does any linear function have an inflection point? Does any quadratic function have an inflection point?

4.7. Find both coordinates of an inflection point on the curve $y = x^2 + (1/x)$.

4.8. Suppose you have a function f and a point c where $f'(c) = 0$ and $f''(c)$ is positive. Have you found a local *minimum* or a local *maximum* of f?

§5. The Rule for Squares

There remain four shortcut rules to be discussed in this chapter. We need to find out how to take derivatives of *squares*, of *square roots*, of *products*, and of *quotients*. One might suspect that we shall therefore have to go through all the details of Fermat's method four more times. Fortunately, things can be arranged so that we have to do Fermat's method only once more, to find a rule for squares. The only algebraic trick we shall need is a simple one. The difference of two squares factors into the product of their sum and difference:

$$a^2 - b^2 = (a+b)(a-b).$$

Suppose we are presented with a function whose rule involves a square. For example, suppose we have a rule $f(x)$ given by $(x^2 + 3)^2$, or by $(5x - (7/x))^2$, or, more generally, by $(g(x))^2$, where g is a function whose derivative we know. By Fermat's method, if $f(x) = (g(x))^2$, then

$$f'(x) = \operatorname*{Limit}_{h \to 0} \frac{f(x+h) - f(x)}{h}$$

$$= \operatorname*{Limit}_{h \to 0} \frac{(g(x+h))^2 - (g(x))^2}{h}.$$

How can this difference quotient be simplified, in order to find its limit? *Answer*: Since the numerator is the difference of two squares, we can factor it into the product of their sum and difference, to get

$$f'(x) = \lim_{h \to 0} \frac{(g(x+h) + g(x))(g(x+h) - g(x))}{h}$$

$$= \lim_{h \to 0} (g(x+h) + g(x))\left(\frac{g(x+h) - g(x)}{h}\right)$$

$$= (g(x) + g(x))g'(x) \quad \text{[by equation (4)]}$$

$$= 2g(x)g'(x),$$

provided, of course, that g has a derivative. We have our rule.

Rule for Squares. *The derivative of $(g(x))^2$ is $2g(x)g'(x)$ if the function g has a derivative.*

By this rule, for example,

the derivative of $(x^2 + 3)^2$ is $2(x^2 + 3)(2x)$,
the derivative of $(5x - (7/x))^2$ is $2(5x - (7/x))(5 + (7/x^2))$.

EXERCISES

5.1. Find the derivative of $(3x + 5)^2$ by using the rule for squares. *Answer*: Here we have $(g(x))^2$, where $g(x) = 3x + 5$. By the rule for squares, its derivative is $2gg'$, which is $2(3x + 5)(3)$, or $18x + 30$.
[*The reader should check that this is the same answer one obtains by first writing $(3x + 5)^2$ as $9x^2 + 30x + 25$, and then taking the derivative by the quadratic rule.*]

5.2. Find the derivative of $(6 - 7x)^2$
(a) by applying the rule for squares.
(b) by first squaring the expression $6 - 7x$ and then applying the rule for quadratics.

5.3. Find the derivative of each of the following functions, expressed as algebraic rules.
(a) $(x^2 - 5x)^2$.
(b) $(7L + (48/L))^2$.
(c) $(5x^2 + (6/(x^2 - 2)))^2$.
(d) $(3x^2 - 5x + \sqrt{2})^2$.
(e) $(2x^2 + (3/x) - 4)^2$.
Answer: (b) $2(7L + (48/L))(7 - (48/L^2))$.

5.4. True or false? *The derivative of a square is equal to the square of the derivative.*

5.5. Find the derivative of x^4 by regarding x^4 as $(x^2)^2$ and using the rule for squares. *Answer*: $4x^3$.

5.6. Find the derivative of $(1/x)^2$
(a) by applying the rule for squares.
(b) by first writing $(1/x)^2 = 1/x^2$ and then using the general reciprocal rule.

5.7. What is the derivative of
 (a) $(x^2 + 5x)^2$?
 (b) $(f(x) + g(x))^2$? *Answer:* $2(f(x) + g(x))(f'(x) + g'(x))$.

§6. The Product Rule and the Square Root Rule

It is *not true* that the derivative of a product is the product of the derivatives. The product rule is a little more complicated than that. It is easy to derive the product rule, though, because *a product can always be expressed in terms of squares*, and we already know the rule for squares.

To see the relationship between products and squares, begin with the familiar identity
$$(a + b)^2 = a^2 + 2ab + b^2,$$
and "solve" this equation for the product ab. You get
$$ab = \frac{1}{2}((a + b)^2 - a^2 - b^2),$$
which expresses the product ab in terms of squares. The same thing holds, of course, for functions. If f and g are functions, then their product can be written as
$$fg = \frac{1}{2}((f + g)^2 - f^2 - g^2).$$
Taking derivatives by using the rule for squares, and also using the result of exercise 5.7, we find that
$$(fg)' = \frac{1}{2}[2(f + g)(f' + g') - 2ff' - 2gg']$$
$$= (f + g)(f' + g') - ff' - gg'$$
$$= ff' + fg' + gf' + gg' - ff' - gg'$$
$$= fg' + gf'.$$

Rule for Products. *The derivative of $f(x)g(x)$ is $f(x)g'(x) + g(x)f'(x)$, provided that f and g have derivatives.*

The reader may find it easier to remember the product rule by reading it in words: *The derivative of a product is equal to the first term times the derivative of the second, plus the "other way around".*

As an example, let us find the derivative of the product $(x + 2)(x - 3)$. By the product rule, it is
$$(x + 2)(x - 3)' + (x - 3)(x + 2)'$$
$$= (x + 2)(1) + (x - 3)(1)$$
$$= 2x - 1.$$

In this example we can "check" our answer by noting that the product $(x + 2)(x - 3)$ is equal to the quadratic expression $x^2 - x - 6$, whose derivative is indeed $2x - 1$. □

Here are some more examples, with the answers left in an unsimplified form.

The derivative of $(x^2 + x)(5x - 2)$ is equal to
$$(x^2 + x)(5) + (5x - 2)(2x + 1).$$
The derivative of $(7L + (48/L))(L^2 - \pi)$ is equal to
$$\left(7L + \frac{48}{L}\right)(2L) + (L^2 - \pi)\left(7 - \frac{48}{L^2}\right). \quad \square$$

Now, what about square roots? How do we get the derivative of f, if the function f is given by $f(x) = \sqrt{1 + x^2}$? Or by $\sqrt{2x - 3}$? Or, more generally, by $\sqrt{g(x)}$, where g is some function whose derivative we already know? We can guess the answer to this question by using the rule for squares: If $f = \sqrt{g}$, then (by squaring both sides) we have

$$f^2 = g,$$
$$2ff' = g' \quad \text{(by rule for squares)},$$
$$f' = \frac{1}{2f}g' \quad \text{(solving for } f'\text{)},$$
$$f' = \frac{1}{2\sqrt{g}}g' \quad (\text{since } f = \sqrt{g}).$$

Square Root Rule. *The derivative of* $\sqrt{g(x)}$ *is* $\dfrac{1}{2\sqrt{g(x)}}g'(x)$, *provided the function g has a derivative.*

The application of this rule is quite straightforward.
The derivative of $\sqrt{1 + x^2}$ is
$$\frac{1}{2\sqrt{1 + x^2}}(2x) = \frac{x}{\sqrt{1 + x^2}}.$$
The derivative of $\sqrt{2x - 3}$ is
$$\frac{1}{2\sqrt{2x - 3}}(2) = \frac{1}{\sqrt{2x - 3}}.$$
The derivative of $\sqrt{x}$ is
$$\frac{1}{2\sqrt{x}}(1) = \frac{1}{2\sqrt{x}}. \quad \square$$

6. The Product Rule and the Square Root Rule

Exercises

6.1. Find the derivative of x^3, by regarding x^3 as the product of x^2 and x.
Answer: $(x^3)' = (x^2 x)' = x^2(1) + x(2x) = 3x^2$.

6.2. Find the derivative of x^4 by regarding x^4 as the product of x^3 and x. Does your answer agree with exercise 5.5?

6.3. Find the derivative of each of the following.
 (a) x^5.
 (b) x^6.
 (c) x^7.
 (d) x^n, where n is a positive integer.

6.4. *Any quotient can be expressed as a product.* Find the derivative of $x/(x + 3)$, by regarding this quotient as the product of x and $1/(x + 3)$.

Answer:
$$\left(\frac{x}{x+3}\right)' = \left(x \cdot \frac{1}{x+3}\right)'$$

$$= x \frac{-1}{(x+3)^2} + \frac{1}{x+3} (1)$$

$$= \frac{-x}{(x+3)^2} + \frac{1}{x+3}.$$

6.5. Find the derivative of $x^3/(5x + 1)$ by regarding this quotient as the product of x^3 and $1/(5x + 1)$.

6.6. Use the square root rule to find the derivatives of the following.
 (a) $\sqrt{9 + x^2}$.
 (b) $17\sqrt{3x^2 - 2x}$.
 (c) $\sqrt{\pi x}$.
 (d) $\sqrt{2}$.

6.7. Find the derivative of $x^4\sqrt{1 + x}$.

Answer:
$$(x^4\sqrt{1+x})' = x^4(\sqrt{1+x})' + (\sqrt{1+x})(x^4)'$$

$$= x^4 \left(\frac{1}{2\sqrt{1+x}}\right) + (\sqrt{1+x})(4x^3)$$

$$= \frac{x^4}{2\sqrt{1+x}} + 4x^3\sqrt{1+x}.$$

6.8. Find the first derivatives of the following.
 (a) $x^2\sqrt{1 + x^2}$.
 (b) $x\sqrt{x}$.
 (c) $x^3\sqrt{2x - 3}$.
 (d) $x^6\sqrt{3x^2 - 2x}$.

6.9. Find the second derivative y'' if $y = \sqrt{2x + 5}$.

§7. The Quotient Rule

Exercises 6.4 and 6.5 give the clue to finding a rule for obtaining the derivative of a quotient f/g: Regard it as the product of $1/g$ and f. Then we have

$$\left(\frac{f}{g}\right)' = \left(\frac{1}{g}f\right)'$$

$$= \left(\frac{1}{g}\right)f' + f\left(\frac{1}{g}\right)'$$

$$= \left(\frac{g}{g^2}\right)f' + f\left(\frac{-1}{g^2}\right)g'$$

$$= \frac{gf' - fg'}{g^2}.$$

Rule for Quotients. *The derivative of $f(x)/g(x)$ is*

$$\frac{g(x)f'(x) - f(x)g'(x)}{(g(x))^2}$$

if the functions f and g have derivatives.

The reader may find it easier to remember the quotient rule by reading it in words: *The derivative of a quotient is equal to the bottom times the derivative of the top,* minus *the other way around, over the bottom squared.*

For example,

the derivative of $x/(x+3)$ is

$$\frac{(x+3)(1) - (x)(1)}{(x+3)^2},$$

the derivative of $x^3/(5x+1)$ is

$$\frac{(5x+1)(3x^2) - (x^3)(5)}{(5x+1)^2}.$$

(The reader should check that these answers, with the help of a little algebra, may be seen to agree with the answers to exercises 6.4 and 6.5.) □

EXERCISES

7.1. If $y = 2x/(x^2 - 3)$, what is y'? *Answer:* $y' = (-2x^2 - 6)/(x^2 - 3)^2$.

7.2. If $y = 2x/(x^2 - 3)$, what is y''? *Hint.* Use the quotient rule to find the derivative of the answer to exercise 7.1. In the course of doing this, the rule for squares will come in handy in finding the derivative of the bottom. Don't take time to simplify your answer.

7.3. Describe the local behavior of the curve $y = 2x/(x^2 - 3)$ near the point $(2, 4)$ and near $(1, -1)$, and find both coordinates of an inflection point.

7.4. Find the first derivatives of the following. Do not simplify.
   (a) $(x + 2)/(x - 2)$.
   (b) $(x^2 - 3)/(x - 2)^2$.
   (c) $x^3/\sqrt{1 + x^2}$.
   (d) $\sqrt{(x + 2)/(x - 2)}$.
   (e) $((3x - 1)/x^2)^2$. Answer: $2((3x - 1)/x^2)((x^2(3) - (3x - 1)(2x))/x^4)$.

7.5. Consider the function given by $f(x) = 5x^2 + (6/(x^2 - 2))$. What does the curve f look like, locally, near the point $(0, -3)$? Near $(2, 23)$?

7.6. What does the curve $y = \sqrt{16 + x^2}$ look like, locally, as it passes through $(0, 4)$?

§8. Solving Optimization Problems

Where are we now? We have just completed an unavoidable digression from our original theme, which was the solution of optimization problems. As we saw in Chapter 1, an optimization problem leads to the problem of finding the highest (or lowest) point on a certain curve. This, in turn, has led to the study of derivatives, because derivatives cast light on the behavior of a curve. And now, at last, we know how to bypass Fermat's method and use the following rules instead.

(1) $(cf)' = c \cdot f'$ (constant multiples).
(2) $(f + g)' = f' + g'$ (sums).
(3) $(1/g)' = (-1/g^2)g'$ (reciprocals).
(4) $(g^2)' = 2gg'$ (squares).
(5) $(\sqrt{g})' = (1/2\sqrt{g})g'$ (square roots).
(6) $(fg)' = fg' + gf'$ (products).
(7) $(f/g)' = (gf' - fg')/g^2$ (quotients).

The reader should practice using these rules until they have been memorized. Then the taking of derivatives will be quite a routine matter, and the most important step in solving an optimization problem will have been mastered.

We can finally come to grips with the topic to which the title of this chapter alludes. What are the steps leading to the solution of an optimization problem? Basically, there are just two steps. First, translate the problem into the geometric problem of finding the highest (or lowest) point on a certain curve f; and second, find f' and use it as an aid in understanding how the curve f behaves.

The critical points to be found in sketching a curve f are those where the tangent line to the curve is horizontal. [That leads to a definition: *To say that x is a **critical point** of f is to say that* $f'(x) = 0$.] Usually, although not always, the function f will attain its optimal value at a critical point.

To verify whether the optimum has been found, make a rough sketch of the curve near each critical point (the second derivative is helpful here) and near each endpoint of the domain.

As we have seen, some curves do not have a highest (or lowest) point. It can be proved, however, that a curve must have such points if it comes from a *continuous* function and if the domain is an interval *containing its endpoints*. This is a deep theorem of analysis, the modern branch of mathematics into which seventeenth-century calculus evolved, and cannot be proved here. The moral for us is to be aware of what a function is doing near the endpoints of its domain, particularly if the domain does not include endpoints. If a continuous curve fails to have a highest (or lowest) point, then by the theorem of analysis the trouble must lie in the behavior of the function near an endpoint missing from its domain.

EXAMPLE 5. Find the highest point on the curve f given by
$$f(x) = 2x + 3,$$
on the domain

(a) $0 \leq x \leq 4$.
(b) $0 < x < 4$.

Let us look first for all critical points in the domain, that is, all values x for which $f'(x) = 0$. Here we have $f'(x) = 2$, which shows that there are no such values. Since f has no critical points, the principle of analysis mentioned above guarantees that the extreme values of f must occur at the endpoints of the domain. At the endpoint 0, the value of f is 3; at the endpoint 4, the value of f is 11. Therefore,

(a) if the domain is $0 \leq x \leq 4$, then $(4, 11)$ is the highest point on the curve f.
(b) if the domain is $0 < x < 4$, then the curve f contains no highest point.

Note that, to draw the conclusions (a) and (b), *we did not have to draw a picture of the curve f!* The reader may wish to draw a picture anyway, to see better what is going on. The expression $2x + 3$ reveals f to be a linear function of slope 2:

(a) Domain: $0 \leq x \leq 4$
    Range: $3 \leq y \leq 11$.

(b) Domain: $0 < x < 4$
    Range: $3 < y < 11$.

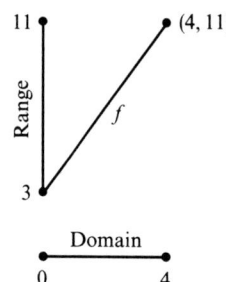
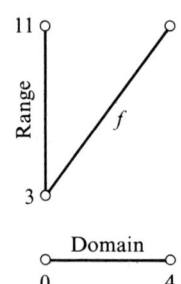

The highest point on the curve is $(4, 11)$. The greatest number in the range is 11; the least is 3.

There is no highest (or lowest) point on the curve, because the range contains no greatest (or least) number.

8. Solving Optimization Problems

EXAMPLE 6. Let $C = 7L + (48/L)$, with domain $0 < L$. Find the least possible value of C.

Let us first find all critical points in the domain, that is, points where C' is zero. Since $C' = 7 - (48/L^2)$, C' is zero when

$$7 - \frac{48}{L^2} = 0,$$

$$7L^2 - 48 = 0 \quad \text{(multiplying through by } L^2\text{)},$$
$$7L^2 = 48,$$

$$L^2 = \frac{48}{7},$$

$$L = \pm\sqrt{\frac{48}{7}}.$$

L	C	C'
?		0

Because $-\sqrt{48/7}$ is not in the domain, the only critical point in the domain is

$$\sqrt{\frac{48}{7}} = 2.619\ldots.$$

At the critical point, the corresponding value of C is given by

$$C = 7\sqrt{\frac{48}{7}} + \frac{48}{\sqrt{\frac{48}{7}}} = 36.661\ldots.$$

We must now show that this is the least possible value of C.

The second derivative helps here. It is given by

$$C'' = \frac{96}{L^3},$$

which is (obviously) *positive* for *all* values of L in the domain $0 < L$. Therefore, the curve $C = 7L + (48/L)$ is *always* bending to its left (or smiling). The point

$$(2.619\ldots, 36.661\ldots),$$

being the point on the curve where a tangent line is horizontal, must be the lowest point on the curve.

We can now answer the question raised in Example 1 of Chapter 1. The least amount of money that will pay for the fencing is $36.66 (rounded off to the nearest cent). □

The preceding example was discussed rather thoroughly without ever drawing the curve. We found the lowest point and we discovered that the curve was always bending to the left. If it is desired to sketch the curve, what additional information is needed? *Answer*: Information about the curve's behavior near the "endpoints" of the domain, i.e., when L is very small and when L is very large.

To get this information, use common sense. Look at the two terms $7L$ and $48/L$, whose sum gives C. What happens to each of them when L is very small? The first term $7L$ is *negligible* (i.e., nearly zero), so the curve behaves essentially like the graph of $48/L$ when L is small. The expression $48/L$ increases without bound as $L \to 0^+$. (See Example 7, near the end of Chapter 1.)

What happens when L is very large? Then the expression $48/L$ is negligible, so the curve behaves essentially like the graph of $7L$ when L is large. The expression $7L$ produces a line of slope 7. For large L, the curve $C = 7L + (48/L)$ approximates a straight line of slope 7.

Putting these facts together produces the following sketch:

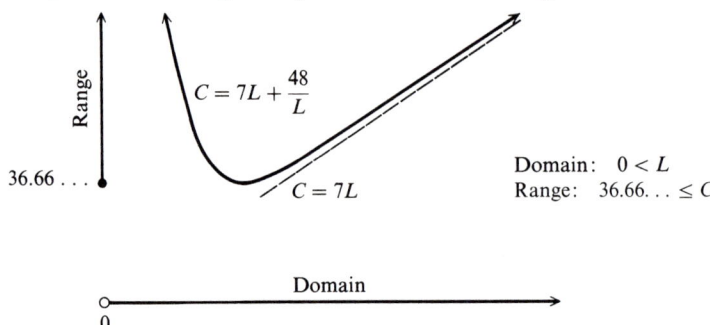

EXERCISES

8.1. For each of the following functions, find its maximum value, if it has one.
   (a) $f(x) = 5 - 2x$, $0 \le x < 3$.
   (b) $F(x) = x^2 - 2x$, $0 \le x \le 4$.
   (c) $g(x) = x - (1/x) + 6$, $0 < x \le 8$.
   (d) $G(x) = x + (1/x) + 6$, $0 < x \le 8$.
   Answers: (b) max F is 8. (d) max G does not exist.

8.2. For each of the functions in exercise 8.1, find its minimum value, if it has one. *Answers*: (a) min f does not exist. (b) min F is -1.

8.3. Sketch the curve $y = 4x + (36/x)$, with domain $0 < x$, indicating both coordinates of the lowest point. Also, indicate how the curve looks when x is very small and when x is very large.

8.4. A rectangular pen containing 16 square meters is to be fenced in. The front will cost $4 per meter of fencing, while each of the other three sides will cost $3 per meter. What is the least amount of money that will pay for the fencing?

8.5. In the problem set at the end of Chapter 1, read again problem 19. Find the greatest possible volume V.

§9. Summary

Here, in detail, are the steps that have been illustrated above.

Step 1. Algebraic formulation:
  (a) See the problem in terms of variables. (The quantity to be optimized is one variable, say y, and you have to find a second variable, say x, on which y depends.)
  (b) Write down an algebraic rule f, giving y in terms of x.
  (c) Specify the domain of the function f.

Step 2. Geometric analysis:
  (a) See the problem as one of finding the highest (or lowest) point on the curve f.
  (b) Find the derivative f'. (And find f'' too, if it can be done without much trouble.)
  (c) Find the critical points, if any, that lie in the domain of f. (That is, find all values of x in the domain of f that satisfy the equation $f'(x) = 0$.)
  (d) Check what happens near the endpoints of the domain.
  (e) Using the information of steps 2(c) and 2(d), find the desired highest (or lowest) point on the curve f.
  [The second derivative may be helpful in steps 2(d) and 2(e).]

Step 3. Back to everyday life:
  (a) Read the problem again, to determine exactly what was called for. (Was it the *first* or *second* coordinate, or *both*, of the *highest* or *lowest* point of the curve that you were seeking?)
  (b) Give a direct answer to the question raised in the problem, by writing a complete, concise sentence.

Step 1(c) is easy to forget, and thus deserves emphasis. The domain must be specified; otherwise, steps 2(c) and 2(d) cannot be carried out. Step 3 is also easy to forget. In concentrating on step 2, you can lose sight of your goal and, as a consequence, do unnecessary work. When a problem takes a long time to work, it is a good idea to remind yourself now and then what you are after.

Here is another example to illustrate these steps.

EXAMPLE 7. An ordinary metal can (shaped like a cylinder) is to be fashioned, using 54π square inches of metal. What choice of radius and height will maximize the volume of the can?

Here, we want to maximize the volume, so let V denote the volume, which is given in terms of the radius r and height h by the formula

$$V = \text{(area of base)(height)}$$
$$= \pi r^2 h. \qquad (7)$$

The rule $V = \pi r^2 h$ gives V in terms of *two* variables. We need to get V in terms of only *one* variable, and this can be done, as follows, by finding a relation between r and h. The picture below shows that the area of the side of the can is given by $2\pi r h$:

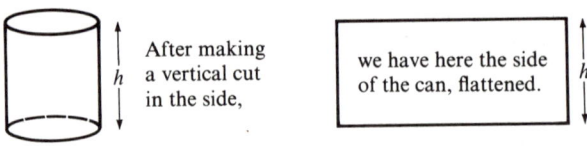

The total amount of metal available, 54π square inches, must equal the amount in the side of the can, plus the amount in the circular top and bottom:

$$54\pi = 2\pi rh + 2\pi r^2.$$

This is a relation between r and h. It is easy to solve for h (the reader is asked to do it), and obtain

$$h = \frac{27 - r^2}{r}. \tag{8}$$

Putting equations (7) and (8) together gives

$$V = \pi r^2 \left(\frac{27 - r^2}{r} \right)$$

$$= \pi r(27 - r^2)$$

$$= 27\pi r - \pi r^3,$$

which expresses V in terms of r alone. The problem now is to find the value of r that yields the maximal volume V, where

$$V = 27\pi r - \pi r^3, \quad 0 < r < \sqrt{27}.$$

[The radius r must be less than $\sqrt{27}$. *Reason*: The height h must be positive, so, by equation (8), $27 - r^2$ must be positive.]

r	V	V'
?		0
r	$27\pi r - \pi r^3$	$27\pi - 3\pi r^2$

Let us find critical points. The derivative is given by

$$V' = 27\pi - 3\pi r^2,$$

which is zero when (dividing by 3π)

$$0 = 9 - r^2,$$
$$r^2 = 9,$$
$$r = \pm 3.$$

Since -3 is not in the domain, the only critical point is 3.

We now show that when r is 3, the volume V is maximal. This is easy to see, for the second derivative is given by

$$V'' = -6\pi r,$$

which is (obviously) negative *throughout the domain*. The curve is therefore always bending to its right (or frowning), and hence it must reach its highest point at the place where it has a horizontal tangent line. (At both endpoints of the domain, V tends to zero.)

To maximize the volume, the radius should be 3 inches, and the corresponding height, by equation (8), should be 6 inches. □

A Final Remark. As in Examples 5, 6, and 7, it is not really necessary to sketch the curve in order to do the problem. If f has a derivative, then the *extreme values* (maximum and minimum) can be located by checking among the endpoints and the critical points. Curve sketching is to be encouraged, because pictures say more than words, but the principles of analysis are valid regardless of how well one draws.

Problem Set for Chapter 4

1. A rectangular pen containing 36 square meters is to be fenced in. The front fence will cost $7 per meter, while each of the other three sides will cost $3 per meter. What is the least amount of money that will pay for the cost of the fence?

2. A book company wants to put 60 square inches of type on a rectangular page, leaving margins of 1 inch on the sides and bottom and of 2 inches at the top. What should be the dimensions of the page in order to minimize the amount of paper used?

3. Write an equation of the tangent line to the curve $y = 5/x^2$ at the point $(1, 5)$.

4. Tell whether the curve $y = 5/x^2$ is bending to the *right* or to the *left* as it passes through
   (a) $(1, 5)$.
   (b) $(-1, 5)$.

5. Consider the function f defined by $f(x) = 5x\sqrt{1 - 3x}$.
   (a) Is the curve f *rising* or *falling* as it passes through the point $(0, 0)$?
   (b) Is the curve f bending to the *left* or to the *right* as it passes through the point $(0, 0)$?

6. Consider the function given by $f(x) = ax^2 + bx + c$, where a, b, and c are constants. Which way does the curve f bend if
   (a) $a > 0$?
   (b) $a < 0$?
   (c) $a = 0$?

7. Consider the curve $C = 3L + (27/L)$, $0 < L$.
   (a) Find the first coordinate of a point on this curve where the tangent line is horizontal.
   (b) This curve always bends the same way on the domain $0 < L$. Which way?
   (c) From your answer to part (b), you know the point found in part (a) must be the *highest* or *lowest* point on the curve?

8. Consider the curve $C = 3L + (27/L)$, $L < 0$.
   (a) This curve always bends the same way on the domain $L < 0$. Which way?
   (b) Find both coordinates of the highest point on the curve.
   (c) Does this curve have a lowest point?

9. What is the range of f, where $f(L) = 3L + (27/L)$, with domain $L \neq 0$? *Hint.* First sketch the curve f, using the information obtained in problems 7 and 8.

10. (*A problem in curve sketching.*) Consider the cubic equation $y = x^3 - 3x + 2$.
    (a) The derivative $y' = 3x^2 - 3$ is a simple quadratic function. Plot the graph of this quadratic, and, on a different coordinate system, plot the graph of the simple linear function $y'' = 6x$.
    (b) The two graphs just sketched give much information about the original cubic. Use these two graphs to specify on what interval(s) the cubic is
       (i) rising.
       (ii) falling.
       (iii) bending to the left.
       (iv) bending to the right.
    (c) Find both coordinates of an inflection point of the cubic.
    (d) There are two points on this cubic where there is a horizontal tangent line. Find both coordinates of both points.
    (e) Sketch the curve $y = x^3 - 3x + 2$, using all the information just obtained.
    (f) Specify the range of the cubic if the domain is
       (i) $0 \leq x \leq 3$.
       (ii) $-2 \leq x < 0$.
       (iii) $-2 < x \leq 0$.
       (iv) unrestricted.

11. Consider the cubic equation $y = x^3 + 3x + 2$.
    (a) Sketch the graph of this cubic, after first investigating its first and second derivatives, as in the preceding problem.
    (b) Find both coordinates of an inflection point.
    (c) Specify the range of this cubic if its domain is given by $-1 \leq x \leq 4$.

12. As in the preceding two problems, carry out an analysis of the cubic $y = x^3 + 2$, sketch its graph, and find its point of inflection. What is its range if its domain is given by $-1 \leq x < 2$?

13. Suppose $y = (t^2 - 3)/(t + 2)$. Find y' and y'' when t is 0, and use this information to sketch the curve locally, near the point $P = (0, -\frac{3}{2})$.

14. Sketch the curve $y = x^2 + (8/x)$, $x \neq 0$, a portion of which has already been sketched in Section 4.

15. Express the number 10 as the sum of two positive numbers in such a way that the sum of the cube of the first and the square of the second is as small as possible.

16. Find the point on the graph of $y^2 = 4x$ that is nearest the point $(2, 1)$.

17. What is the smallest slope that a tangent line to the curve $y = x^3 + 3x + 2$ could possibly have?

18. Identical squares are to be cut out of each corner of a piece of metal that is shaped like a rectangle of dimensions 5 feet by 8 feet. The four squares are then discarded, and the sides folded upwards to make a large box, with open top. Let x be the length of the sides of the squares cut out, and let V be the corresponding volume of the box. Find the value of x that maximizes V.

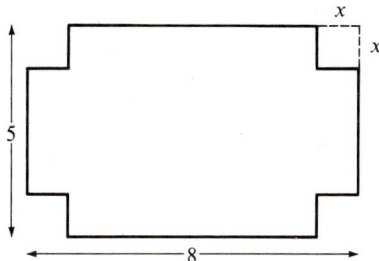

Discard squares, fold up sides:

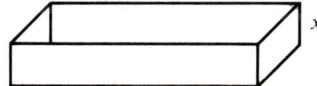

19. Identical squares are to be cut out of each corner of a rectangular piece of metal measuring 10 meters by 4 meters. Then the squares are to be discarded, and the sides folded up to make a water trough for thirsty horses. What size squares should be cut out in order to maximize the volume of the trough?

20. Out of 100 square centimeters of metal, the sides, top, and bottom of a cylindrical can are to be fashioned. What should be the radius of the base of the can in order to maximize the amount of chicken soup that the can will hold?

21. In exercise 1.13 of Chapter 2, a certain cost C was given in terms of a length x by the equation

$$C = 7\sqrt{9 + x^2} - 2x + 26, \qquad 0 \leq x \leq 13.$$

(a) Find C', then fill in the table below.

x	C	C'
0	47	?
4	53	?

(b) Tell whether the curve is *rising* or *falling* as it passes through $(0, 47)$ and through $(4, 53)$. Can you conclude that the lowest point on the curve lies somewhere between these two points?
(c) Find the value of x that yields the least cost C.
(d) Read again exercise 1.13 of Chapter 2. Then draw a picture of how the cable should be built in order to minimize the cost of the cable.

22. A lighthouse is located 4 miles offshore. The nearest town is 5 miles downshore. Whenever she goes into town, the lighthouse keeper must take a motorboat containing her motorcycle, dock at a point somewhere downshore, then ride the rest of the way by motorcycle. Where should the boat be docked in order to minimize the time of the trip to town if
    (a) the motorboat goes 20 miles per hour and the motorcycle goes 40 miles per hour?
    (b) the motorboat and the motorcycle travel at the same speed?
    (c) the motorboat goes A miles per hour and the motorcycle goes B miles per hour?

23. In problem 18 at the end of Chapter 1, a certain cost C was given in terms of a length L by the equation

$$C = 3L^2 + \frac{96}{L}, \qquad 0 < L.$$

   (a) Find the value of L that minimizes the cost.
   (b) Read again problem 18 of Chapter 1, and draw a picture indicating the dimensions of the metal container that will minimize its cost.

24. Find the dimensions of the cheapest possible trash can with square base and rectangular sides, subject to the following specifications. The volume of the can is to be 3 cubic meters, the material for the sides costs $0.30 per square meter, and the material for the base costs $0.50 per square meter.

25. In the preceding problem, suppose it is decided to add a top to the can, made out of light metal costing only $0.10 per square meter. With this addition, what are the dimensions of the cheapest can?

26. The metal used in making the top and bottom of a *cylindrical* can will cost $0.03 per square centimeter, while the metal used in the side of the can will cost $0.02 per square centimeter. If the volume of the can is to be 100 cubic centimeters, what should be the dimensions of the can in order to minimize the cost?

27. Out of 160 square feet of material, a container is to be made. What dimensions will maximize the volume of the container if the container is to be shaped like
    (a) a rectangular figure with square base and open top?
    (b) a rectangular figure with square base and with a top?
    (c) a cylindrical can without a top?
    (d) a cylindrical can with a top?

28. A Norman window is in the shape of a rectangle surmounted by a semicircle. Find the dimensions of the window that will allow the most light to pass, provided that the perimeter of the window is 8 meters.

29. A wire is to be cut in two. The first part is to be bent into the circumference of a circle, and the second part into the perimeter of a square. How should the wire be cut in order to minimize the combined area of the circle and square if
    (a) the wire is 100 centimeters long?
    (b) the wire is A centimeters long?

30. The definition of *concave upward* is as follows: A curve f is **concave upward** *if it lies above each tangent line (with the obvious exception of the point of tangency).*
    (a) If f'' is always positive, is the curve f concave upward?

(b) If the area in the plane lying up above the curve f forms a concave figure, is the curve f concave upward?

31. Find the first derivatives of the following. Do not simplify your answers.
    (a) $(x/(x-6))^2$.
    (b) $\sqrt{x/(x-6)}$.
    (c) $(4x^5 - 3x)(x^2 - x + \sqrt{2})^2$.
    (d) $x^4/(x^3 + x - 3)^2$.
    (e) $(x^5 - x)\sqrt{2 + 7x}$.
    (f) $\sqrt{x^5}$.

32. Mathematics shares some characteristics with experimental science. One notices a pattern developing, and then one tries to guess a general rule. The rule must then be tested for its applicability to new situations. One hopes that a widely applicable rule can be derived logically from simpler principles that are already accepted. Consider the rule

$$(x^n)' = nx^{n-1},$$

which ought to have been guessed in exercise 6.3 of this chapter. This rule applies where n is a *positive integer*. Let us test this rule for wider applicability.
    (a) Apply the rule above to find $(x^{-1})'$. Does it result in the correct answer? (We already know that the derivative of x^{-1} is $-1/x^2$, from our work in Section 1.)
    (b) Apply the rule to find $(x^{-2})'$. Does it give the derivative of $1/x^2$?
    (c) Apply the rule to find $(x^{1/2})'$. Does it give the derivative of $\sqrt{x}$?
    (d) Apply the rule to find $(x^{3/2})'$. Does it give the derivative of $x\sqrt{x}$? (The derivative of $x\sqrt{x}$ can be taken by the product rule.)
    (e) Apply your rule to find $(x^{5/2})'$. Does it give the derivative of $\sqrt{x^5}$?
    (f) Apply your rule to find $(x^{-7/2})'$. Does it give the derivative of $1/\sqrt{x^7}$?
    (g) If it made any sense to speak of raising a number to the power π, what would you guess is the derivative of x^π?
    (The widely applicable rule that suggests itself here will be derived logically in Chapter 9.)

33. Suppose that a function g has a derivative at a point x. Does it necessarily follow that g is *continuous* at x? That is, if $g'(x)$ exists, does it follow that $g(x) = \text{Lim } g$ at x? *Hint.* In Section 2 we saw that if $g'(x)$ exists, then equation (4) necessarily follows.

34. Suppose that a function g is continuous at a point x. Does it necessarily follow that g has a derivative at x? *Hint.* Consider a function whose curve has a "corner", as pictured below, at the point $(x, g(x))$.

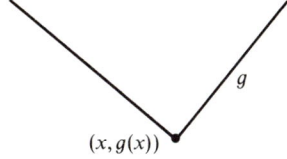

35. Use Fermat's method to show that the derivative of a constant function is zero. [This is so easy that it is easy to miss. You are to show that if $g(x) = c$, where c is a constant, then $g'(x) = 0$ for all x.]

36. Give two proofs of the rule for constant multiples, which states that $(c \cdot f)' = c \cdot f'$.
    (a) First, by applying the product rule to the product $f \cdot g$, where $g(x) = c$, and using the result of problem 35.
    (b) Secondly, by applying Fermat's method to the situation pictured below:

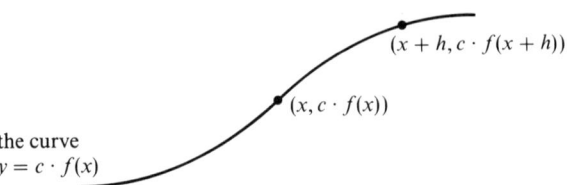

the curve
$y = c \cdot f(x)$

37. By applying Fermat's method to the situation pictured below, prove the rule for sums, which states that $(f + g)' = f' + g'$.

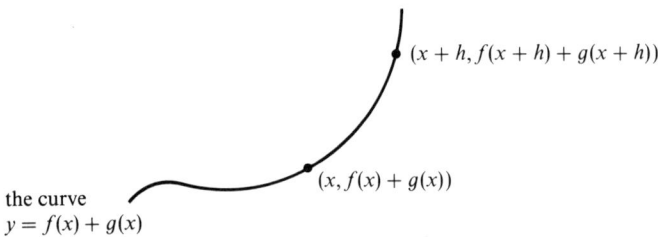

the curve
$y = f(x) + g(x)$

38. Use Fermat's method (not a shortcut rule) to show that if $f(x) = \sqrt{x}$, then $f'(x) = 1/2\sqrt{x}$. *Hint.* Simplify the difference quotient of f by multiplying both the top and the bottom by the expression $\sqrt{x+h} + \sqrt{x}$.

39. Use Fermat's method (not a shortcut rule) to prove directly the product rule, which states that $(fg)' = fg' + gf'$. *Hint.* First find the difference quotient of the product function given by $y = f(x)g(x)$. Simplify it by inserting the expression $f(x + h)g(x) - f(x + h)g(x)$ into the numerator to get

$$f(x+h)\frac{g(x+h) - g(x)}{h} + g(x)\frac{f(x+h) - f(x)}{h}.$$

Then find the limit as h tends to 0.

40. Derive the reciprocal rule from the product rule, by proceeding as follows. Assuming that the expression $1/g$ has a derivative, begin with the obvious equality

$$g\left(\frac{1}{g}\right) = 1,$$

and use the product rule, together with the result of problem 35, to write

$$g\left(\frac{1}{g}\right)' + \left(\frac{1}{g}\right)g' = 0,$$

then solve for the derivative of $1/g$.

41. Match each of the following functions (a) through (j) with its derivative. [The derivatives of (k) and (l) are not pictured.]

(a)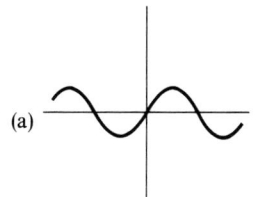

(b)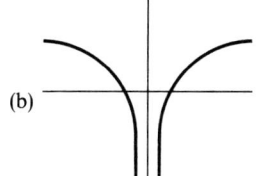

(c)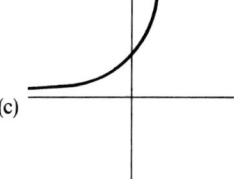

(d)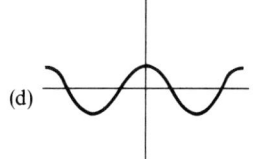

(e)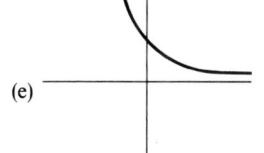

(f)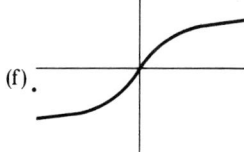

(g)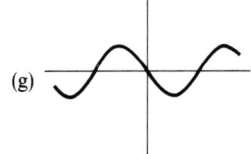

(h)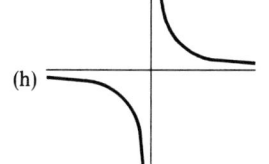

(i)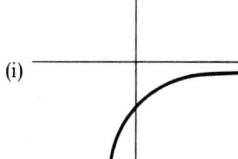

(j)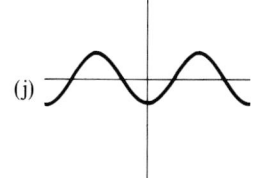

(k)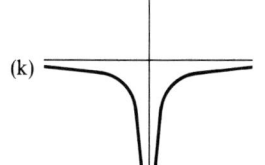

(l)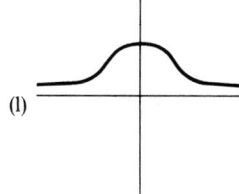

5 Chains and Change

Things change. The world is in flux. How can one understand a world in which change plays so great a role? The seventeenth-century answer given by Leibniz and Newton is simplicity itself:

Study change.

To study change is to study the way things vary. We have done a little of this in the preceding chapters, but we have not yet taken up this study in earnest. The derivative has a remarkable ability to capture the dynamics of change. A main point of this chapter is that *the derivative may be viewed as measuring the instantaneous rate of change.*

How can this be? Before answering, we need to develop symbolism that is suggestive of the ideas involved. The symbolism of *primes* (as in y' or f') to denote the derivative offers no aid to our new endeavor. In fact, the main advantage of denoting the derivative by f' is that this notation suggests that the derivative is a *function*. Once this important fact has been hammered home, the use of primes to denote derivatives offers no special advantage, and may be discarded in the presence of a superior system of symbolism.

§1. Leibniz's Notation: Mathematics and Poetry

A superior system of notation for the calculus was developed by Leibniz. If y is a function of x, Leibniz denoted the derivative by

$$\frac{dy}{dx},$$

1. Leibniz's Notation: Mathematics and Poetry

instead of by y'. Or, if A is a function of s, Leibniz called the derivative

$$\frac{dA}{ds},$$

instead of A'.

At this point the reader is doubtless mystified as to why this symbolism is supposed to be more helpful than the perfectly good notation already developed. It is helpful only if one views the derivative the way Leibniz did. Let us illustrate how Leibniz would go about showing that *the derivative of x^2 is $2x$*.

Consider a fixed point (x, y) on the curve $y = x^2$.

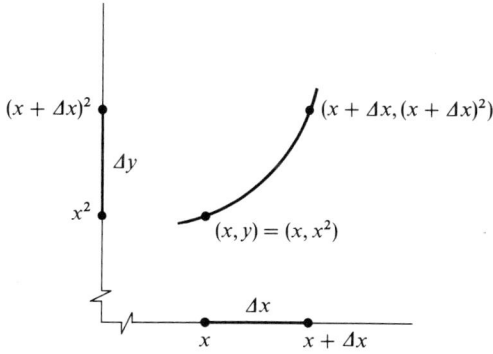

A portion of the curve $y = x^2$

Let Δx be a small change in the variable x. ("Δ" is the Greek letter *delta*. The expression Δx is to be taken as a whole, and not to be confused with a product. The change Δx may be either positive or negative.) *What is the corresponding change Δy in the variable y?* From the figure, it is clearly given by

$$\Delta y = (x + \Delta x)^2 - x^2$$
$$= x^2 + 2x(\Delta x) + (\Delta x)^2 - x^2$$
$$= 2x(\Delta x) + (\Delta x)^2.$$

Therefore, the ratio of the change (or *increase*) in y to the increase in x which caused it is given by

$$\frac{\Delta y}{\Delta x} = 2x + \Delta x. \tag{1}$$

As mentioned above, the increase Δx may be either *positive* or *negative* (a *negative increase* of course represents a decrease), but may not be zero. What happens in equation (1) as Δx tends to 0? Then the fraction $\Delta y/\Delta x$ approaches what might be termed the *instantaneous rate of increase of y with respect to x,* which, using equation (1), is equal to

$$\underset{\Delta x \to 0}{\text{Limit}} \frac{\Delta y}{\Delta x} = \underset{\Delta x \to 0}{\text{Limit}} (2x + \Delta x) = 2x.$$

(The reason that we get the derivative of x^2 should be plain. As seen in the figure above, the ratio of changes $\Delta y/\Delta x$ is also the slope of a line that approaches the tangent line at (x, y) when Δx approaches zero.)

The ratio $\Delta y/\Delta x$ of changes taking place over an interval of length Δx is *not* of primary interest here. Leibniz wanted the "ultimate ratio", or the *instantaneous* rate of increase taking place at the point x. This is what happens as the length Δx shrinks to zero, and this Leibniz called dy/dx. That is, the symbol dy/dx is defined as follows:

$$\frac{dy}{dx} = \lim_{\Delta x \to 0} \frac{\Delta y}{\Delta x}.$$

Why did Leibniz choose to denote the derivative this way? What is in the symbol dy/dx that is not in the name *derivative*? A lot, as it turns out. First, dy/dx reminds us that the derivative is the limit of ratios of changes. Secondly, because the symbol dy/dx looks like a fraction, it reminds us that the derivative is a limit of fractions, of "quotients of differences". The symbol dy/dx, by its very form, gives a hint that *the derivative might be expected to exhibit some of the familiar properties of fractions*. In a lighter vein, the reason Leibniz chose this symbolism is that, by the seventeenth century, the ancient Greek letter Δ had evolved "in the limit" to the modern d. What could be more natural than to denote the limit of $\Delta y/\Delta x$ by dy/dx?

It is hard to overestimate the value of appropriate symbolism. Of all creatures, only human beings have much ability to name things and to coin phrases. Poets do this best of all.

> ... as imagination bodies forth
> The forms of things unknown, the poet's pen
> Turns them to shapes and gives to airy nothing
> A local habitation and a name.
>
> Shakespeare

It can be contended that Leibniz's way of writing the calculus approaches the poetic. One can be borne up and carried along purely by his symbolism, while his symbols themselves may appear to take on a life all their own. Mathematics and poetry are different, but they are not so far apart as one might think.

The reader who is skeptical of the remarks just made is asked to suspend a final judgment until this chapter and the next are completed. Any skepticism that still remains may be eliminated by Chapter 10, which can be read immediately following Chapter 6. In the meantime, just to show that the remarks above are not especially radical, here is a well-known quotation from a man who won the Nobel Prize in literature:

> Mathematics, rightly viewed, possesses not only truth, but supreme beauty—a beauty cold and austere, like that of sculpture, without appeal to

1. Leibniz's Notation: Mathematics and Poetry

any part of our weaker nature, without the gorgeous trappings of painting or music, yet sublimely pure, and capable of a stern perfection such as only the greatest art can show. The true spirit of delight, the exaltation, the sense of being more than man, which is the touchstone of the highest excellence, is to be found in mathematics as surely as in poetry.

Bertrand Russell

Before we can see anything in Leibniz's notation, we must learn how to use it. There is no poetry in the examples that follow, only illustrations of how things are said in the language of Leibniz.

$$\text{If } y = x^2, \text{ then } \frac{dy}{dx} = \frac{d}{dx}(x^2) = 2x.$$

$$\text{If } A = s^2, \text{ then } \frac{dA}{ds} = \frac{d}{ds}(s^2) = 2s.$$

$$\text{If } f(x) = x^3, \text{ then } \frac{df}{dx} = 3x^2.$$

$$\frac{d}{dt}(\sqrt{1 + 3t^2}) = \frac{6t}{2\sqrt{1 + 3t^2}}.$$

$$\text{If } y = f \cdot g, \text{ then } \frac{dy}{dx} = f\frac{dg}{dx} + g\frac{df}{dx} \quad \text{(the product rule).}$$

$$\frac{d}{dx}\left(\frac{1}{g}\right) = \frac{-1}{g^2}\frac{dg}{dx} \quad \text{(the reciprocal rule).}$$

EXERCISES

1.1. Write the *square root rule* in Leibniz's notation: If $y = \sqrt{g}$, then $dy/dx = ?$

1.2. Write the *quotient rule*: $d(f/g)/dx = ?$

1.3. Find dC/dL if $C = 7L + (48/L)$.

1.4. What is $d(t^2 + 3t + \pi)/dt$?

1.5. Use the product rule to find $d(w^4\sqrt{3 + w})/dw$.

1.6. Find the derivatives of each of the following, expressing your answer in Leibniz's notation.
   (a) $L = 12/W$.
   (b) $C = (84/W) + 4W$.
   (c) $C = 2\pi r$.
   (d) $A = \pi r^2$.
   Answers: (b) $dC/dW = (-84/W^2) + 4$. (c) $dC/dr = 2\pi$.

1.7. (*This question is not entirely frivolous, as will be seen in Chapter 6.*) The ancient Greek letter Δ has by the seventeenth century evolved "in the limit" to the letter d. What about the Greek letter Σ (sigma)? What is the "seventeenth-century limit" of Σ?

1.8. Is dy/dx the quotient of "dy" and "dx"? *Answer*: No. The derivative is denoted by the entire symbol dy/dx. Just as one understands the word *rainbow* without feeling any need to know what *ra* and *inbow* might mean, so one can understand dy/dx without ascribing meaning to dy and dx.

1.9. A familiar rule for fractions is $(A/B)(B/C) = A/C$. If derivatives behaved like fractions, what would the product

$$\frac{dy}{dx}\frac{dx}{dt}$$

be equal to? *Answer*: dy/dt.

1.10. If derivatives behaved like fractions, what would the following products of derivatives be equal to?
   (a) $(dC/dL)(dL/dW)$.
   (b) $(dA/dr)(dr/dt)$.
   (c) $(dL/dW)(dW/dL)$.

§2. The Derivative as Instantaneous Speed

Suppose a rock is thrown directly upward, and suppose that, at time t seconds after it is released, its height h (in feet) is given by the equation

$$h = -16t^2 + 64t.$$

To illustrate the ideas just introduced concerning change, let us try to answer the following questions.

(a) During its first second of flight, what is the rock's *average* speed?
(b) What is the rock's *instantaneous* speed when $t = 1$ (i.e., 1 second after release)?
(c) When $t = 3$ (i.e., 3 seconds after release), is the rock going *up* or *down*?
(d) When does the rock attain its maximum height?
(e) What is the rock's *initial* velocity, i.e., what speed was given the rock at the instant of release?

To get hold of this situation, let us set up the usual table. Since $h = -16t^2 + 64t$, the derivative is given by

$$\frac{dh}{dt} = -32t + 64.$$

Plugging in a few numbers gives rise to this table.

2. The Derivative as Instantaneous Speed

t (in seconds)	h (in feet)	dh/dt (in feet per second)
$\left.\begin{array}{c}0\\1\end{array}\right]\Delta t$	$\left.\begin{array}{c}0\\48\end{array}\right]\Delta h$	32
2	64	0
3	48	-32
t	$-16t^2 + 64t$	$-32t + 64$

Why is dh/dt in the units of feet per second (ft/sec)? Because the change Δh in h is in feet and the change Δt is in seconds, so $\Delta h/\Delta t$ is in feet per second, showing that dh/dt is the limit of numbers $\Delta h/\Delta t$ that are in units of feet per second. Let us answer the questions raised in turn.

(a) During the initial 1-second interval, we have $\Delta t = 1$. The corresponding distance traveled is the change in height from 0 to 48 feet. That is, $\Delta h = 48$. The *average* speed for the first second is then given by

$$\frac{\text{distance traveled}}{\text{time taken}} = \frac{\Delta h}{\Delta t}$$

$$= \frac{48 \text{ feet}}{1 \text{ second}}$$

$$= 48 \text{ ft/sec.}$$

(b) The derivative dh/dt gives the instantaneous rate of increase of height with respect to time. When $t = 1$, $dh/dt = 32$ ft/sec. If the rock had a speedometer inside it to measure the upward speed, the speedometer should read 32 when t is 1.

(c) When t is 3, then dh/dt is -32, so that dh/dt, which measures the rate of *increase* of height, is *negative*. Since the rate of increase of height is negative, the rock is *falling* when t is 3. (The instantaneous speed is 32 ft/sec *downward* when t is 3.)

(d) The rock is going up when the upward speed dh/dt is positive and is going down when dh/dt is negative. The rock must therefore attain its maximum height when dh/dt is zero. This occurs when t is 2.

(e) It is easy to be confused about speeds at the moment of release and at the moment of impact with the ground. But there is no question that at any intermediate time t, the speed is given by the expression $-32t + 64$. To avoid confusion, let us agree that the initial speed is the limit of this expression as t tends to zero from the right:

$$\underset{t \to 0^+}{\text{Limit}} (-32t + 64) = 64 \text{ ft/sec.} \qquad \square$$

It is easy to sketch the quadratic curve $h = -16t^2 + 64t$, and thus it is easy to picture the situation described above. *Avoid the mistake of thinking* that the rock travels along the curve, however. *The rock moves straight up (until $t = 2$), then straight down, along the vertical axis.*

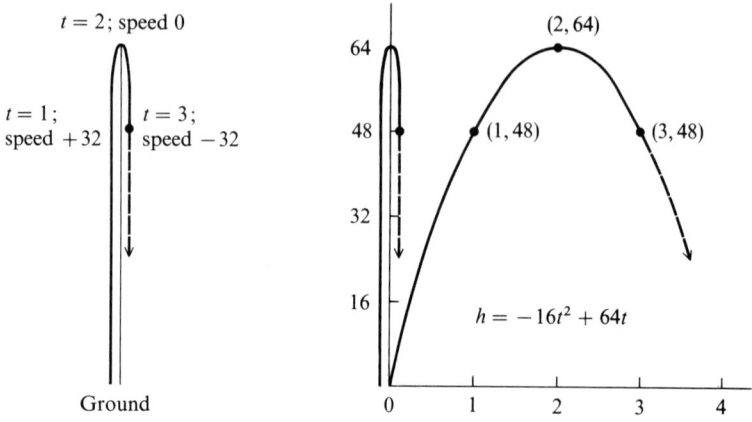

The rock hits the ground when $t = 4$. *What is its speed at the moment of impact?* God only knows. We can know if we interpret the question as requiring us to find the speed the rock is *approaching* as t tends to the moment of impact. Then the answer is easy. The upward speed at the moment of impact is approaching

$$\text{Limit}_{t \to 4^-} (-32t + 64) = -64 \text{ ft/sec.}$$

(The negative sign occurs because $-32t + 64$ gives the *upward* speed.) At the moment of impact, the rock is approaching a *downward* speed of 64 ft/sec.

Exercises

2.1. A rock is thrown directly upward. Its height h (in feet) at time t seconds after release is given by

$$h = -16t^2 + 128t.$$

(a) What is the rock's *average* speed during its first second of flight?
(b) What is the rock's *instantaneous* speed when $t = 1$?
(c) Is the rock going *up* or *down* when $t = 3$?
(d) When is the maximum height attained?
(e) What is the average speed of the rock during the time interval between $t = 1$ and $t = 3$?
(f) What is the instantaneous speed when $t = 2$?
(g) What is the rock's initial speed?
(h) When does the rock hit the ground, i.e., when is the height h equal to zero? *Answer:* When $t = 8$.
(i) What is the speed of the rock when it hits the ground? (See the discussion above for a proper interpretation of this question.)

2. The Derivative as Instantaneous Speed   115

2.2. A rocket travels directly upward. At time t seconds after it is launched, its height h in feet is given by
$$h = 50t^3 + 80t.$$

(a) What is the rocket's average speed during its first 2 seconds of flight? *Answer*: 280 ft/sec.
(b) What should the rocket's speedometer read when $t = 1$?
(c) Let v stand for the rocket's speedometer reading (so that $v = dh/dt$). Then v, like h, is a function of t. Fill in the question marks appropriately in the following table.

t	h	v	dv/dt
1	?	230	?
2	560	680	?
π	?	?	?
t	$50t^3 + 80t$	?	?

(d) Think about the units in which things are measured. Here we have t in seconds, h in feet, so v is in feet per second. What units is dv/dt measured in? *Answer*: ft/sec per second.
(e) When t is 2, is the speedometer reading v *increasing* or *decreasing*? *Hint*. This is the same question as, "Is dv/dt *positive* or *negative*?" It is also the same question as, "Is the rocket *accelerating* or *decelerating* in its upward movement?"
(f) Is the rocket *accelerating* or *decelerating* when $t = 1$?
(g) **Acceleration** *is defined as the rate of increase of speed*. What is the rocket's instantaneous acceleration when $t = 1$? *Answer*: The rate of increase of speed, dv/dt, is equal to 300 ft/sec per second, when $t = 1$.
(h) What is the rocket's instantaneous acceleration when $t = 2$?

2.3. Go back to the situation described in exercise 2.1.
(a) Fill in the following table.

t	h	v	dv/dt
1	112	96	?
4	?	?	?
6	?	?	?
t	$-16t^2 + 128t$	?	?

(b) Since v is the *upward* speed, and since dv/dt measures the rate of *increase* of v, it follows that dv/dt measures the *upward* acceleration. In this case the upward acceleration is constant. What is it? *Answer*: It is -32 ft/sec per second. (This is what gravity does, near the earth's surface. Each second the effect of gravity is to decrease the upward speed of a freely falling body by 32 ft/sec.)
(c) If a freely falling body is given an initial speed of $+128$ ft/sec, how many seconds will gravity take to change the speed to
   (i) 64 ft/sec?
   (ii) -64 ft/sec?
(d) In the rocket problem of exercise 2.2, why isn't the acceleration -32 ft/sec per second, since this is the acceleration due to gravity?

§3. Continuity and Nature

Laws that govern nature command our interest. Leibniz believed that all such laws are subject to the following basic principle.

Leibniz's Principle of Continuity. *Nature must behave in a continuous fashion.*

What does this mean? It is best to look at a concrete example, so consider again the motion of the rock discussed at length in the preceding section. This could be regarded as a simple experiment in physics, out of which arise the variables h and t, related by the function f pictured here.

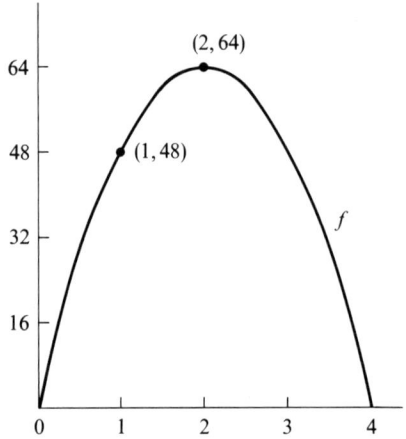

Notice that f is continuous at each point in its domain $0 \leq t \leq 4$. According to Leibniz's principle, it could be no other way, for f describes a process that actually takes place in nature. It would be impossible, for example, for the rock to behave as described by the function g pictured below.

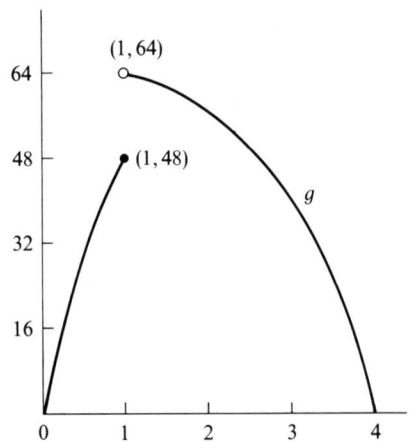

3. Continuity and Nature

This describes the rock climbing steadily to 48 feet, then instantly leaping to a height of almost 64 feet. Only a miracle (i.e., something that disregards the laws of nature) could accomplish this! Leibniz's principle says that nature simply cannot allow the discontinuity of the function g at the point $t = 1$. If there are laws of nature, then these laws determine an underlying purpose, and the action of nature must agree with that purpose. Thus only *continuous* functions can arise out of this experiment, or any experiment, in physics. Or so the philosopher thought.

> Nothing happens all at once, and it is one of my great maxims, and among the most completely verified, that *nature never makes leaps*: which I called the *Law of Continuity*....
>
> Leibniz

Let us go into this a bit further. Consider the instant when t is 1. What happens *naturally* (i.e., in the course of nature) is supposed to be continuously related both to the past and to the future. What does this mean in terms of *change*? Does continuity mean that a small change Δt in time will produce only a small change Δh in height? Certainly *not*, because anyone can think of occasions where nature allows large changes in little time. Instead, continuity means that, as Δt is taken nearer and nearer to zero, then the corresponding change Δh must also tend to zero:

$$\Delta t \to 0 \quad \text{implies} \quad \Delta h \to 0.$$

In other words, to say that h is a continuous function of t is to say

$$\underset{\Delta t \to 0}{\text{Limit}} \, \Delta h = 0. \tag{2}$$

To illustrate this, notice the difference in the behavior of f and g near the point $t = 1$.

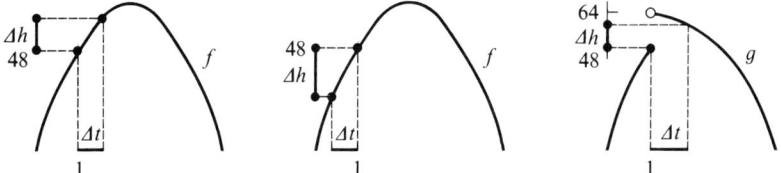

The first two figures show that as $\Delta t \to 0$, either through positive or negative values, $\Delta h \to 0$. Thus, $\text{Limit}_{\Delta t \to 0} \, \Delta h = 0$ and f is continuous at 1. The third figure shows that as $\Delta t \to 0$ through positive values, $\Delta h \to 16$. Thus, $\text{Limit}_{\Delta t \to 1^+} \, \Delta h \neq 0$ and g is discontinuous at 1.

Condition (2) expresses the definition of continuity in terms of change. We should check to see that the definition of continuity by condition (2) agrees with the definition of continuity given in Chapter 1. To see this, assume that $h = f(t)$ satisfies condition (2). Because $\Delta h = f(t + \Delta t) - f(t)$, this condition tells us that

$$\underset{\Delta t \to 0}{\text{Limit}} \, f(t + \Delta t) - f(t) = 0,$$

which means

$$\underset{\Delta t \to 0}{\text{Limit}} \, f(t + \Delta t) = f(t),$$

which says that
$$f(t) = \text{Lim } f \text{ at } t,$$
showing that f satisfies the definition of continuity given in Chapter 1.

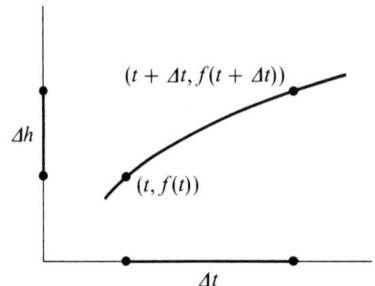

Think of t as being fixed, with Δt tending to zero

EXERCISES

3.1. Suppose that, at a certain point, the derivative dh/dt exists. Prove that condition (2) must then be satisfied, showing that continuity follows from the existence of the derivative. *Hint.* As $\Delta t \to 0$, $\Delta h/\Delta t$ tends to the limit dh/dt. Why does this imply that $\Delta h \to 0$?

3.2. Does the existence of a derivative follow from continuity? That is, if $\Delta t \to 0$ implies $\Delta h \to 0$, does it automatically follow that the limit of $\Delta h/\Delta t$ exists?

3.3. If y is a function of x, which of the following conditions mean that the function is continuous?
   (a) $\Delta y \to 0$ as $\Delta x \to 0$.
   (b) $\Delta x \to 0$ as $\Delta y \to 0$.
   (c) $\text{Limit}_{\Delta y \to 0} \Delta x = 0$.
   (d) $\text{Limit}_{\Delta x \to 0} \Delta y = 0$.

3.4. If derivatives behaved like fractions, what would you expect the following products of derivatives to be equal to?
   (a) $(dA/dC)(dC/dr)$.
   (b) $(dy/dx)(dx/dy)$.
   (c) $(dV/dh)(dh/dt)$.

§4. A Chain Rule?

Believe it or not, there is still something to be learned from the example given on the first page of Chapter 1. Three variables arise from that example, related in the following way.

$$C = \text{cost of } \boxed{} W, \quad \text{where } W \cdot L = 12.$$
$$\phantom{C = \text{cost of }} L$$

4. A Chain Rule?

The variable C can be expressed either in terms of L alone or of W alone, while the variables W and L are themselves related by the fact that their product must be 12, the area of the rectangle. This leads to the following relations.

$$C = 7L + \frac{48}{L}, \tag{3}$$

$$L = \frac{12}{W}, \tag{4}$$

$$C = \frac{84}{W} + 4W. \tag{5}$$

Equations (3) and (4) might be thought of as links in a *chain* of relations which together produce equation (5). That is, the first two equations show how C is a function of L and L is a function of W. This chain of relations forces C to be a function of W, namely, the function specified in equation (5). One feels that there ought to be a rule governing derivatives in the presence of such a chain. The derivatives that arise from (3), (4), and (5) are as follows.

$$\frac{dC}{dL} = 7 - \frac{48}{L^2}. \tag{3'}$$

$$\frac{dL}{dW} = \frac{-12}{W^2}. \tag{4'}$$

$$\frac{dC}{dW} = \frac{-84}{W^2} + 4. \tag{5'}$$

Is there a "chain rule", as one feels there ought to be? Leibniz's notation suggests one to us. The notation suggests that derivatives might act like fractions, in which case we might expect that the product $(dC/dL)(dL/dW)$ is equal to dC/dW. Let us see if this is so.

$$\frac{dC}{dL}\frac{dL}{dW} = \left(7 - \frac{48}{L^2}\right)\left(\frac{-12}{W^2}\right) \quad \text{[from (3') and (4')]}$$

$$= \frac{-84}{W^2} + \frac{48 \cdot 12}{L^2 W^2}$$

$$= \frac{-84}{W^2} + \frac{48 \cdot 12}{(LW)^2}$$

$$= \frac{-84}{W^2} + 4 \quad \text{(since } LW = 12\text{)}$$

$$= \frac{dC}{dW} \quad \text{[from (5')].}$$

It is so! Leibniz's notation has suggested a *chain rule* for derivatives. Has any magician's trick ever been so delightful as this?

EXERCISES

4.1. From equation (4) we get $W = 12/L$.
   (a) Find dW/dL.
   (b) If derivatives behaved like fractions, one might expect that the product of dW/dL and dL/dW is 1. Is it?

4.2. Suppose $y = 1/x$.
   (a) Find dy/dx.
   (b) Suppose, in addition, that $x = t^2 + 3t$. Find dx/dt.
   (c) The chain of relations $y = 1/x$ and $x = t^2 + 3t$ tells us that $y = 1/(t^2 + 3t)$. Find dy/dt by using the general rule for reciprocals.
   (d) Using your answers to parts (a) and (b), find the product of dy/dx and dx/dt. Does your answer agree with dy/dt as found in part (c)? *Hint*. After finding the product, get your answer entirely in terms of t by replacing x with $t^2 + 3t$.

4.3. Suppose $y = \sqrt{u}$.
   (a) Find dy/du.
   (b) Suppose, in addition, that $u = x^2 + 9$. Find du/dx.
   (c) The chain of relations $y = \sqrt{u}$ and $u = x^2 + 9$ tells us that $y = \sqrt{x^2 + 9}$. Find dy/dx by using the square root rule.
   (d) Using your answers to parts (a) and (b), find the product of dy/du and du/dx. Does your answer agree with dy/dx as found in part (c)?

4.4. Consider the chain of relations $y = u^5$ and $u = 3x^2 + 7x$. What does this tell us about the dependence of y upon x? *Answer*: The dependence of y upon x is expressed by the rule $y = (3x^2 + 7x)^5$.

4.5. Consider each of the following chains of relations. What does it tell us about the dependence of y upon x?
   (a) $y = u^3$, $u = 7x - 13$.
   (b) $y = 5/(2 - t)$, $t = 2 - x$. *Answer*: $y = 5/x$.
   (c) $y = u^2$, $u = x^3 - 3x + \pi$.

4.6. A complicated dependence can often be regarded as made up of a chain of simpler dependences. For each of the following, specify such a chain.
   (a) $y = (4x^2 - 6x)^7$. *Answer*: This can be regarded as the result of the chain $y = u^7$, $u = 4x^2 - 6x$.
   (b) $y = \sqrt{3 - 2x + x^2}$.
   (c) $y = (19x - 4)^5$.
   (d) $y = (5x + (1/x))^4$. *Answer*: This is $y = u^4$, where $u = 5x + (1/x)$.

§5. The Chain Rule

Suppose that we have two functions that form a chain of relations, and suppose that each has a derivative. That is the setting for the chain rule.

5. The Chain Rule

Chain Rule. *If y is a function of u and u is a function of x, then*
$$\frac{dy}{dx} = \frac{dy}{du}\frac{du}{dx}.$$

This is the rule that Leibniz's notation enabled us to guess. It is also a rule that Leibniz's notation enables us to remember, since it says, essentially, that derivatives multiply just like fractions, provided that Leibniz's notation is used.

Why is this rule true? First note that u is a continuous function of x, since du/dx exists. This means that

$$\Delta u \to 0 \quad \text{as } \Delta x \to 0. \tag{6}$$

(See exercise 3.1.) This fact will be useful in a moment.

To see the plausibility of the chain rule, consider what is produced by a nonzero change Δx in x. First, a change Δu in u occurs (since u is a function of x), and then the change Δu in turn produces a change Δy in y (since y is a function of u). By ordinary multiplication of fractions,

$$\frac{\Delta y}{\Delta x} = \frac{\Delta y}{\Delta u}\frac{\Delta u}{\Delta x}, \tag{7}$$

provided $\Delta u \neq 0$. As $\Delta x \to 0$, equation (7) becomes "in the limit"

$$\frac{dy}{dx} = \frac{dy}{du}\frac{du}{dx},$$

by virtue of condition (6). □

What has just been given is more of a "plausibility argument" than a real proof to justify the chain rule. The trouble is that equation (7) does not hold if $\Delta u = 0$, i.e., if a nonzero Δx should produce no change in u. A more careful proof, taking account of this troublesome case, will be outlined in a problem at the end of Chapter 8, for those rare readers blessed with both skepticism and patience. Let us for the time being accept the chain rule as true, and learn how to use it. It is the most important rule governing derivatives.

EXAMPLE 1. What is the derivative of $(3x^2 + 7x)^5$?

Here we want dy/dx, where $y = (3x^2 + 7x)^5$. As in exercise 4.4, we may regard y as being given in terms of x by the chain of relations
$$y = u^5 \quad \text{and} \quad u = 3x^2 + 7x.$$
By the chain rule,
$$\frac{dy}{dx} = \frac{dy}{du}\frac{du}{dx}$$
$$= 5u^4(6x + 7)$$
$$= 5(3x^2 + 7x)^4(6x + 7). \quad \square$$

EXAMPLE 2. What is the derivative of $(x^3 - 3x + \pi)^2$?

Here we want dy/dx, where $y = (x^3 - 3x + \pi)^2 = u^2$, if we set u equal to $x^3 - 3x + \pi$. We then have the chain

$$y = u^2 \quad \text{and} \quad u = x^3 - 3x + \pi.$$

By the chain rule,

$$\frac{dy}{dx} = \frac{dy}{du}\frac{du}{dx}$$

$$= 2u(3x^2 - 3)$$
$$= 2(x^3 - 3x + \pi)(3x^2 - 3).$$

(Note that Example 2 can be done by the rule for squares, to get the same answer. The rule for squares is simply the special case of the chain rule that arises when a chain of relations involves a square.) □

EXERCISES

5.1. What is the derivative of $(7x - 13)^3$? *Hint.* The chain of relations in exercise 4.5(a) arises here.

5.2. Find the derivative of $(4x^2 - 6x)^7$. *Hint.* We want dy/dx, where $y = (4x^2 - 6x)^7 = u^7$, if we set u equal to $4x^2 - 6x$.

5.3. Regard each of the following as being given by an appropriate chain of relations, and use the chain rule to obtain the derivative.
   (a) $(5x + (1/x))^4$.
   (b) $(x^2 - 2x + 1)^3$.
   (c) $(t^2 + t)^3$.
   (d) $(5L - 16\pi L^2)^4$.

5.4. The area, radius, and circumference of a circle are related by the chain $A = \pi r^2$ and $r = (1/2\pi)C$. Find dA/dC by the chain rule. *Answer:* $dA/dC = (1/2\pi)C = r$.

5.5. On the basis of the answer to exercise 5.4, and with nothing but Leibniz's notation to guide your intuition, guess what dC/dA is.

5.6. From the equations $A = \pi r^2$ and $C = 2\pi r$,
   (a) find an algebraic rule giving C in terms of A.
   (b) find dC/dA from your answer to part (a), and see if it agrees with the guess made in exercise 5.5.
   Partial answer: $C = \sqrt{4\pi A}$. (Find dC/dA by the square root rule.)

5.7. If $y = x^2$, $x > 0$, then it follows that $x = \sqrt{y}$, $y > 0$.
   (a) Find dy/dx from the equation $y = x^2$.
   (b) Find dx/dy from the equation $x = \sqrt{y}$.
   (c) The expression dy/dx "looks like" the reciprocal of dx/dy. Is it?

§6. Related Rates

Problems involving *related rates* are tailor-made for the chain rule. Let us do an easy example, then a harder example, in order to make some observations on how such problems may be handled.

EXAMPLE 3. A pebble is dropped in still water, forming a circular ripple whose radius is expanding at a rate of 3 inches per second. When the radius is 7 inches, how fast is the area A of the ripple increasing?

In rate problems it is important to get straight exactly what we are required to find, as well as what we are given to start off with. One must remember that the derivative measures the instantaneous rate of increase. The rate of increase of area A (with respect to time) is then dA/dt. The goal of Example 3 is then to find dA/dt when r is 7. This may be abbreviated by

$$\left. \frac{dA}{dt} \right|_{r=7} \tag{8}$$

(read "dA/dt, evaluated when r is 7"). The expression (8) is the rate we are required to find.

What are we given to work with? The first sentence of Example 3 tells us a related rate:

$$\frac{dr}{dt} = 3, \tag{9}$$

and we know that there is a chain of relations connecting the variables A, r, and t:

$$A = \pi r^2 \quad \text{and} \quad r \text{ is a function of } t.$$

By the chain rule, using (9),

$$\frac{dA}{dt} = \frac{dA}{dr}\frac{dr}{dt} = 2\pi r(3).$$

Therefore,

$$\frac{dA}{dt} = 6\pi r.$$

To evaluate expression (8), plug in $r = 7$:

$$\left. \frac{dA}{dt} \right|_{r=7} = 6\pi r \Big|_{r=7}$$
$$= 6\pi(7)$$
$$= 42\pi \text{ in}^2/\text{sec}.$$

(The expression "in²/sec" abbreviates the phrase "square inches per second". Why must dA/dt come out in these units?) □

EXAMPLE 4. The bottom end of a 10-foot ladder resting against a wall is pulled away from the wall at a rate of 2 ft/sec. At what rate is the top end falling at the instant when the bottom end is 6 feet from the wall?

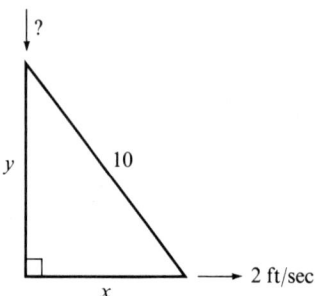

There are two ways to work this problem, and we shall look at both of them. As with virtually every calculus problem, the first step is to *see the problem in terms of variables*. Time is certainly one variable, and the others are x and y, the legs of a right triangle formed by the ladder, the wall, and the floor. As time t increases, it is evident that x increases and y decreases. The derivative dy/dt gives the rate of *increase* of y. The rate at which the top of the ladder *falls* is the rate of *decrease* of y, which is the negative of dy/dt. Thus we are required to find

$$-\frac{dy}{dt}\bigg|_{x=6}. \qquad (10)$$

What are we given to work with? We know a related rate:

$$\frac{dx}{dt} = 2, \qquad (11)$$

and we know an age-old relation that connects the variables x and y:

$$x^2 + y^2 = 100. \qquad (12)$$

Solving this equation for y shows that we have the following chain of relations connecting y, x, and t:

$$y = \sqrt{100 - x^2} \quad \text{and} \quad x \text{ is a function of } t.$$

By the chain rule, with help from equation (11),

$$\frac{dy}{dt} = \frac{dy}{dx}\frac{dx}{dt} = \frac{-2x}{2\sqrt{100-x^2}}(2),$$

so that

$$-\frac{dy}{dt} = \frac{2x}{\sqrt{100-x^2}}.$$

6. Related Rates

To find (10), as desired, plug in $x = 6$:

$$-\frac{dy}{dt}\bigg|_{x=6} = \frac{12}{\sqrt{100-36}}$$
$$= \frac{3}{2} \text{ ft/sec.} \qquad \square$$

An alternate way to finish this problem is as follows. In equation (12), x and y both depend upon t. Taking the derivative with respect to t yields

$$\frac{d}{dt}(x^2 + y^2) = \frac{d}{dt}(100),$$

$$2x\frac{dx}{dt} + 2y\frac{dy}{dt} = 0 \quad \text{(by rule for squares)},$$

$$4x + 2y\frac{dy}{dt} = 0 \quad [\text{by (11)}],$$

so that

$$-\frac{dy}{dt} = \frac{2x}{y}.$$

Therefore,

$$-\frac{dy}{dt}\bigg|_{x=6} = \frac{2x}{y}\bigg|_{x=6} = \frac{2(6)}{8} = \frac{3}{2} \text{ ft/sec},$$

since (by the Pythagorean theorem) $y = 8$ when $x = 6$. $\qquad \square$

Related rates problems may seem difficult at first because everything in them seems to be changing at once. But this is only an invitation to see the problem in terms of variables and to use the derivative's magic power to measure change. Then adopt the philosopher's point of view. Seek that which does not change, that "holds sway above the flux". Search for a relation between the variables that always holds. This relation may be as simple as the Pythagorean theorem (in Example 4) or the formula for the area of a circle (in Example 3). Finally, express yourself in the language of Leibniz. It will lead you to the truth.

EXERCISES

6.1. A pebble is dropped in still water, forming a circular ripple whose radius is increasing at a rate of 5 inches per second. When the radius is 3 inches, how fast is the area of the ripple increasing?

6.2. The bottom end of a 13-foot ladder resting against a wall is pulled away from the wall at a rate of 3 feet per second. How fast is the top end falling when
   (a) the bottom end is 5 feet from the wall?
   (b) the top end is 5 feet from the floor?
   Answers: (a) $\frac{5}{4}$ ft/sec. (b) $\frac{36}{5}$ ft/sec.

6.3. An airplane is flying horizontally at 5000 feet, with speed 600 ft/sec, and an observer is on the ground. Let s be the distance from the observer to the airplane.

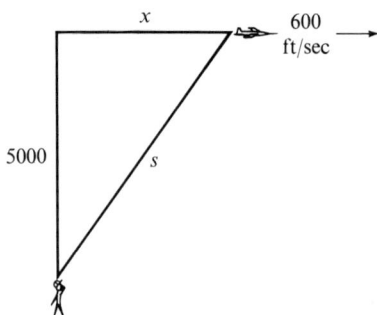

Find ds/dt, the rate of increase of s, at each of the following instants.
(a) Two seconds after the plane passes directly above the observer.
(b) One second after the plane is directly overhead.
(c) At the instant the plane is overhead.
(d) Three seconds before the plane is directly overhead.
Hint. (c) At this instant the plane is closest to the observer, so s assumes its minimum. What value does the derivative take when a minimum is attained?
Hint. (d) You want $(ds/dt)|_{x=-1800}$, and you should expect a negative answer, since the distance s is decreasing at this instant.

6.4. Consider again the rock whose motion is described at length in Section 2. Suppose there is an observer at ground level, 36 feet from the point where the rock is released. Let s be the distance from the observer to the rock.

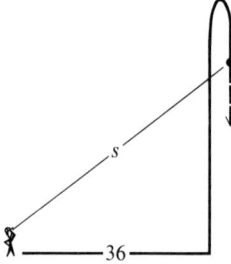

Find ds/dt when
(a) $t = 1$ (and $h = 48$).
(b) $t = 2$ (and $h = 64$).
(c) $t = 3$ (and $h = 48$).
Hint. From the table in Section 2, dh/dt is 32, 0, and -32 at times $t = 1, 2,$ and 3, respectively.

6.5. A child 4 feet tall walks directly away from a street light that is 10 feet above the ground. She walks at a rate of 5 feet per second. How fast does the tip of her shadow move?
Hint. You want dL/dt, knowing that $dx/dt = 5$, and knowing, from similar triangles, that the relation $L/10 = (L - x)/4$ always holds. Proceed as in the alternate solution to Example 4.

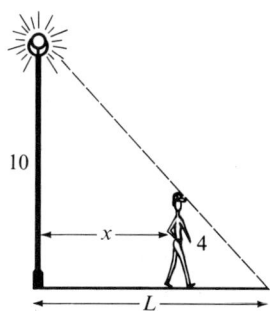

§7. Antiderivatives

We have so far been mainly concerned with the following operation: getting a function, forming with it a quotient of differences, and taking a limit in order to get its derivative. This operation is called *differentiation*. To *differentiate* a function is to take its derivative. For example, we get

$$
\begin{aligned}
&1 \quad \text{by differentiating} \quad t, \\
&t \quad \text{by differentiating} \quad \frac{t^2}{2}, \\
&t^2 \quad \text{by differentiating} \quad \frac{t^3}{3}, \\
&t^3 \quad \text{by differentiating} \quad \frac{t^4}{4}.
\end{aligned} \tag{13}
$$

In general, we get

$$t^n \quad \text{by differentiating} \quad \frac{t^{n+1}}{n+1}.$$

What we have been studying so far is called the *differential* calculus. The name is due to Leibniz who, writing in Latin, spoke of "calculus differentialis". An important concern of the differential calculus is simply to fill in the question mark given the following table.

t	y	dy/dt
t	$f(t)$	?

The differential calculus is concerned with how to get from the second column above, to the third. It is done, of course, by means of Fermat's method. We shall now consider the reverse problem: *how to get from the third column back to the second*. This is a principal concern of "calculus integralis", as Leibniz called it, writing in 1696.

t	y	dy/dt
t	?	$f(t)$

This is the problem of finding an *antiderivative*. Fortunately, we already know a little about antiderivatives. From the formulas (13) it is completely obvious that

t is an antiderivative of 1,

$\dfrac{t^2}{2}$ is an antiderivative of t,

$\dfrac{t^3}{3}$ is an antiderivative of t^2,

$\dfrac{t^4}{4}$ is an antiderivative of t^3.

In general,

$\dfrac{t^{n+1}}{n+1}$ *is an antiderivative of* t^n.

Knowing this enables one to find easily antiderivatives of many functions whose rules involve powers only:

An antiderivative of -32 *is* $-32t$.
An antiderivative of $-32t$ *is* $-16t^2$.
An antiderivative of $64 - 32t$ *is* $64t - 16t^2$.
An antiderivative of $1 + 4t - 9t^2$ *is* $t + 2t^2 - 3t^3$.

We generally say "an", rather than "the", in speaking of antiderivatives, because there is generally more than one antiderivative of a given function. Having found an antiderivative, you can easily find another, simply by adding any constant to the one already in hand. *Reason*: If F is an antiderivative of f (i.e., if $F' = f$), then $F + C$ is too [since $(F + C)' = F'$, the derivative of a constant C being 0]. Thus, for example,

$$-32t, \qquad -32t - 7, \qquad -32t + \pi, \qquad -32t + C,$$

where C can be any constant, are all antiderivatives of -32. We cannot speak of "the" antiderivative of -32, unless we specify, by giving additional information, exactly which antiderivative we mean.

EXAMPLE 5. Consider the function given by $f(t) = -32$, with domain $0 \leq t$. Find

(a) *an* antiderivative F of f.
(b) *the* antiderivative F of f that takes the value 64 when t is 0.
(c) *the* antiderivative F of f that takes the value -40 when t is 5.

We have already answered part (a). Any function whose rule is of the form $-32t + C$ will do, where C can be any constant (including, of course, 0). □

To answer (b), note that what is required is to fill in properly the following table.

7. Antiderivatives

t	$F(t)$	$f(t)$
0	64	
t	?	-32

The first line of the table gives enough information (we hope) to make the antiderivative unique. By part (a) we expect

$$F(t) = -32t + C, \qquad (14)$$

and by the first line of the table we must have

$$F(0) = 64. \qquad (15)$$

From (14), with $t = 0$, we get

$$F(0) = -32(0) + C = C.$$

This equation, together with (15), shows that C must be 64. Thus, in equation (14), not just any constant C will do; we must have $F(t) = -32t + 64$. □

To answer (c) we must satisfy the condition

$$F(5) = -40, \qquad (16)$$

in addition to equation (14), which implies

$$F(5) = -32(5) + C. \qquad (17)$$

Putting (16) and (17) together determines C:

$$-40 = -160 + C,$$
$$C = 120.$$

The answer to part (c) is then given by $F(t) = -32t + 120$. □

An antiderivative is generally determined, not uniquely, but only "up to an additive constant". Additional information, as in parts (b) and (c) of Example 5, is required to specify a unique antiderivative. How can we be *sure* of uniqueness, though? Our procedure here will be justified in the next section.

EXERCISES

7.1. In each of the following, specify an antiderivative F of the given function f.
  (a) $f(t) = 3t + 2$. Answer: $F(t) = \frac{3}{2}t^2 + 2t + C$.
  (b) $f(t) = -32t + 96$.
  (c) $f(t) = 1 + t + t^2$.
  (d) $f(t) = \pi t^3$. Answer: $F(t) = \frac{1}{4}\pi t^4 + C$.

7.2. In each of parts (a) through (d) of exercise 7.1, find the antiderivative F that takes the value 6 when $t = 2$.
Answers: (a) $F(t) = \frac{3}{2}t^2 + 2t - 4$. (d) $F(t) = \frac{1}{4}\pi t^4 + 6 - 4\pi$.

7.3. In each of the following the derivative of h is specified. Use antiderivatives to find h itself.
  (a) $dh/dt = 96 - 32t$.
  (b) $dh/dt = -40 - 32t$.
  (c) $dh/dt = -32t$.

7.4. In each of parts (a) through (c) of exercise 7.3, find the antiderivative h that takes the value 100 when $t = 0$. *Answer:* (a) $h = 96t - 16t^2 + 100$.

7.5. In each of parts (a) through (c) of exercise 7.3, find the antiderivative h that takes the value 100 when $t = 1$. *Answer:* (a) $h = 96t - 16t^2 + 20$.

§8. A Fundamental Principle and Freely Falling Bodies

Taking antiderivatives points us in a direction exactly opposite the direction of differential calculus, and leads to the study of *integral* calculus. The reason for the use of the word *integral* will be explained in Chapter 6.

Let us examine more carefully the notion of an antiderivative. Because we get 0 by differentiating a constant function $F(t) = C$, it seems plausible that

$$\text{any antiderivative of 0 is a constant function.} \tag{18}$$

Can we be sure of this? Another way of saying the same thing is as follows:

$$\text{If } F'(t) = 0, \text{ then } F(t) = C \text{ for some constant } C. \tag{19}$$

Statements (18) and (19) are true, but only with the additional understanding that the domain in question is *connected*, that is, has no holes in it. The example pictured below shows that statement (19) can fail with holes in the domain. In this example $F'(t) = 0$, yet the function F is not constant.

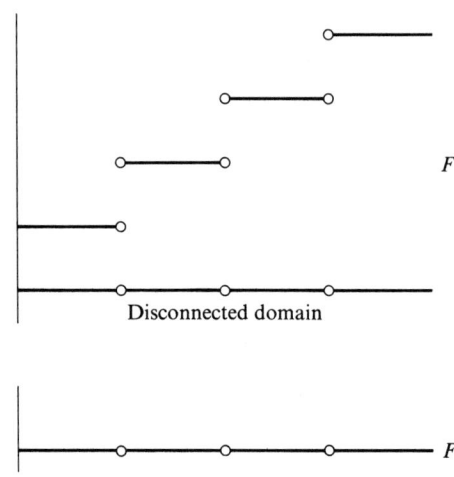

Disconnected domain

8. A Fundamental Principle and Freely Falling Bodies 131

In analysis it is shown that (18) and (19) are true, *provided the domain is connected*. The reader is asked to accept this as intuitively obvious. It has an important consequence, which will provide the basis for much that will follow.

A Fundamental Principle of Integral Calculus. *Let F and A be functions defined on the same connected domain, and assume that $dA/dt = dF/dt$. Then,*
$$A(t) = F(t) + C$$
for some constant C.

PROOF. First note that
$$\frac{d}{dt}(A - F) = \frac{dA}{dt} - \frac{dF}{dt} = 0, \qquad (20)$$

since $dA/dt = dF/dt$, by assumption. Therefore,

$$A - F = \text{an antiderivative of } 0 \quad [\text{by (20)}]$$
$$= \text{a constant function} \quad [\text{by (18)}],$$

since the domain is assumed connected. Thus, for some constant C,

$$A(t) - F(t) = C,$$
$$A(t) = F(t) + C. \qquad \square$$

The fundamental principle just established is sometimes phrased this way: *Two antiderivatives of the same function differ by a constant*. The reader is cautioned to remember that zero is a perfectly good constant.

Intuition often runs ahead of reason. A good example of this is found in Section 7, where we expected equation (14) to hold and to justify what followed there. Now we know that we were right. The fundamental principle guarantees equation (14), for it says that any antiderivative *whatsoever* of -32 differs by a constant from $-32t$, on a connected domain. Thus, parts (b) and (c) of Example 5 do indeed have unique answers, for the domain of Example 5 is connected.

Though the fundamental principle may appear abstract, it has quite practical uses. It comes into play whenever we know the rate of change of a quantity and want to know the quantity itself. An example of this is furnished by the study of *freely falling bodies*. This refers to the vertical movement of objects thrown in the air near the earth's surface. If gravity is the only force acting on the body (which means that the body is not self-propelled and that the effect of air friction is ignored), the body is said to be *freely falling*.

Near the earth's surface, the effect of gravity is very simple to describe. Each second gravity *decreases* the upward speed of a freely falling body by

32 ft/sec. That is, if v is the upward speed of a freely falling body, then the effect of gravity is specified by the equation

$$\frac{dv}{dt} = -32. \tag{21}$$

This equation gives the rate of increase of v. If we want to know v itself, the fundamental principle says that

$$v = -32t + C$$

for some constant C. We need additional information to determine C. If, for example, the initial speed was known to be 64 ft/sec, then $v = -32t + 64$, as in part (b) of Example 5. If, as in part (c) of Example 5, the speed is known to be downwards at 40 ft/sec when $t = 5$, then we must have $v = -32t + 120$.

EXAMPLE 6. A rock is thrown upward from ground level with an initial speed of 64 ft/sec. Treating the rock as a freely falling body, answer the following:

(a) What is the maximum height attained by the rock?
(b) Where is the rock 3 seconds after it is released?
(c) When, and with what speed, will it hit the ground?

Here, we know that the upward speed v is given by $v = -32t + 64$, by the remarks preceding the example. But the upward speed v is equal to dh/dt, the rate of increase of height. Hence.

$$\frac{dh}{dt} = -32t + 64.$$

Therefore, by the fundamental principle,

$$h = -16t^2 + 64t + C \tag{22}$$

for some constant C. What is C? Since the rock is thrown from ground level, we must have $h = 0$ when $t = 0$, so that, from (22),

$$0 = -16(0)^2 + 64(0) + C.$$

Therefore, $C = 0$ and (22) becomes

$$h = -16t^2 + 64t.$$

But this is the height formula that was discussed at length in Section 2. From that section we know that the maximum height of the rock is 64 feet, the rock is at 48 feet and falling when $t = 3$, and it hits the ground when $t = 4$ with a downward speed approaching 64 ft/sec. □

8. A Fundamental Principle and Freely Falling Bodies

The way the preceding example was begun involved two steps that can be schematized as follows:

t (time since release)	h (height)	$v\ (=dh/dt)$ (upward speed)	dv/dt (upward acceleration due to gravity)
0	0	+64	
t	(Step 2)	(Step 1)	−32

Example 6 involves two antiderivatives. In step 1, when we "pull back" from the fourth column to the third, we must adjust the constant so that the initial speed is 64, as required. When we pull back from the third column to the second, another constant must be adjusted to be in accordance with the given initial height.

EXAMPLE 7. From a building 200 feet high, a ball is thrown downward at an initial speed of 50 ft/sec. Find an algebraic expression for the height of the ball in terms of the time since release, treating the ball as a freely falling object.

Here, we begin with the following information, and we want to fill in the question mark giving h in terms of t. To do this we must first fill in the other question mark properly.

t	h	$v = dh/dt$	dv/dt
0	200	−50	
t	?	?	−32

We must simply pull back twice by taking antiderivatives, each time adjusting the constant in accordance with the initial conditions. Pulling back to the third column yields $v = -32t - 50$, and thus in the second column we must have $h = -16t^2 - 50t + 200$. This equation gives h in terms of t, so long as gravity is the only force that acts upon the ball, that is, until $h = 0$ when the ball hits the ground. □

The method outlined in Examples 6 and 7 develops a mathematical "model" predicting the motion of a freely falling body. Given the initial speed and the initial location, we can find the formulas governing the speed and the height at *any* time, so long as gravity acts. Thus the dynamics of a freely falling body can be worked out with ease, through the help of our fundamental principle of integral calculus.

Our model is hardly perfect, however, for the notion of a *freely falling body* is an idealization of what actually happens when a rock is tossed up in the air. Air friction has its effect, particularly at high speeds, and a more complex model is required to account for this and other factors.

EXERCISES

8.1. A ball is thrown vertically from a cliff. Find its upward speed v in terms of the time t since release if
(a) it is initially thrown upward at 96 ft/sec.
(b) it is initially thrown downward at 60 ft/sec.

8.2. A rock is thrown upward from ground level at an initial speed of 96 ft/sec.
(a) What is its maximum height?
(b) Where is the rock 4 seconds after release?
(c) When will the rock hit the ground?
(d) What is the speed of the rock at the moment of impact? *Answer*: $\text{Limit}_{t \to 6^-} v = -96$ ft/sec.

8.3. From a tower 256 feet high, a ball is thrown upward at an initial speed of 96 ft/sec. When, and with what speed, will it hit the ground? *Hint.*

t	h	v	dv/dt
0	256	96	
t	(Step 2)	(Step 1)	-32
?	0	?	

Partial answer: At impact the downward speed approaches 160 ft/sec.

8.4. Suppose, in exercise 8.3, the ball is thrown *downward* initially at 96 ft/sec. When, and with what speed, will it hit the ground? *Hint.* This is done like exercise 8.3, except the initial speed is -96 instead of 96.

8.5. A rifle is supposed to have a muzzle velocity of 1000 ft/sec. If it is fired straight up, how high will the bullet go?

8.6. A certain rifle, when fired straight up, will send a bullet to a height of 2000 feet. What is the muzzle velocity of the rifle? *Hint.* Letting v_0 be the muzzle velocity, we have

t	h	v
0	0	v_0
	2000	0

Find v_0, beginning with equation (21).

8.7. A boy hurls a ball directly upwards. It hits the ground 8 seconds later. What was the ball's initial speed? *Hint.*

t	h	v
0	0	v_0
8	0	

Find v_0.

§9. Antiderivatives and Distance

The method of freely falling bodies does not apply, of course, to self-propelled objects like motorcycles, cars, and rockets. Nevertheless, antiderivatives come into play with self-propelled objects when it is desired to convert speedometer readings into distance traveled. Suppose a navigator charts his speedometer reading as it varies over the span of an hour. How can the navigator determine from his chart the distance traveled during this hour? The answer involves antiderivatives.

EXAMPLE 8. A rocket ship blasts through the firmament on a journey directly away from the earth. At noon on a certain day the navigator becomes interested in the ship's speedometer reading as a function of time, and finds that it is given by $100t^3 - 400t^2 + 800t$, where t is the time in hours since noon. If this function f gives the speedometer reading in km/hr (kilometers per hour), find the distance traveled by the rocket

(a) between noon and two o'clock.
(b) between one and four o'clock.

The speedometer reading is the instantaneous rate of change of distance from the earth. If we let s be the distance from the earth, we then have

$$\frac{ds}{dt} = 100t^3 - 400t^2 + 800t = f(t).$$

The distance s must be in kilometers, since the speedometer reading is given in km/hr. Schematically, the situation we are faced with can be pictured as follows.

t (hours since noon)	s (distance from earth)	ds/dt (speed)
0	?	
1	?	
2	?	
4	?	
t	$F(t)$	$f(t)$

We know the expression for $f(t)$ and we need to fill in the question marks correctly to answer (a) and (b) above.

We first find an antiderivative F of f:

$$s = F(t) = \frac{100t^4}{4} - \frac{400t^3}{3} + \frac{800t^2}{2} + C.$$

We know the *position function* F must be of this form by the fundamental principle, but we are not given enough information to determine C. Nevertheless, by plugging in the values 0, 1, 2, and 4 into this expression for F, we can easily figure out what was required:

(a) The distance traveled between time $t = 0$ and $t = 2$ is equal to

$$\text{(position at } t = 2\text{) minus (position at } t = 0\text{)}$$
$$= F(2) - F(0)$$
$$= 933.7 + C - C$$
$$= 933.7 \text{ km.}$$

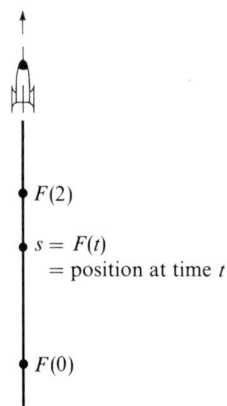

(b) The distance traveled between time $t = 1$ and $t = 4$ is equal to

$$F(4) - F(1) = 4266.7 + C - (291.7 + C)$$
$$= 3975 \text{ km.} \qquad \square$$

To get the distance traveled, given the speed function f, is then a job for antiderivatives. If F is an antiderivative of f, then F gives the position at time t, so that the initial and final positions are readily determined. The distance traveled is simply the distance between the initial and final position, if the ship does not reverse course. (In Example 8 the speed function is always positive, so the direction of travel is always away from the earth.)

What happens if the speed function changes sign in the midst of the journey, so that the course of travel is reversed? Then the distance traveled must be calculated in two steps, as illustrated in the next example.

EXAMPLE 9. A rock is thrown upward at an initial speed of 64 ft/sec. How far does the rock travel during the first 3 seconds of its flight?

Here the speed function f is given by $f(t) = 64 - 32t$, because of the influence of gravity. When $t = 2$, the sign of the speed function changes from positive to negative, showing that the rock's motion changes from up to down. We must calculate separately the distance traveled by the rock

9. Antiderivatives and Distance

during its upward and downward journey. An antiderivative F is given by

$$F(t) = -16t^2 + 64t + C$$

for some constant C. We are not given enough information to determine C, for we do not know the initial height of the rock.

Nevertheless, the distance traveled upward is

$$F(2) - F(0) = 64 + C - C = 64 \text{ ft.}$$

The distance traveled downward from $t = 2$ to $t = 3$ is

$$F(2) - F(3) = 64 + C - (48 + C)$$
$$= 16 \text{ ft.}$$

The total distance traveled is then 80 feet, even though the distance between the rock's final and initial positions, given by $F(3) - F(0)$, is only 48 feet. (Example 9 is, of course, essentially the same situation that we have met twice before, in Section 2 and in Example 6, Section 7.) □

EXERCISES

9.1. In Example 8, find the distance traveled by the rocket ship between
 (a) $t = 0$ and $t = 3$.
 (b) $t = 1$ and $t = 3$.

9.2. If ds/dt is the upward speed, then its rate of increase, which is $d(ds/dt)/dt$, is the upward acceleration. In Leibniz's notation, the symbol $d(ds/dt)/dt$ is abbreviated to d^2s/dt^2. Find the upward acceleration in Example 8, and then answer the following:
 (a) Is the rocket *accelerating* or *decelerating* in its upward movement when $t = 3$?
 (b) Is the rocket *accelerating* or *decelerating* when $t = 0$? *Answer*: Since $d^2s/dt^2|_{t=0} = 800$ km/hr per hour, which is *positive*, the rocket is accelerating.

9.3. In Example 9, how far does the rock travel between
 (a) $t = 1$ and $t = 3$?
 (b) $t = 0$ and $t = 4$? *Answer*: 128 feet.

9.4. A stone is thrown upward from a tower window at an initial speed of 48 ft/sec. Find the distance traveled by the stone during its first 3 seconds of flight, treating it as a freely falling body.

9.5. Do exercise 9.4 with the modification that the stone is thrown downward instead of upward.

9.6. The speed function f of a ship stays constant at 30 km/hr, i.e., $f(t) = 30$. Find how far the ship travels between $t = 1$ and $t = 4$,
 (a) by the method of antiderivatives, as in Example 8.
 (b) by common sense.

§10. A Token

There is a lot to be learned in the simple pastime of contemplating a circle. (Chapter 7, in fact, will be entirely devoted to this.) Through a problem at the end of Chapter 2, it was established that the area A of a circle is given by

$$A = \pi r^2, \tag{23}$$

where r is the radius. As also noted in Chapter 2, problem 16, Archimedes established the fact that the two figures below have the same area.

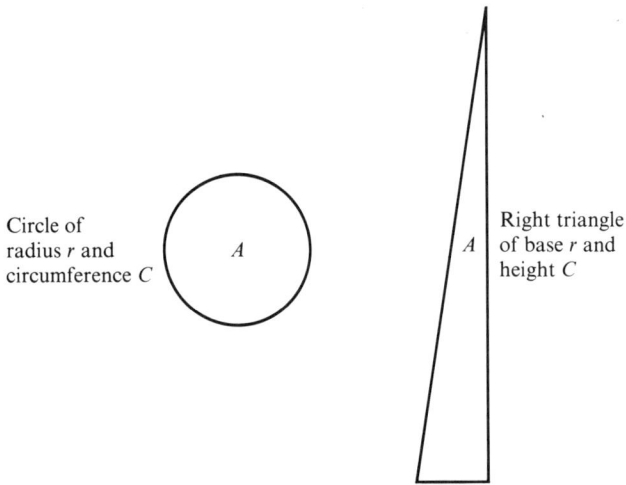

The equality of areas produces the equation $\pi r^2 = \frac{1}{2}Cr$, from which we get the formula for the circumference,

$$C = 2\pi r. \tag{24}$$

From (23) we derive the equation $dA/dr = 2\pi r$, so that by (24) we have

$$\frac{dA}{dr} = C. \tag{25}$$

Thus the derivative (with respect to r) of the area of a circle is equal to the circumference! It takes only a little sensitivity to recognize that there must be here some sort of underlying harmony that has so far gone unnoticed. Equation (25) is a token from the gods. It is up to us to figure out what it really means. Remember the words of Xenophanes and Heraclitus!

Is equation (25) just an accident? Or should we have realized, by adopting the proper point of view, that this equation was bound to be true? *Let us set about trying to derive equation (25) directly from fundamental considerations.* We may discover something worth knowing in the process.

The equation $C = 2\pi r$ defines, of course, a straight line of slope 2π, passing through the origin. The variables r, C, and A are then related as indicated in the figure.

10. A Token

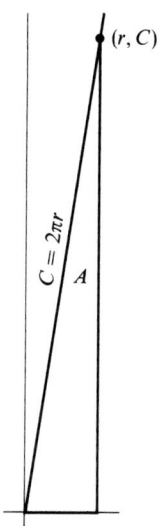

If r is changed by a small amount Δr, what are the corresponding changes ΔC and ΔA? They are as indicated in the figure below.

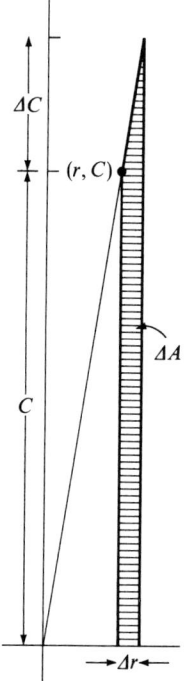

To calculate the change ΔA in area, regard it as being made up of a rectangle surmounted by a triangle. The rectangle has base Δr and height C, and the triangle has base Δr and height ΔC. Therefore,

$$\Delta A = C(\Delta r) + \frac{1}{2}(\Delta r)(\Delta C).$$

Dividing by Δr produces

$$\frac{\Delta A}{\Delta r} = C + \frac{1}{2}(\Delta C).$$

As $\Delta r \to 0$, we must have $\Delta C \to 0$ also, because C is a continuous function of r. Therefore,

$$\frac{dA}{dr} = \underset{\Delta r \to 0}{\text{Limit}} \frac{\Delta A}{\Delta r}$$

$$= \underset{\Delta r \to 0}{\text{Limit}} \left[C + \frac{1}{2}(\Delta C) \right]$$

$$= C.$$

What have we learned? We have learned that the equation $dA/dr = C$ is simply a consequence of the fact that A is the area beneath the curve giving C as a continous function of r. That is, given the picture

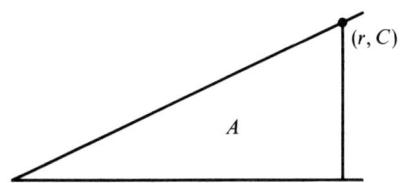

it must follow that $dA/dr = C$.

Are areas beneath continuous functions always related to the functions in this way? What is the secret that is still eluding us?

It turns out that what is behind all this is the fundamental theorem of calculus. Leibniz guessed it, probably sometime in the 1670s. Actually, Isaac Newton had come upon it in 1666 (at the age of twenty-three), but kept it a secret.

The fundamental theorem is discussed in Chapter 6. The reader may possibly be able to guess what it says beforehand, however, after doing the following exercises. The many uses of this theorem will still bring surprise.

EXERCISES

10.1. Make a guess about a relation between the three variables that occur in each of the following pictures.

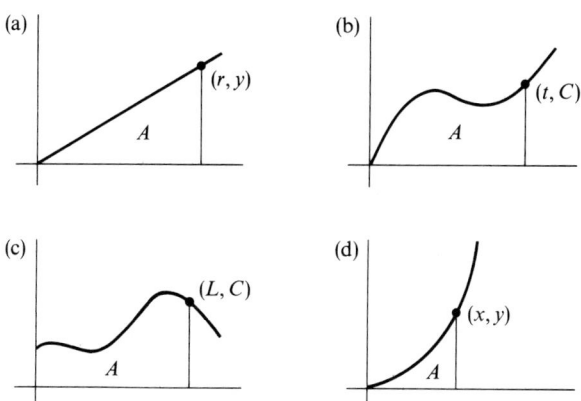

Answer: (c) $dA/dL = C$. (d) $dA/dx = y$.

10.2. Guess again, as in exercise 10.1, but utilize the fact that equations for the curves are furnished.

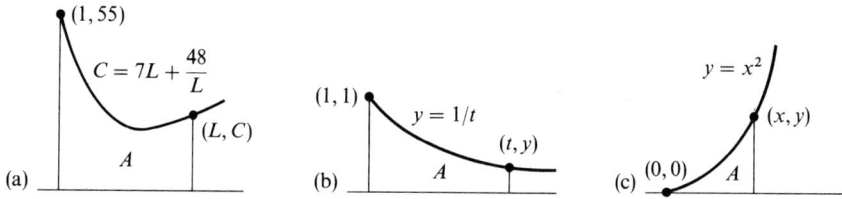

Answer: (c) $dA/dx = x^2$.

§11. Leibniz

Leibniz, like Descartes, is one of several mathematicians who were also distinguished philosophers. He said that we live in "the best of all possible worlds". Voltaire, who admired Leibniz, still could not tolerate his unrestrained optimism, and satirized it in *Candide*. Relatively recent developments in physics have shown, however, that profound truth can be found in Leibniz's seemingly naive belief:

> Ours, according to Leibniz, is the best of all possible worlds, and the laws of nature can therefore be described in terms of extremal principles.
>
> C. L. Siegel and J. K. Moser*

* Siegel/Moser, *Lectures on Celestial Mechanics* (New York: Springer-Verlag, 1971) p.1.

MENSIS OCTOBRIS A. M DC LXXXIV.

NOVA METHODUS PRO MAXIMIS ET MI-
nimis, itemque tangentibus, quæ nec fractas, nec irrationales quantitates moratur, & singulare pro illis calculi genus, per G. G. L.

SIt axis AX, & curvæ plures, ut VV, WW, YY, ZZ, quarum ordi- TAB.XII natæ, ad axem normales, VX, WX, YX, ZX, quæ vocentur respe- ctive, v, w, y, z; & ipsa AX abscissa ab axe, vocetur x. Tangentes sint VB, WC, YD, ZE axi occurrentes respective in punctis B, C, D, E. Jam recta aliqua pro arbitrio assumta vocetur dx, & recta quæ sit ad dx, ut v (vel w, vel y, vel z) est ad VB (vel WC, vel YD, vel ZE) vocetur d v (vel d w, vel dy vel dz) sive differentia ipsarum v (vel ipsarum w, aut y, aut z) His positis calculi regulæ erunt tales:

Sit a quantitas data constans, erit da æqualis o, & d ax erit æqu. a dx: si sit y æqu v (seu ordinata quævis curvæ YY, æqualis cuivis ordinatæ respondenti curvæ VV) erit dy æqu. dv. Jam *Additio & Subtractio*: si sit z — y + w + x æqu. v, erit d z — y + w + x seu dv, æqu dz — d y + d w + d x. *Multiplicatio*, d x v æqu. x dv + v d x, seu posito y æqu. xv, fiet d y æqu x dv + v d x. In arbitrio enim est vel formulam, ut xv, vel compendio pro ea literam, ut y, adhibere. Notandum & x & d x eodem modo in hoc calculo tractari, ut y & dy, vel aliam literam indeterminatam cum sua differentiali. Notandum etiam non dari semper regressum a differentiali Æquatione, nisi cum quadam cautione, de quo alibi. Porro *Divisio*, d $\frac{v}{y}$ vel (posito z æqu. $\frac{v}{y}$) dz æqu. $\frac{\pm v dy \mp y dv}{yy}$

Quoad *Signa* hoc probe notandum, cum in calculo pro litera substituitur simpliciter ejus differentialis, servari quidem eadem signa, & pro + z scribi + dz, pro — z scribi — dz, ut ex additione & subtractione paulo ante posita apparet; sed quando ad exegesin valorum venitur, seu cum consideratur ipsius z relatio ad x, tunc apparere, an valor ipsius dz sit quantitas affirmativa, an nihilo minor seu negativa: quod posterius cum fit, tunc tangens ZE ducitur a puncto Z non versus A, sed in partes contrarias seu infra X, id est tunc cum ipsæ ordinatæ

Figure 1. First page of the first paper published on the calculus. Leibniz wrote this short account—only six pages long—in 1684. The long title reads "A new method for maxima and minima, as well as for tangents, which is not obstructed by fractional and irrational quantities, and a unique calculus for it".

Modern books on celestial mechanics show that the course actually chosen as the path of a heavenly body is the optimum among all possible courses. It should be added that the *optimum path* must be defined quite carefully, and in a way that Fermat was more likely to have foreseen than Leibniz. Nevertheless, the optimization techniques of Leibniz's calculus enter the picture in an essential way.

Few men have been more gifted than Leibniz. He invented a calculating machine that could multiply, divide, and take roots. He organized the Berlin Academy of Sciences and was its first president. He knew many languages, was an historian and a diplomat, with interests in economics, and a pioneer in the field of international law.

But when he died in 1716, little notice was taken. Only one mourner attended the funeral of Gottfried Wilhelm von Leibniz, and an observer said that "he was buried more like a robber than what he really was, the ornament of his country."

Problem Set for Chapter 5

1. A motorcycle travels on a straight road leading directly away from a city. At time t hours past noon its distance from the city is $10t^3 - 40t^2 + 80t$ miles.
   (a) How far does the motorcycle go between one o'clock and three o'clock?
   (b) What is its average speed over the time interval between one o'clock and three o'clock?
   (c) What is the speedometer reading at two o'clock?
   (d) At two o'clock, is the motorcycle *accelerating* or *decelerating*?
   (e) At one o'clock, is the motorcycle *accelerating* or *decelerating*?

2. The height of a rock at time t is given by $h = -4.9t^2 + 20t$, where h is in meters and t is in seconds.
   (a) Is the rock *rising* or *falling* when $t = 3$?
   (b) How fast is the rock going when $t = 3$?
   (c) When does the rock attain its maximal height?
   (d) What is the acceleration of the rock?

3. Suppose x and y are each functions of t. Let A denote their product. (If x and y are positive, A can be pictured as the area of a rectangle whose sides vary in length as t increases.)

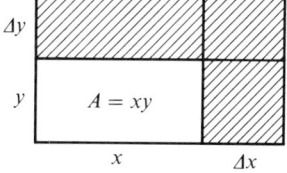

The shaded area is ΔA

(a) In time Δt, x becomes $x + \Delta x$ and y becomes $y + \Delta y$. Hence $A + \Delta A = (x + \Delta x)(y + \Delta y)$. Use this equation to find ΔA in terms of x, y, Δx, and Δy.
(b) Suppose dx/dt and dy/dt exist. What must happen to Δx and Δy as $\Delta t \to 0$?
(c) Derive the *product rule*, by carrying out the following steps.
   (i) Take your answer to (a), and divide both sides of the equation by Δt.
   (ii) Take the limit as $\Delta t \to 0$, using your answer to part (b), to show that

$$\frac{dA}{dt} = x\frac{dy}{dt} + y\frac{dx}{dt}.$$

4. Back in Chapter 1, we encountered this situation linking the three variables A, s, and w:

$A = $ area of [rectangle with width w and height s] w, where $2w + s = 1200$.

This leads to the chain

$$A = 600s - \frac{1}{2}s^2, \quad \text{where } s = 1200 - 2w,$$

which produces the equation $A = 1200w - 2w^2$.
(a) Using the three equations above, find dA/ds, ds/dw, and dA/dw.
(b) Multiply dA/ds by ds/dw. Is your answer equal to dA/dw?
(c) Find dw/ds. Is it equal to the reciprocal of ds/dw?

5. Use the chain rule to find the derivative of each of the following:
(a) $(x^2 + 7x)^4$.
(b) $(x^3 - (1/x))^6$.
(c) $((x-2)/(x+2))^3$.
(d) $(x^4 - 3x + \pi)^5$.

6. A pebble dropped in still water causes a ripple to form, whose radius is increasing at a rate of 7 in/sec. At what rate is the area of the ripple increasing, when the radius is 5 inches?

7. A ladder 20 feet long leans against a wall. If its bottom end is pulled away from the wall at a constant rate of 5 ft/sec, how fast is the top of the ladder descending
(a) when the bottom end is 12 feet from the wall?
(b) when the top end is 12 feet from the floor?

8. A man 5 feet tall walks directly away from the base of a street light. He walks at a rate of 3 ft/sec. How fast does the tip of his shadow move if the street light is 12 feet above the ground?

9. In problem 8, how fast does the *length* of his shadow change?

10. An observer is 80 feet from a railroad track when a train passes at a rate of 50 ft/sec. How fast is the train's engineer moving away from the observer at the instant they are
(a) 80 feet apart?
(b) 100 feet apart?

Problem Set for Chapter 5

11. Molasses is poured on a pancake at a constant rate so that the circular area covered by the molasses is increasing at a rate of 3 in²/sec. How fast is the radius of this circular area increasing at the instant when the radius is 2 inches?

12. Find antiderivatives of each of the following:
    (a) $3t^2 + 12t + \pi$.
    (b) $1/t^2$.
    (c) $t^3 - 5t + 3$.
    (d) $t^5 + 4t^3 - 16t^2$.

13. In each of parts (a) through (d) of problem 12, find an antiderivative that takes the value 0 when $t = 1$. Is there a unique answer in each case?

14. From a window 276 feet high, a rock is thrown upward at an initial speed of 50 ft/sec. Answer the following questions, treating the rock as a freely falling body.
    (a) When will the rock attain its maximal height?
    (b) When will it hit the ground?
    (c) What will be the speed of the rock when it hits the ground?

15. A baseball is thrown straight up. What was its initial speed if
    (a) it reaches a maximum height of 100 feet?
    (b) it hits the ground 5 seconds after it is released?
    (c) it is at a height of 60 feet 2 seconds after it is released?

16. Since 32 feet is about 9.8 meters, equation (21) of Section 8 becomes $dv/dt = -9.8$ m/sec per sec. By taking antiderivatives twice, show that the height h in meters of a freely falling body is $-4.9t^2 + v_0 t + h_0$, where v_0 is the initial upward speed in m/sec, h_0 is the initial height in meters, and t is the time in seconds after the body is released.

17. In Example 7 of Section 8, find the speed at which the ball hits the ground.

18. The derivative of a certain function f is given by $f'(x) = 10 - 6x$. It is also known that $f(2) = 3$. Find the largest number in the range of f.

19. Suppose that the least number in the range of a certain function g is 2. Suppose also that $g'(x) = 2x - 4$. Find $g(3)$.

20. Think about a tennis ball just as it lands on the ground after being dropped. It bounces up. The upward speed is negative just before impact, and positive just after. Does this mean that the speed function must be discontinuous at the instant of impact? Is Leibniz's principle of continuity violated? Or can you see a way to save this principle by a more careful examination of what actually happens at the moment of impact?

21. (*For ambitious students only*) Although we know the derivative of the reciprocal function, we do not yet know an *antiderivative* of it. Nevertheless, suppose that we have somehow found the antiderivative A of the reciprocal function that takes the value 0 at the point 1. That is, we have a function A satisfying the following:

t	$A(t)$	$A'(t)$
1	0	
t	$A(t)$	$1/t$

Although we do not yet have any sort of formula by which to express the rule for the function A, we can nevertheless deduce some interesting things about it.
(a) To begin with, we know that if $L = A(t)$, then $dL/dt = 1/t$. This makes it unlikely that the domain of the function A includes the point 0. Why?
(b) Let $y = A(\pi t)$. This may be regarded as the chain $y = A(u)$, where $u = \pi t$. Use the chain rule to find dy/dt. Hint. $dy/du = 1/u$.
(c) The work in parts (a) and (b) shows that $dL/dt = dy/dt$. By the fundamental principle of integral calculus, there must be some constant C such that $L = y + C$, i.e., $A(\pi t) = A(t) + C$, on a connected domain. Show that the constant C must be $A(\pi)$. Hint. $A(1) = 0$.
(d) We now know that $A(\pi t) = A(\pi) + A(t)$, since $C = A(\pi)$. Assuming that the domain of A is the connected set of all positive numbers, show that, for any $s > 0$ and $t > 0$, we have
$$A(st) = A(s) + A(t).$$
Hint. Use the same reasoning as before. Just consider s instead of π.
(e) The equation in (d) shows that the function A "converts multiplication into addition" in a sense. That is, the action of A on a *product* st is the *sum* of the action on each term. By letting $t = s$ in this equation, prove that $A(s^2) = 2A(s)$ if $s > 0$.
(f) Prove that $A(s^3) = 3A(s)$ if $s > 0$.
(g) In the equation in (d), let $t = 1/s$, and show that $A(1/s) = -A(s)$ if $s > 0$.
(h) In the equation in (d), let $s = t = \sqrt{x}$ and prove that $A(\sqrt{x}) = \frac{1}{2}A(x)$.
(i) If $L = A(f)$, then $dL/df = 1/f$. What would you guess is the formula for df/dL? (*This has been a preview of the logarithmic function that will be discussed in Chapter 9.*)

22. Consider the area A as indicated below:

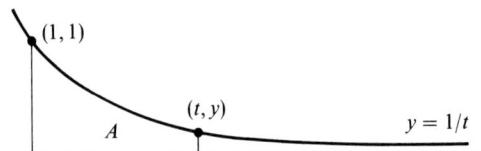

(a) What is A when $t = 1$?
(b) What would you guess dA/dt to be?
(c) Here, A is a function of t. Does it satisfy the table set up at the beginning of problem 21?

23. A rock is thrown up at an initial speed of 96 ft/sec. How far does the rock travel during
(a) the first 2 seconds of flight?
(b) the first 5 seconds of flight?

24. A small, tired bug is climbing up the y-axis. At time $t = 1$, the bug is at the origin and, from that time on, her speed is given by $f(t) = 4/t^2$.
(a) How far does the bug go between times $t = 1$ and $t = 2$?
(b) At what time t will the bug be at position $y = 3$?
(c) At what time t will the bug be at position $y = 3.75$?
(d) How far does the bug go between times $t = 1$ and $t = 1000$?
(e) Will the bug ever reach the position $y = 4$?

Problem Set for Chapter 5   147

25. Match each of the following functions (a) through (g) with its derivative. [The derivative of (h) is not pictured.]

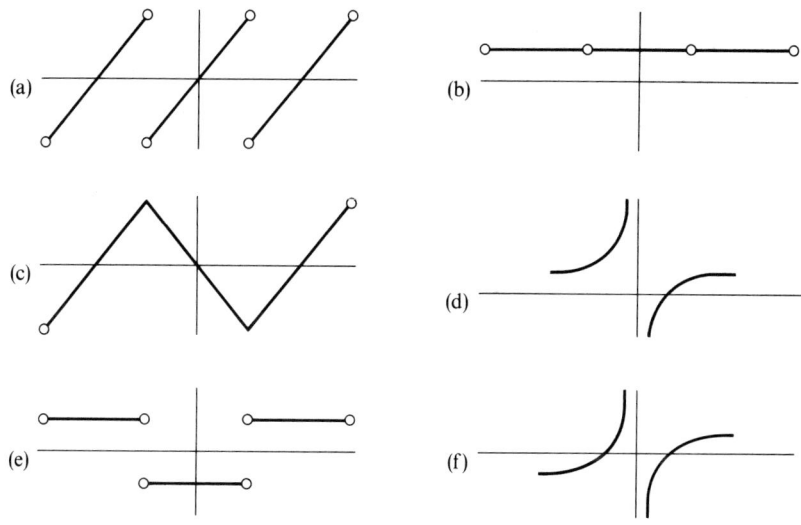

[The curve in (g) lies on the horizontal axis, but has holes in it. The left branch of the curve in (f) is identical with that of (d), but translated downwards.]

26. In problem 25, the curves (d) and (f) have the same derivative, but do not differ by a constant. Doesn't this contradict the fundamental principle of integral calculus?

6

The Integrity of Ancient and Modern Mathematics

When minds of first order meet, sparks fly, even across the centuries. The fundamental theorem of calculus, to be discussed in this chapter, is the result of such a pyrotechnic fusion of ideas. When Leibniz and Newton met Eudoxus and Archimedes, the calculus was rounded out into a whole. By the end of the seventeenth century it was becoming evident that calculus was not a bag of unrelated tricks but was an entity complete unto itself.

The point of this chapter is to see our subject as a unified whole, and the fundamental theorem is what really ties it together. Before coming to this theorem, let us recall briefly what we have seen so far. Calculus is largely the study of the interplay between a function and its derivative. In Chapters 3 and 4 we saw the *geometric* aspect of this interplay, which gives insight into the study of curves lying in a plane. As a by-product, the solution of *optimization* problems was effected. In Chapter 5, a *dynamic* aspect of this interplay revealed itself, throwing light upon the study of *change*. Previously vague terms, like *instantaneous velocity*, *acceleration*, and *rate of growth*, were seen to have natural and precise meanings couched in calculus. And, most importantly, the fundamental notion of *continuity* has been clarified in terms of *limits*.

We have seen by now that the interplay between a function and its antiderivative is signally rich. In this chapter we study still another aspect of this interplay. Calculus permits the easy calculation of the *area* of a figure bounded by curves in the plane.

§1. Areas and Antiderivatives?

Why should there be any connection between the calculus and the calculation of area? Isaac Newton saw the connection at an early age, having learned something, no doubt, from studying mathematics at Cambridge

1. Areas and Antiderivatives?

under the tutelage of Isaac Barrow. While Newton was keeping his secrets to himself, the light came to Leibniz upon studying a mathematics paper by Pascal. The connection is a secret no longer.

Let us try to *guess* the connection first and put off until later an attempt to prove that our guess is correct. The key is to work through several simple examples and to observe that two seemingly different approaches yield the same result.

To see the landscape clearly, a motorcycle ride will help, if the reader will put up with just one more trip. Suppose you are watching the speedometer and therefore know the function f giving the speed of the motorcycle in terms of time. What method(s) can be applied to the speed function f, in order to calculate the distance traveled between, say, the times $t = 1$ and $t = 4$?

EXAMPLE 1. Suppose the speed is constant at 50 km/hr, i.e., the speed function is given by $f(t) = 50$. What is the distance traveled between $t = 1$ and $t = 4$?

One way the distance traveled can be found is by the antiderivative method illustrated in Section 9 of Chapter 5. Since the speed is always positive in this example, the distance traveled is just the distance between the motorcycle's initial and final positions. The position function F is an antiderivative of the speed function f, so

$$F(t) = 50t + C,$$

where C is some constant. The distance traveled is then

$$F(4) - F(1) = 200 + C - (50 + C)$$
$$= 150 \text{ kilometers.}$$

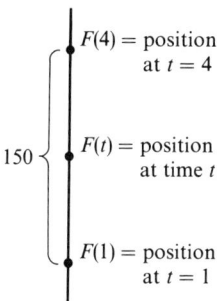

Common sense reveals a simpler way to do this problem, however, for the speed is *constant* at 50 km/hr. Traveling at 50 km/hr for 3 hours, the motorcycle covers a distance of

$$50 \cdot 3 = 150 \text{ kilometers.}$$

The product $50 \cdot 3$ has a striking significance if we look at the graph of the speed function f, which is simply a horizontal line. One cannot help but notice that *the distance traveled between times $t = 1$ and $t = 4$ is numerically equal to the area beneath the curve f, between $t = 1$ and $t = 4$*:

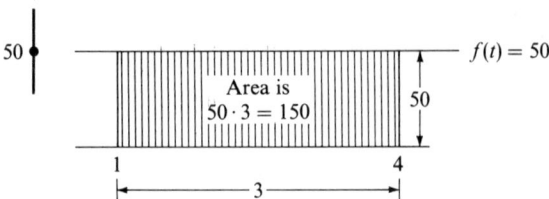

Area beneath f, from $t = 1$ to $t = 4$, is $150 = F(4) - F(1)$,

where F is an antiderivative of f. Could it be that the area beneath *any* curve is so simply related to an antiderivative? □

EXAMPLE 2. Suppose the speed is given by $f(t) = 2t$. What is the distance traveled between $t = 1$ and $t = 4$?

An antiderivative F is given by $F(t) = t^2 + C$. Since the speed $2t$ is always positive between $t = 1$ and $t = 4$, the distance traveled is

$$F(4) - F(1) = 16 + C - (1 + C) = 15 \text{ units.}$$

Let us check to see if this is equal to the area beneath the curve f. Since the graph of $f(t) = 2t$ is simply a line of slope 2, the area in question looks like this:

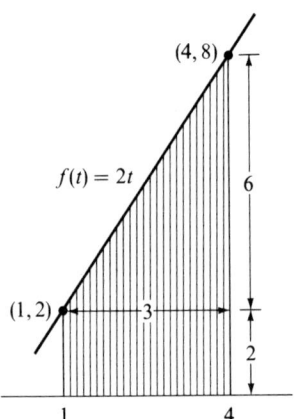

The area is made up of a rectangle of area $3 \cdot 2 = 6$, surmounted by a right triangle of area $\frac{1}{2}(3)(6) = 9$. The area beneath the curve f is then

$$6 + 9 = 15.$$ □

The two methods agree once again! The area beneath the graph of the positive function f again turns out to be the same number as that calculated

1. Areas and Antiderivatives?

by the antiderivative method, i.e.,

$$F(4) - F(1). \qquad (1)$$

We shall see a lot of such expressions as (1), and it will be convenient to have an abbreviation for them. The notation $F|_a^b$ or $[F(t)]_a^b$ is defined to do this:

$$F|_a^b = [F(t)]_a^b = F(b) - F(a).$$

For example,

$$50t|_1^4 = 50(4) - 50(1) = 150,$$
$$[t^2]_1^4 = 4^2 - 1^2 = 15,$$
$$[t^2 - 2t]_1^4 = (16 - 8) - (1 - 2) = 9. \qquad (2)$$

EXAMPLE 3. Consider the area beneath the curve given by $f(t) = 2t - 2$, between $t = 1$ and $t = 4$. Sketch this area and see if it is equal to that calculated by the antiderivative method.

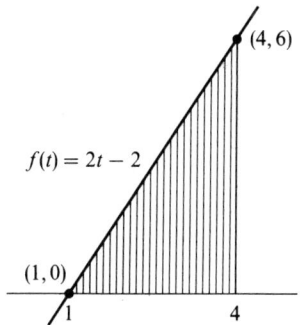

The area is easily seen to be a right triangle of base 3 and height 6, having an area of

$$\frac{1}{2}(3)(6) = 9,$$

which agrees with the number calculated by the antiderivative method in equation (2) preceding the example. □

EXERCISES

(*Remember that the phrase* "beneath the curve" *means* "below the curve and above the horizontal axis".

1.1. Sketch the graphs of each of the following linear functions f and find the area beneath f, between $t = 1$ and $t = 4$, by splitting the area into a rectangle surmounted by a triangle.
   (a) $f(t) = 10 - 2t$.
   (b) $f(t) = t$.
   (c) $f(t) = 4t - 3$.

1.2. For each of the three linear functions of exercise 1.1, apply the antiderivative method. That is, find an antiderivative F and calculate the expression (1). *Answer:* (b) $\frac{1}{2}t^2\big|_1^4 = \frac{1}{2}(16) - \frac{1}{2}(1) = \frac{15}{2}$.

1.3. Apply the method of antiderivatives to each of the following.
 (a) $f(t) = 4t + 2$, from $t = 2$ to $t = 5$.
 (b) $f(t) = 4t + 2$, from $t = 1$ to $t = 4$.
 (c) $f(t) = t$, from $t = 0$ to $t = 1$.
 (d) $f(t) = 5 - t$, from $t = 0$ to $t = 2$.
 Answer: (a) $[2t^2 + 2t]_2^5 = 60 - 12 = 48$.

1.4. The answer to each of the four parts of exercise 1.3 ought to be equal to a certain area. In each case, sketch the area. *Answer:* (a) The area of 48 is that lying beneath the curve $f(t) = 4t + 2$, $2 \leq t \leq 5$.

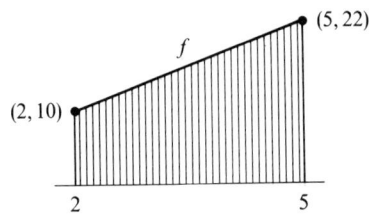

1.5. Apply the method of antiderivatives to each of the following.
 (a) $f(t) = 1/t^2$, from $t = 1$ to $t = 4$.
 (b) $f(t) = 1/t^2$, from $t = 2$ to $t = 6$.
 (c) $f(t) = t^2$, from $t = 0$ to $t = 1$.
 (d) $f(t) = t^2 - 4t + 5$, from $t = 1$ to $t = 4$.
 Answer: (a) $[-1/t]_1^4 = -\frac{1}{4} - (-1) = \frac{3}{4}$.

§2. Areas Bounded by Curves

Consider the area beneath the quadratic curve given by
$$f(t) = t^2 - 4t + 5, \qquad 1 \leq t \leq 4.$$

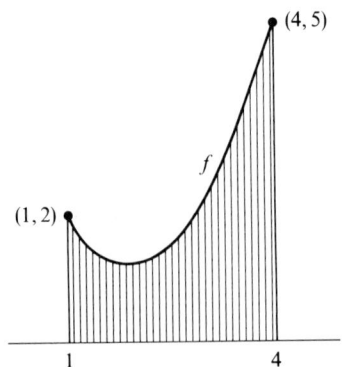

2. Areas Bounded by Curves   153

An antiderivative F of f is given by

$$F(t) = \frac{1}{3}t^3 - 2t^2 + 5t.$$

In view of the way things have turned out up to now, one might *guess* that this area is equal to

$$\left[\frac{1}{3}t^3 - 2t^2 + 5t\right]_1^4 = \left(\frac{64}{3} - 32 + 20\right) - \left(\frac{1}{3} - 2 + 5\right)$$
$$= 6.$$

The answer of 6 square units is surely easy to calculate by the antiderivative method. But how can we be sure that this method gives the correct area? We must first have a clear definition of *area*.

The importance of the role of definitions (in any subject, but particularly in mathematics) is not often noticed. At first we generally have only an intuitive conception of some notion that seems of interest. However, we can deal with intuitive notions, like *tangent line* and *area*, only in a superficial way until we assign these notions a precise significance, showing how they are related to ideas with which we are quite at home. Even more important (in any subject) is the choice of *what* terms to define, for that choice will determine one's language and consequently will ease—or hinder—one's way. When Fermat chose to speak in terms of the intuitive notion of a *limit*, he rendered invaluable service to all who would enter mathematics.

Fermat gave a definition that clarified the idea of a *tangent line* and enabled us to travel in this book as far as we have. To travel much further with security, we must seek clarification of the notion of *area*. What does it mean to assert that the area pictured above is 6 square units? The figure is bounded by a curve on one side! Is it nonsense to speak of the "area" inside a curved figure?

This question was profoundly considered long ago by Archimedes, who became the master of a method introduced still earlier by Eudoxus. Archimedes, of course, had no notion of antiderivatives, but he could calculate areas (and volumes!) enclosed by curved figures. He used the method of Eudoxus, coupled with his own awesome technique.

The exercises below may suggest the essence of Eudoxus' method, but the discussion in depth of this method is postponed until Section 5. There we shall again seek out Eudoxus and Archimedes, who knew what they were talking about.

EXERCISES

2.1. Review problems 8 through 16 in the problem set at the end of Chapter 2.

2.2. Consider the two "stairstep" figures superimposed on the curve $f(t) = t^2 - 4t + 5$, $1 \le t \le 4$.

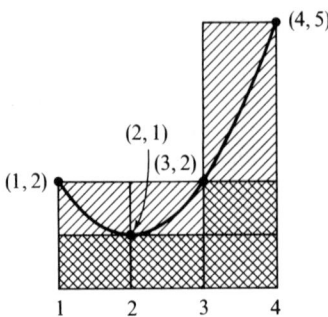

Use them to convince yourself that the area beneath the curve exceeds 4 square units but is less than 9 square units.

2.3. What can you deduce by considering twice as many steps?

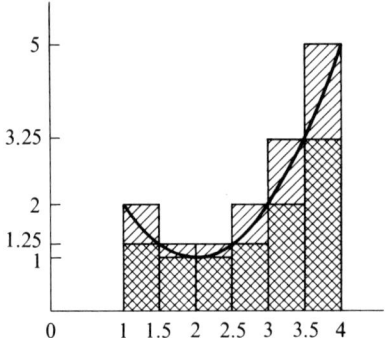

Answer: The area beneath the curve exceeds 4.875 square units but is less than 7.375 square units.

2.4. By putting in a few more steps, convince yourself that the area beneath the curve exceeds 5 square units but is less than 7 square units.

2.5. (*A question for speculation*) Make up a definition of the *area* enclosed by a curved figure lying in the plane. There are several ways this might be defined. Can you think of a way to define the area as a number that is the limit of other numbers that approximate it ever so closely?

§3. Areas and Antiderivatives

The exercises in Section 2 point the way toward a definition of the notion of *area*. The definition will be stated precisely in Section 5. Right now, let us take for granted the fact that the notion of area dates from antiquity, and ask a seventeenth-century question: *What have areas got to do with antiderivatives?*

3. Areas and Antiderivatives

The answer to this question was given independently by Newton and Leibniz, and runs somewhat as follows. The key step in most calculus problems is to see the problem in terms of variables. How can we see the problem of calculating this area, for example, as a problem involving variables?

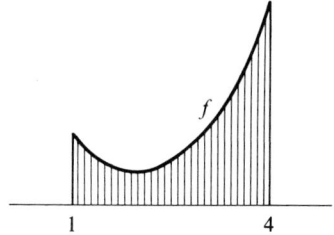

The answer is to consider the way the indicated area A varies in terms of t in the picture below.

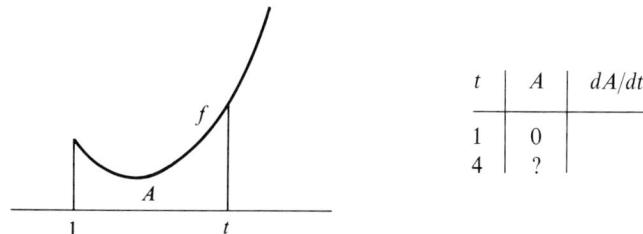

(Do you see why we set $A = 0$ when $t = 1$?) We want to find the area A when $t = 4$. We have made a guess that the antiderivative method will probably give it to us:

$$\text{When } t = 4, \text{ then } A = F(4) - F(1), \qquad (3)$$

where F is an antiderivative of f.

So far, equation (3) is only an educated guess. To prove that it is correct, let us try to find a formula expressing A in terms of t, in order to plug in $t = 4$. From the picture given above, we may expect that

$$\frac{dA}{dt} = f(t). \qquad (4)$$

(See Section 10 of Chapter 5.) A proof of (4) will be forthcoming shortly, but first note that (4) says that A, like F, is an antiderivative of f. By the fundamental principle of integral calculus, A and F differ by some constant C, i.e.,

$$A = F(t) + C. \qquad (5)$$

What is C? Since $A = 0$ when $t = 1$, equation (5) shows

$$0 = F(1) + C,$$

156 6 The Integrity of Ancient and Modern Mathematics

so that $C = -F(1)$ and (5) becomes

$$A = F(t) - F(1). \tag{6}$$

Statement (3), which we were trying to prove, is now an obvious consequence of (6)! □

A proof of (4), on which the preceding argument hangs, will be given below, but the style of argument just seen will be valuable later and ought to be remembered. It consists of three steps, culminating in a proof of (3):

Step 1. By (4), we have

t	A	dA/dt
1	0	
t		$f(t)$

Step 2. By the fundamental principle, since $F' = f$, we have

t	A	dA/dt
1	0	
t	$F(t) + C$	$f(t)$

Step 3. Adjusting C so that $A = 0$ when $t = 1$ yields this information from which (3) follows easily.

t	A	dA/dt
1	0	
t	$F(t) - F(1)$	$f(t)$
4	?	

To make things complete, we must prove (4), which shows the connection between areas and antiderivatives. Note that here we are dealing with a function whose graph lies above the axis, i.e., a *positive* function. A more general case is treated in Section 4.

Theorem on Areas and Antiderivatives. *Let f be a positive, continuous function, and let A be the area beneath the curve f from $x = a$ to $x = t$.*

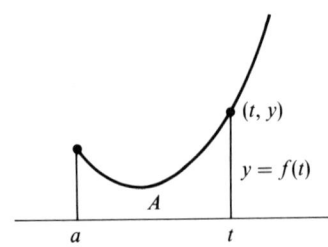

3. Areas and Antiderivatives

Then A is an antiderivative of f:

$$\frac{dA}{dt} = f(t).$$

PROOF. Let t be fixed, and let $y = f(t)$. To find dA/dt, we first consider the change ΔA in area produced by a nonzero change Δt:

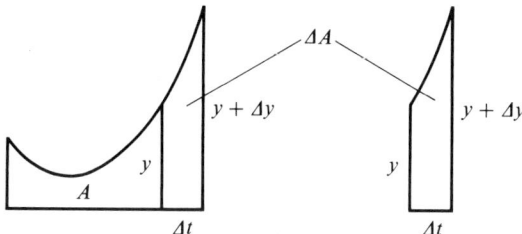

Compare the size of ΔA with that of rectangles built upon the same base of length Δt.

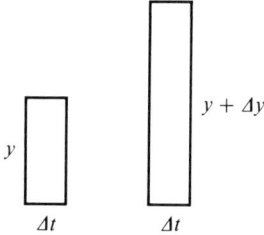

Since the areas of these rectangles are equal, respectively, to $y \cdot \Delta t$ and $(y + \Delta y) \cdot \Delta t$, it follows that ΔA lies between $y \cdot \Delta t$ and $(y + \Delta y) \cdot \Delta t$. Dividing by Δt then shows that

$$\frac{\Delta A}{\Delta t} \text{ lies between } y \text{ and } y + \Delta y. \tag{7}$$

From (7) it is easy to determine the limit of $\Delta A/\Delta t$ as $\Delta t \to 0$, for Δy must tend to zero as well (since y is a continuous function of t). Thus $\Delta A/\Delta t$, being sandwiched between y and $y + \Delta y$, must tend to y, i.e.,

$$\frac{dA}{dt} = \operatorname*{Limit}_{\Delta t \to 0} \frac{\Delta A}{\Delta t} = y = f(t). \qquad \square$$

The proof just given may seem to rely on the picture that shows the curve f rising as it passes through (t, y) and also shows the change Δt as being positive. If the curve is falling, or if Δt is negative, the pictures have to be redrawn, but the proof has been worded in such a way as to require no change. If the function "wiggles" violently near (t, y) so that the curve is neither rising nor falling there, our proof is invalid, but the theorem is still true, as shown in a more careful demonstration better deferred to a course in analysis.

158　　　　　　　　　　　6 The Integrity of Ancient and Modern Mathematics

EXERCISES.

3.1. Find the area beneath the graph of each of the following equations, from 1 to 4.
   (a) $f(t) = t^2 - 2t + 6$.
   (b) $f(t) = 1/t^2$.
   (c) $f(x) = x^2$.
   (d) $f(x) = x^3$.
   (e) $f(x) = x^2 + x^3$.
   (f) $y = 3t^2 + 5$.
   (g) $y = 4x^3 - 3x^2$.
   (h) $y = \pi$.
   (i) $h = -16t^2 + 64t$.
   (j) $g(s) = 600s - \frac{1}{2}s^2$.
   Answers: (b) $\frac{3}{4}$ square units. [See exercise 1.5(a).] (d) $x^4/4|_1^4 = (256/4) - (1/4) = 255/4 = 63\frac{3}{4}$ square units. (f) 78 square units. (h) 3π square units.

3.2. In Section 10 of Chapter 5 you were asked to do some problems by guesswork. With the aid of the theorem on areas and antiderivatives, go back and do these problems without guessing.

3.3. With the aid of the theorem on areas and antiderivatives, find dA/dt in each of the following situations. *Specify your answer in terms of t.*

(a)

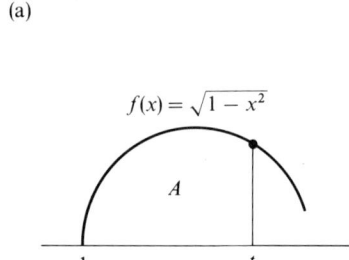

(b)

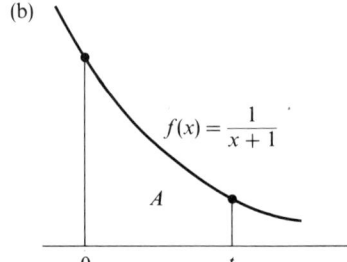

Answer: (a) Since $f(x) = \sqrt{1-x^2}$, we have $f(t) = \sqrt{1-t^2}$. By the theorem, $dA/dt = f(t) = \sqrt{1-t^2}$. (Note that this problem really has nothing to do with "x". The answer would be the same if the function f had been expressed by writing $f(s) = \sqrt{1-s^2}$, or by writing the equation $y = \sqrt{1-L^2}$ to specify the curve f. In this problem x is a *dummy variable*, in the sense that the answer is unchanged if "x" is renamed as "s" or "L".)

3.4. The algebraic rule $\sqrt{1-t^2}$ has domain $-1 \le t \le 1$. Find an antiderivative of this function. *Answer*: Let A be the function of t specified by the picture in exercise 3.3(a). (This function is specified in words, not as an algebraic rule, but it is a perfectly good function, and the theorem on areas and antiderivatives shows that it answers this question.)

3.5. Find an antiderivative of each of the following functions, expressed as algebraic rules.
   (a) $1/(t+1), 0 \le t$.
   (b) $1/t, 1 \le t$.
   (c) $\sqrt{4-t^2}, -2 \le t \le 2$.
   (d) $1/(t^2+1), 0 \le t$.
   Answers: (a) Let A be the function of t specified in the picture in exercise 3.3(b). (b) Let A be the function of t specified in the picture in problem 22 at the end of Chapter 5.

§4. Areas between Curves

The preceding section studied how to find the area between a curve and a certain straight line (the horizontal axis). It is just as easy to find the area between a curve and another curve.

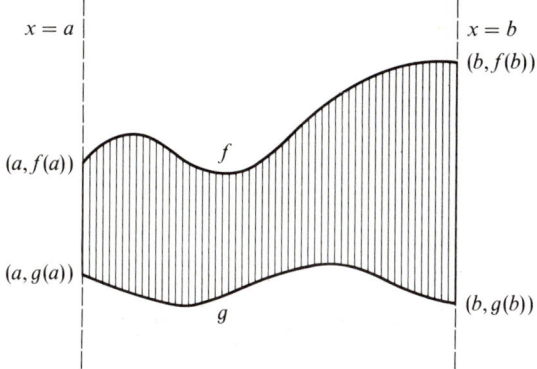

Finding the area between two curves is a problem that can be approached by the method of Newton and Leibniz outlined in Section 3. The key is to see the problem in terms of variables. Let A be the area indicated below, so that A is a function of t.

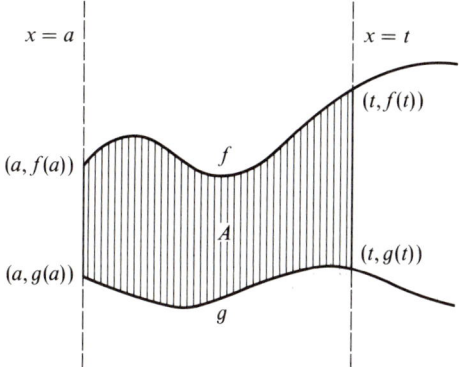

If both f and g are continuous, with f lying above g, it follows that

$$\frac{dA}{dt} = f(t) - g(t). \tag{8}$$

The proof of (8) will not be given, because the idea of the proof is so similar to that of the theorem of Section 3. [The only basic difference is this. In the theorem of Section 3, ΔA was seen to be roughly equal to the product $f(t)\Delta t$; whereas here ΔA is roughly equal to $(f(t) - g(t))\Delta t$.] From (8) it is easy to deduce, as explained below, a more general area principle.

General Area Principle. *Let f and g be continuous curves, with f lying above g. Then the area between f and g, from $x = a$ to $x = b$, is given by*

$$[F - G]_a^b,$$

where F is an antiderivative of f and G is an antiderivative of g.

PROOF. The proof follows exactly the pattern of the three steps described in Section 3. Using (8), we have the following information about the dependence of A upon t:

t	A	dA/dt
a	0	
t	?	$f(t) - g(t)$
b	?	

Equation (8) says that A is an antiderivative of $f - g$. Since $F - G$ is too (*why?*), the fundamental principle of integral calculus says that

$$A = (F(t) - G(t)) + C \tag{9}$$

for some constant C. What is C? Because $A = 0$ when $t = a$, we get from equation (9) that

$$0 = (F(a) - G(a)) + C.$$

This shows that $C = -(F(a) - G(a))$, so that (9) becomes

$$A = (F(t) - G(t)) - (F(a) - G(a)) = [F - G]_a^t.$$

Thus the formula $A = [F - G]_a^t$ expresses A in terms of t. When $t = b$, then A becomes the desired area, as pictured at the beginning of this section. From the formula, when $t = b$,

$$A = [F - G]_a^b. \qquad \square$$

In applying the general area principle to areas bounded by curves, it is essential to note which curve is on top. If the curves cross one or more times, several applications of the area principle may be required. (See Example 8.)

EXAMPLE 4. Find the indicated area.

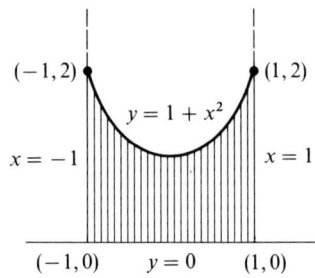

4. Areas between Curves

This is the area between the curves f and g, where $f(x) = 1 + x^2$, $g(x) = 0$, $a = -1$, $b = 1$. Antiderivatives of f and g are given by $F(x) = x + x^3/3$ and $G(x) = 0$. By the general area principle, the area is

$$[F - G]^1_{-1} = \left[x + \frac{x^3}{3} \right]^1_{-1}$$

$$= \frac{4}{3} - \left(\frac{-4}{3} \right) = \frac{8}{3} \text{ square units.} \qquad \square$$

EXAMPLE 5. Find the indicated area.

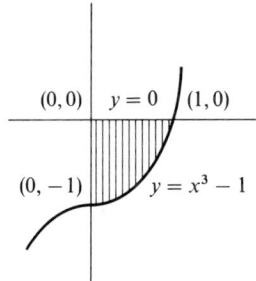

The curve lying on top here is given by $f(x) = 0$, while the bottom curve's equation is $g(x) = x^3 - 1$. Here, $a = 0$ and $b = 1$. Antiderivatives are given by $F(x) = 0$ and $G(x) = x^4/4 - x$. By the general area principle, the area is equal to

$$[F - G]^1_0 = \left[0 - \left(\frac{x^4}{4} - x \right) \right]^1_0$$

$$= \frac{3}{4} \text{ square units.} \qquad \square$$

EXAMPLE 6. Find the indicated area.

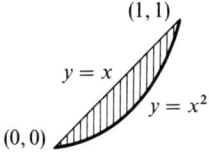

Let $f(x) = x$, $g(x) = x^2$, $a = 0$, $b = 1$. The area is equal to

$$[F - G]^1_0 = \left[\frac{x^2}{2} - \frac{x^3}{3} \right]^1_0$$

$$= \frac{1}{2} - \frac{1}{3} = \frac{1}{6} \text{ square units.} \qquad \square$$

EXAMPLE 7. Find the indicated area.

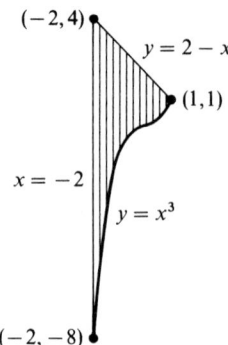

Let $f(x) = 2 - x$, $g(x) = x^3$, $a = -2$, $b = 1$. The area is given by

$$\left[2x - \frac{x^2}{2} - \frac{x^4}{4}\right]_{-2}^{1},$$

which, when evaluated, is seen to be $\frac{45}{4}$ square units. □

EXAMPLE 8. Consider the curve given by $f(x) = x^2 - 1$, with domain $-2 \leq x \leq 4$.

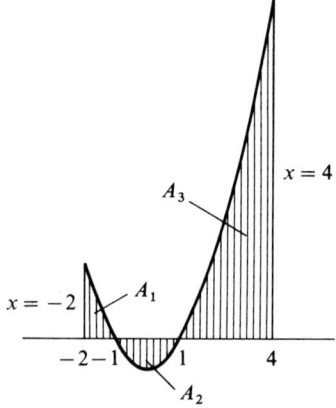

Find the area between the curve f and the x-axis.

Since the curve crosses the x-axis twice, the required area splits into three pieces, A_1, A_2, and A_3, as indicated. In each piece the area principle may be applied, taking account as to which of the curves $y = x^2 - 1$ and $y = 0$ is on top. We get

$$A_1 = \left[\frac{x^3}{3} - x\right]_{-2}^{-1} = \frac{4}{3},$$

$$A_2 = \left[0 - \left(\frac{x^3}{3} - x\right)\right]_{-1}^{1} = \left[x - \frac{x^3}{3}\right]_{-1}^{1} = \frac{4}{3},$$

$$A_3 = \left[\frac{x^3}{3} - x\right]_{1}^{4} = 18.$$

4. Areas between Curves

The total area between the curve f and the x-axis is then

$$A_1 + A_2 + A_3 = 20\frac{2}{3} \text{ square units.} \qquad \square$$

EXERCISES

4.1. In each of the following, find the indicated area. *Hint for* A_4. First find A_3. $A_4 = A_2 - A_3$ (why?).

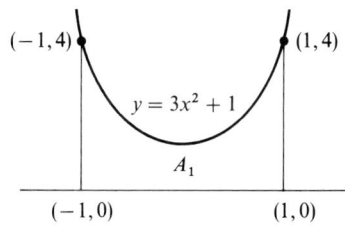
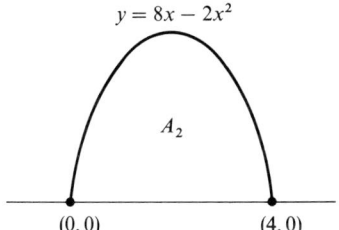
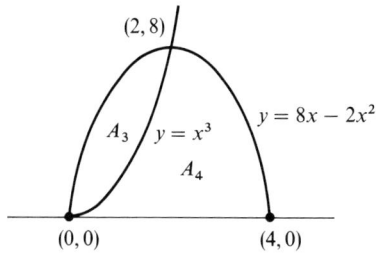
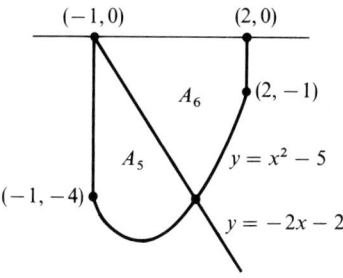
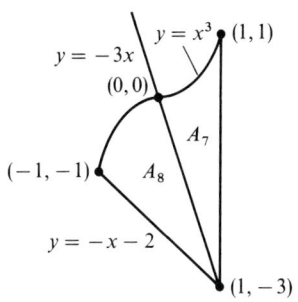
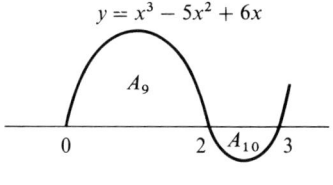

Partial answer: $A_2 = \frac{64}{3}$. $A_5 = \frac{16}{3}$. $A_7 = \frac{7}{4}$. $A_9 = \frac{8}{3}$.

4.2. Find the area between each of the following curves and the x-axis, as illustrated in Example 8.
   (a) $f(x) = 4 - x^2$, $-3 \le x \le 4$.
   (b) $f(x) = x^3 - 5x^2 + 6x$, $-2 \le x \le 4$.
   (c) $f(x) = 1 - (4/x^2)$, $1 \le x \le 3$.

Answer: (c) $\frac{4}{3}$ square units:

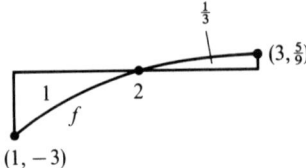

§5. Eudoxus' Method and the Integral

Integrity, integer, integration, integral—these words have the same root meaning, that of "wholeness". To *integrate* is to collect into a whole. What we are now studying is called *integral calculus,* and it is high time to explain why Leibniz chose to call it that. The reason may stem from an observation made by Leibniz [and, before him, by Cavalieri (1598–1647) and others], an observation that may be confirmed by the general area principle. *If, when figures in the plane are set one above the other, they are seen to be made up of equal "vertical segments", then the areas of the figures must be equal.* This delightful observation is generally known as **Cavalieri's principle**. A concrete illustration is below.

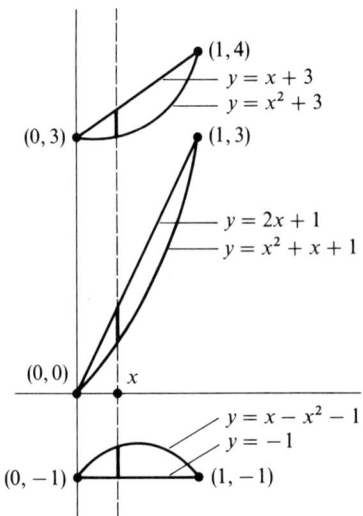

Here, a vertical line through any point x on the horizontal axis shows a vertical segment of length $x - x^2$ in all three figures. Cavalieri's principle says the three figures must have the same area. This is confirmed by the general area

5. Eudoxus' Method and the Integral 165

principle, which says that all three figures have an area of $[(x^2/2) - (x^3/3)]_0^1 = \frac{1}{6}$ *square units.*

A rough* statement of Cavalieri's principle is that the area of a figure is determined by the vertical line segments that make it up. The area of a figure would thus seem to be the result of collecting into a whole, or *integrating*, all its vertical line segments. Leibniz toyed with the idea of regarding any area as an *integral* of (infinitely many) line segments.

This idea raises serious questions. The area of each vertical line segment is of course zero, since a line segment has no width. Yet somehow Leibniz would have us believe that *infinitely many* zeros integrate into a nonzero total area! The paradoxical nature of this idea was recognized by Leibniz, who nevertheless persisted in believing the idea valuable, at least on an intuitive level. Leibniz was never able to describe clearly this intuitive perception, and it has generally been regarded with suspicion.

Nonetheless, by thinking on this intuitive level Leibnize was able to make important discoveries. Justification for some of these discoveries often had to wait for later mathematicians, as Leibniz sometimes had difficulty in saying what he meant. The difficulty is understandable, for it is related to one of the old paradoxes of Zeno (ca. 495–435 B.C.), but further discussion of this is postponed until Chapter 10. □

The point of the preceding discussion was to explain how the word *integral* entered the calculus. Leibniz wanted to refer to an area as an integral, and out of respect for Leibniz we shall do likewise. However, we discard his fuzzy notion about an "integral of zeros" collecting together to yield a nonzero number. We seek a slight modification of Leibniz's notion of an *integral* to bring things into clear focus. How can this be done?

Once again we turn for help to the notion of a limit, which has already done more than its share to clarify the idea of a tangent to a curve and the idea of continuity. As we shall see, the integral has a natural definition in terms of a limit, by means of a modification of a method introduced by Eudoxus over 2000 years ago.

Eudoxus, of course, never spoke of limits, nor did Archimedes. The Greeks never called limits by name, but could sometimes manage to get the same job done by using the method of elimination. (In modern terms, this amounts to finding an area A by somehow eliminating all numbers larger than A, together with all numbers smaller, leaving the desired number A as the only number left.[†]) Our experience in Chapter 3 suggests that the use of limits may be preferable to the use of the principle of elimination.

Here, then, is **Eudoxus' method** (in modern dress), defining the **integral** of a function f on the domain $a \leq x \leq b$.

---
* And, as it stands, quite inaccurate. See problem 21 at the end of this chapter.

† An example of this method may be found in the appendix on Archimedes.

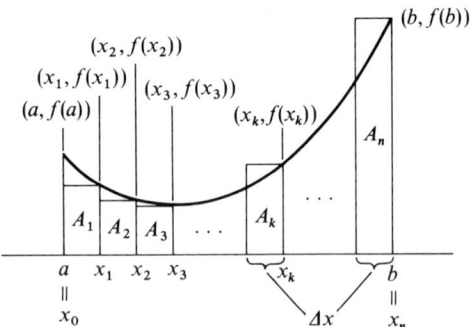

Idea: *As n gets larger, the area beneath the staircase with n steps approaches the area beneath the curve f.*

Consider any large positive integer n, and divide the interval $a \leq x \leq b$ into n subintervals, each of the same length Δx, so that

$$\Delta x = \frac{b-a}{n}. \tag{10}$$

Thus (see the picture) we have $x_0 = a$, $x_1 = a + \Delta x$, $x_2 = a + 2\Delta x$, $x_3 = a + 3\Delta x$, and so on. Finally, at the last, we have

$$\begin{aligned} x_n &= a + n\Delta x \\ &= a + \frac{n(b-a)}{n} = b. \end{aligned}$$

There is a convenient way to abbreviate the preceding two sentences, namely,

$$x_k = a + k\Delta x, \quad \text{for } k = 0, 1, 2, \ldots, n. \tag{11}$$

The area A_k of the k-th rectangle (see the picture) is simply the product of its height and width:

$$A_k = f(x_k)\,\Delta x. \tag{12}$$

Therefore, the total area S_n beneath the staircase figure with n steps is the sum $A_1 + A_2 + \cdots + A_n$. We abbreviate this by writing

$$S_n = \sum_{k=1}^{n} A_k$$

(read "S_n equals the sum, as k runs from 1 to n, of A_k"). Substituting the expression (12) for A_k shows

$$S_n = \sum_{k=1}^{n} f(x_k)\,\Delta x. \tag{13}$$

Now S_n is the area beneath the staircase figure with n steps, and it is *not* likely to be equal to the area beneath the curve f. However, as n is taken larger and larger, S_n clearly approximates the area beneath f to great

5. Eudoxus' Method and the Integral

accuracy. The area beneath f is the *limit* of S_n as n increases without bound, so we define the integral of f to be this limit. That is, the integral of f from a to b is defined as

$$\text{Limit } S_n = \underset{\Delta x \to 0}{\text{Limit}} \sum_{k=1}^{n} f(x_k) \Delta x \quad [\text{by (13)}], \tag{14}$$

since (10) shows that $\Delta x \to 0$ as n increases without bound.

The right-hand side of (14) suggests that the integral of f from a to b might be denoted by

$$\int_a^b f(x)\,dx, \tag{15}$$

since this is the symbol that results from replacing the Greek Δ by the letter d, the Greek Σ by the letter $\int$ (a seventeenth-century S) and replacing the discrete points x_k by the continuous variable x, which runs from a to b.

Definition. Let f be a function with domain $a \leq x \leq b$. **The integral of f from a to b** is denoted by

$$\int_a^b f(x)\,dx \quad \left(\text{or, for short, by } \int_a^b f \right)$$

and is defined to be the number calculated by Eudoxus' method:

$$\int_a^b f(x)\,dx = \underset{\Delta x \to 0}{\text{Limit}} \sum_{k=1}^{n} f(x_k) \Delta x = \text{Limit } S_n,$$

where S_n is defined by equation (13).

The idea of Eudoxus' method is not unlike the idea behind Fermat's method. The "right answer" for the integral $\int_a^b f$ is the limit of "wrong answers" S_n that come quite close to the integral when n is quite large. In integral calculus Eudoxus' method assumes a role of importance parallel to the role played by Fermat's method in differential calculus. It defines the basic notion to be studied.

Just as we have found shortcuts to Fermat's method, so we can find shortcuts to the method of Eudoxus. We can sometimes guess the value of an integral by interpreting the integral as an area. For instance, from exercise 3.1(d) we may expect that

$$\int_1^4 f(x)\,dx = 63\frac{3}{4}, \quad \text{where } f(x) = x^3,$$

or, more briefly,

$$\int_1^4 x^3\,dx = 63\frac{3}{4}.$$

If the variable is called t instead of x, we modify our notation accordingly. From exercise 1.1, without using Eudoxus' method, we expect to have

$$\int_1^4 (10 - 2t) \, dt = 15, \qquad \int_1^4 t \, dt = \frac{15}{2}, \qquad \int_1^4 (4t - 3) \, dt = 21.$$

These examples should suggest that integrals, defined by Eudoxus' method, can be calculated by the method of antiderivatives. This is true, and it is essentially the content of the fundamental theorem. Before the fundamental theorem can be appreciated, however, we must learn to be at home with

(a) interpreting integrals as areas (being careful, because an integral is not always an area).
(b) calculating integrals by Eudoxus' method (as a limit of sums).

The first of these should be accomplished by the following exercises. Section 6 deals with the second.

EXERCISES

5.1. In exercise 3.1, ten areas were found. Express each of these areas as integrals, and express the answers you found in exercise 3.1 in integral notation. *Answers*:
(b) $\int_1^4 (1/t^2) \, dt = \frac{3}{4}$. (c) $\int_1^4 x^2 \, dx = 21$. (h) $\int_1^4 \pi \, dt = \int_1^4 \pi \, dx = \int_1^4 \pi \, ds = 3\pi$.

5.2. Interpret each of the following integrals as an area.
(a) $\int_0^4 (8x - 2x^2) \, dx$.
(b) $\int_{-1}^1 (3x^2 + 1) \, dx$.
(c) $\int_0^2 (x^3 - 5x^2 + 6x) \, dx$.
(d) $\int_0^5 \pi \, dt$.
(e) $\int_{-3}^3 \sqrt{9 - t^2} \, dt$.
(f) $\int_{-2}^2 \sqrt{4 - t^2} \, dt$.
(g) $\int_0^5 \sqrt{25 - t^2} \, dt$.
(h) $\int_{-3}^0 \sqrt{9 - x^2} \, dx$.

Answers: (a) This integral is equal to the area A_2 of exercise 4.1. (d) This integral is equal to the area beneath the curve $y = \pi$, from $t = 0$ to $t = 5$. (e) This integral is equal to the area beneath the curve $y = \sqrt{9 - t^2}$, a semicircle (why?), from $t = -3$ to $t = 3$.

5.3. Evaluate each of the integrals in 5.2 by some means *other than* Eudoxus' method. *Answer*: (e) The area of a semicircle of radius 3 is equal to half the area of the full circle, or $9\pi/2$. Therefore, $\int_{-3}^3 \sqrt{9 - t^2} \, dt = 9\pi/2$.

5.4. If a function f is negative, i.e., if its graph lies below the horizontal axis, then all the A_k's of equation (12) are negative. Use this to explain why, *when f is negative*, then $\int_a^b f(x) \, dx$ will *not* be an area. *Hint*. No area is negative.

§6. The Integral as a Limit of Sums

It takes time and patience to carry out Eudoxus' method of calculating an integral. Since the antiderivative method is shorter, one may ask why time should be spent studying Eudoxus. There are several reasons.

6. The Integral as a Limit of Sums

(a) $\int_a^b f$ cannot be calculated by antiderivatives unless an antiderivative of f is known. [There are many functions, such as $1/x$ and $1/(1 + x^2)$, whose antiderivatives we do not yet know.]
(b) Eudoxus' method leads to a clear understanding of what is meant by the *area beneath a curve*.
(c) Eudoxus' method emphasizes that an integral is a limit of sums. Areas are not the only quantities that are limits of sums. As we shall see, *volumes* can also be regarded as limits of sums, and they can be expressed by integrals. Integrals are of use in expressing other quantities as well, such as the quantity of *work* required to put a satellite in orbit.

Eudoxus' method involves the sum of n numbers, where n is a large integer. Such a sum must be simplified before the limit can be taken in order to find the integral. There is one such sum which, thanks to the Pythagoreans, we know how to calculate already.

$$1 + 2 + 3 + \cdots + n = \frac{n(n + 1)}{2}. \tag{16}$$

(See Chapter 2, problem 2.) The formula for the sum of the *squares* of the first n positive integers has also been known since antiquity:

$$1 + 4 + 9 + \cdots + n^2 = \frac{n(n + 1)(2n + 1)}{6}. \tag{17}$$

Let us take formula (17) for granted here. It is discussed in an appendix on sums.

To deal efficiently with sums, an efficient system of notation must be developed. The symbol

$$\sum_{k=1}^{n}$$

is used as an indication to sum up n numbers that are to be indexed by k. We refer to k as the *index of summation*. For instance,

$$\sum_{k=1}^{3} k = 1 + 2 + 3 = 6,$$

because $\sum_{k=1}^{3} k$ indicates the sum of 3 numbers, the numbers being expressed by k, where k runs from 1 to 3. Similarly,

$$\sum_{k=1}^{3} 5k = 5 \cdot 1 + 5 \cdot 2 + 5 \cdot 3 = 5(1 + 2 + 3). \tag{18}$$

This simplifies, of course, to $5 \cdot 6 = 30$, but it is more important to note that equation (18) shows that

$$\sum_{k=1}^{3} 5k = 5 \cdot \sum_{k=1}^{3} k. \tag{19}$$

What is the "real reason" that the number 5 can be brought out in front of the summation sign, as in (19)? It is simply because 5 is an expression that is independent of the index k. It therefore occurs in each of the summands and can be factored out in front, as seen in (18).

What has just been illustrated in the simple example given above is most important to remember when trying to simplify sums. *Note when an expression can be brought out in front of the summation sign.* We can see, for example, that

$$\sum_{k=1}^{n} \left(\frac{1}{n^2}\right) k = \left(\frac{1}{n^2}\right) \cdot \sum_{k=1}^{n} k, \qquad (20)$$

simply because the expression $1/n^2$ is independent of the index k. In fact, beginning with equation (20), we can carry out a complete simplification as follows. We already know how to simplify the sum that occurs in the right-hand side of (20). Equation (16) says

$$\sum_{k=1}^{n} k = \frac{n(n+1)}{2}. \qquad (21)$$

[*Do you see why equation (16) says exactly this?*] When this is used in (20) we get

$$\sum_{k=1}^{n} \left(\frac{1}{n^2}\right) k = \left(\frac{1}{n^2}\right) \frac{n(n+1)}{2} = \frac{n+1}{2n} = \frac{1}{2} + \frac{1}{2n}. \qquad (22)$$

As a first example of Eudoxus' method, let us calculate an integral whose value we already know by other means.

EXAMPLE 9. Calculate $\int_0^1 x\,dx$ directly from its definition by applying the method of Eudoxus.

Here, we must apply Eudoxus' method to the function given by $f(x) = x$ on the domain $0 \leq x \leq 1$. Thus we have $a = 0$, $b = 1$, and

$$\Delta x = \frac{b-a}{n} = \frac{1-0}{n} = \frac{1}{n} \quad [\text{from (10)}],$$

$$x_k = a + k\,\Delta x = 0 + k\left(\frac{1}{n}\right) = \frac{k}{n} \quad [\text{from (11)}],$$

$$f(x_k) = x_k = \frac{k}{n}.$$

Using these, we first find an approximation S_n to the desired integral. From (13),

$$S_n = \sum_{k=1}^{n} f(x_k)\,\Delta x = \sum_{k=1}^{n} \left(\frac{k}{n}\right)\left(\frac{1}{n}\right) = \sum_{k=1}^{n} \left(\frac{1}{n^2}\right) k.$$

The integral, by definition, is the limit of S_n as n increases without bound. In order to find that limit we must first simplify the expression for S_n. This

6. The Integral as a Limit of Sums

has already been carried out in (22), so we have

$$S_n = \frac{1}{2} + \frac{1}{2n}. \tag{23}$$

From (23), it is easy to find Limit S_n, for it is obvious that, as n grows increasingly larger, $1/2n \to 0$. Therefore,

$$\int_0^1 x\, dx = \text{Limit } S_n = \text{Limit}\left(\frac{1}{2} + \frac{1}{2n}\right) = \frac{1}{2}. \quad \square$$

Example 9 shows that Eudoxus' method, like Fermat's method, can be carried out without ever drawing a geometric picture to describe what goes on. A picture aids the understanding, however, so let us draw one. What was shown in (23) is that the area S_n of the staircase figure with n steps is equal to $\frac{1}{2} + 1/2n$ square units. As n gets larger (or, equivalently, as $\Delta x \to 0$), the jagged figure on the left approximates more and more the area on the right.

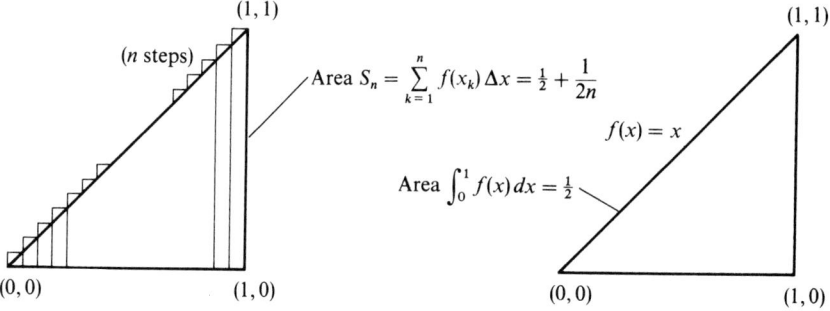

The integral of Example 9 is, of course, calculated much more quickly by simply using the formula for the area of a triangle. Or, by the antiderivative method,

$$\int_0^1 x\, dx = \left.\frac{x^2}{2}\right|_0^1 = \frac{1}{2} - 0 = \frac{1}{2}. \quad \square$$

Before doing a second example it might be well to make a small point about summations. Here is a question that is easy to miss because it is too simple. *What is*

$$\sum_{k=1}^3 1$$

equal to? To answer this question, remember that "$\sum_{k=1}^3$" indicates that 3 numbers are to be summed. For instance,

$$\sum_{k=1}^3 A_k = A_1 + A_2 + A_3.$$

172   6 The Integrity of Ancient and Modern Mathematics

If $A_k = 1$ (that is, if $A_1 = 1, A_2 = 1, A_3 = 1$), this becomes

$$\sum_{k=1}^{3} 1 = 1 + 1 + 1 = 3,$$

answering our question. By the same token we see that

$$\sum_{k=1}^{n} 1 = \underbrace{1 + 1 + 1 + \cdots + 1}_{n \text{ summands}} = n. \tag{24}$$

EXAMPLE 10. Calculate $\int_1^4 2x\, dx$ directly from its definition by applying the method of Eudoxus.

We apply Eudoxus' method to the function given by $f(x) = 2x$ on the domain $1 \le x \le 4$. Thus we have $a = 1$, $b = 4$, and

$$\Delta x = \frac{b - a}{n} = \frac{4 - 1}{n} = \frac{3}{n} \quad [\text{from (10)}],$$

$$x_k = a + k\Delta x = 1 + k\left(\frac{3}{n}\right) = 1 + \frac{3k}{n} \quad [\text{from (11)}],$$

$$f(x_k) = 2x_k = 2 + \frac{6k}{n}.$$

Hence,

$$S_n = \sum_{k=1}^{n} f(x_k)\Delta x = \sum_{k=1}^{n} \left(2 + \frac{6k}{n}\right)\left(\frac{3}{n}\right)$$

$$= \sum_{k=1}^{n} \left(\frac{6}{n} + \frac{18k}{n^2}\right)$$

$$= \sum_{k=1}^{n} \frac{6}{n} + \sum_{k=1}^{n} \frac{18k}{n^2} \quad (\text{why?})$$

$$= \frac{6}{n}\sum_{k=1}^{n} 1 + \frac{18}{n^2}\sum_{k=1}^{n} k \quad (\text{why?})$$

$$= \frac{6}{n}(n) + \frac{18}{n^2}\cdot\frac{n(n+1)}{2} \quad [\text{by (21) and (24)}]$$

$$= 6 + 18\frac{n+1}{2n} = 15 + \frac{9}{n}.$$

Therefore,

$$\int_1^4 2x\, dx = \text{Limit } S_n = \text{Limit}\left(15 + \frac{9}{n}\right) = 15. \qquad \square$$

6. The Integral as a Limit of Sums

By comparison with Eudoxus' method, antiderivatives evaluate integrals like lightning:

$$\int_1^4 2x\,dx = x^2\big|_1^4 = 16 - 1 = 15.$$

The point of these examples, however, has nothing to do with speed of calculation. Only an electronic computer would regard Eudoxus' method as speedy. The point is to emphasize that the integral is a limit of sums and can be calculated without reference to any geometric figure and without any knowledge whatever of derivatives or antiderivatives.

The integral $\int_a^b f$ does have a geometric interpretation, however, as the area beneath the curve f, *if f is not negative*. By another stroke of good fortune, the integral enjoys a connection with antiderivatives, to be stated precisely in the fundamental theorem. Since such delightful connections can be proved to be true, the most intellectual of minds might regard them as unsurprising, being merely part of the nature of things. Some of the rest of us, who know the meaning of serendipity, happily find it here.

EXERCISES

(Be willing to put in a little time practicing the use of summation notation. It is quite efficient, once learned. The appendix on sums and limits may be helpful.)

6.1. Go through the following steps to calculate the integral $\int_0^7 (3x+2)\,dx$.
  (a) What is Δx?
  (b) Find x_k in terms of k.
  (c) Find $f(x_k)$ in terms of k and n, where $f(x) = 3x + 2$.
  (d) Using formulas (21) and (24) to simplify the expression S_n, show that $S_n = (147/2) + (147/2n) + 14$.
  (e) Find the desired integral by taking the limit, as n increases without bound, of your answer to part (d).
  (f) Interpret the integral as an area and calculate it by the antiderivative method. Do you get the same answer as in part (e)?

6.2. Just as in exercise 6.1, calculate each of the following integrals directly from its definition by Eudoxus' method.
  (a) $\int_0^4 (5x+1)\,dx$.
  (b) $\int_0^2 (5x+1)\,dx$.
  (c) $\int_2^4 (5x+1)\,dx$.
  (d) $\int_2^4 (1-5x)\,dx$.
  Answer: (d) -28. (The integral here is not an area, since the function $1-5x$ is negative on the domain $2 \leq x \leq 4$.)

6.3. Explain why, in exercise 6.2, it is to be expected that the sum of your answers to parts (b) and (c) is equal to the answer to part (a).

6.4. Write formula (17) in summation notation. *Answer:* $\sum_{k=1}^n k^2 = n(n+1)(2n+1)/6$.

6.5. Use your answer to exercise 6.4 to help calculate $\int_0^1 x^2\,dx$ directly from its definition by Eudoxus' method. *Answer:* $\frac{1}{3}$. (See Appendix 2, Section 2.)

§7. Some Properties of the Integral

In Section 2 we guessed that a certain area was equal to 6 square units without knowing, at that time, what was meant by the area within a figure bounded by a curve. We now have Eudoxus' method of determining such an area, and we can therefore check our guess of Section 2. Let us do that, with an eye out for noticing some properties of the integral.

Consider, then, the function given by
$$f(x) = x^2 - 4x + 5, \qquad 1 \le x \le 4.$$
Applying the method of Eudoxus to find the area beneath f, we have $\Delta x = 3/n$, $x_k = 1 + 3k/n$, and

$$S_n = \sum_{k=1}^{n} f(x_k) \Delta x = \sum_{k=1}^{n} [x_k^2 - 4x_k + 5] \Delta x \qquad (25)$$

$$= \sum_{k=1}^{n} \left[\left(1 + \frac{3k}{n}\right)^2 - 4\left(1 + \frac{3k}{n}\right) + 5\right]\left(\frac{3}{n}\right)$$

$$= \sum_{k=1}^{n} \frac{27k^2}{n^3} - \frac{18k}{n^2} + \frac{6}{n} \quad \text{(by collecting terms)}$$

$$= \frac{27}{n^3} \sum_{k=1}^{n} k^2 - \frac{18}{n^2} \sum_{k=1}^{n} k + \frac{6}{n} \sum_{k=1}^{n} 1$$

$$= \frac{27}{6}\left(1 + \frac{1}{n}\right)\left(2 + \frac{1}{n}\right) - 9\left(1 + \frac{1}{n}\right) + 6,$$

by (16), (17), and (24). (The reader is asked to fill in the missing steps in this calculation.) Since $1/n \to 0$ as n gets larger, it is easy to take the limit of S_n, which gives the area beneath f. The area beneath f is then equal to

$$\text{Limit } S_n = \text{Limit}\left[\frac{27}{6}\left(1 + \frac{1}{n}\right)\left(2 + \frac{1}{n}\right) - 9\left(1 + \frac{1}{n}\right) + 6\right]$$

$$= \frac{27}{6}(1)(2) - 9(1) + 6$$

$$= 6 \text{ square units}.$$

This confirms our guess, and shows that
$$\int_1^4 (x^2 - 4x + 5)\,dx = 6.$$

Looking back over the calculation given above, we can notice an important property of the integral. From line (25), we see that

$$S_n = \sum_{k=1}^{n} x_k^2 \Delta x - \sum_{k=1}^{n} 4x_k \Delta x + \sum_{k=1}^{n} 5 \Delta x. \qquad (26)$$

What happens to this equation "in the limit"? As n increases without bound, equation (26) becomes (*do you see why?*)

$$\int_1^4 (x^2 - 4x + 5)\,dx = \int_1^4 x^2\,dx - \int_1^4 4x\,dx + \int_1^4 5\,dx.$$

7. Some Properties of the Integral

This suggests that *the integral of a sum of functions is equal to the sum of their integrals.* This is true:

Sum Rule for Integrals. *If the functions f and g have integrals on the domain $a \leq x \leq b$, then*

$$\int_a^b (f(x) + g(x))\,dx = \int_a^b f(x)\,dx + \int_a^b g(x)\,dx.$$

PROOF.

$$\int_a^b (f(x) + g(x))\,dx = \text{Limit} \sum_{k=1}^n (f(x_k) + g(x_k))\,\Delta x$$

$$= \text{Limit} \sum_{k=1}^n f(x_k)\,\Delta x + \text{Limit} \sum_{k=1}^n g(x_k)\,\Delta x$$

$$= \int_a^b f(x)\,dx + \int_a^b g(x)\,dx. \qquad \square$$

What about a rule for constant multiples? Is it true that $\int_1^4 4x\,dx$ is equal to $4\int_1^4 x\,dx$? Sure it is:

Constant-Multiple Rule for Integrals. *If f has an integral on the domain $a \leq x \leq b$, then for any constant c,*

$$\int_a^b c \cdot f(x)\,dx = c \int_a^b f(x)\,dx.$$

PROOF. We know that

$$\sum_{k=1}^n c \cdot f(x_k)\,\Delta x = c \sum_{k=1}^n f(x_k)\,\Delta x, \qquad (27)$$

since the constant c is independent of the index of summation. Since equation (27) holds for each n, no matter how large, we get, in the limit,

$$\int_a^b c \cdot f(x)\,dx = c \int_a^b f(x)\,dx. \qquad \square$$

Another property of the integral is suggested by this figure.

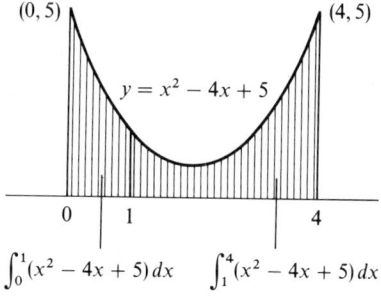

$$\int_0^1 (x^2 - 4x + 5)\,dx \qquad \int_1^4 (x^2 - 4x + 5)\,dx$$

Since the total shaded area is equal to $\int_0^4 (x^2 - 4x + 5)\,dx$, we know that $\int_0^1 (x^2 - 4x + 5)\,dx + \int_1^4 (x^2 - 4x + 5)\,dx = \int_0^4 (x^2 - 4x + 5)\,dx$. We are led to suspect that, in general, if $\int_0^4 f$ exists, then so does $\int_0^1 f$ and $\int_1^4 f$, and we have the *additivity property*

$$\int_0^1 f + \int_1^4 f = \int_0^4 f.$$

We can incorporate this into an existence theorem.

Existence Theorem for Integrals. *If f is a continuous function throughout the domain $a \leq x \leq b$, then the integral*

$$\int_a^b f = \int_a^b f(x)\, dx$$

exists. Moreover, if t is between a and b, then

$$\int_a^b f = \int_a^t f + \int_t^b f.$$

What does it mean to say that an integral "exists"? It means, simply, that Limit S_n exists, where S_n is the approximating sum from Eudoxus' method. The limit will exist, according to the theorem given above, if f is continuous. But the proof of the existence theorem is better left to a course in analysis. Let us take it for granted that f has an integral from a to b if f is continuous on $a \leq x \leq b$.

EXERCISES

7.1. Fill in the missing steps in the calculation of S_n that begins with equation (25).

7.2. Consider the areas A_1 and A_2 in the figure.

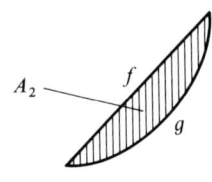

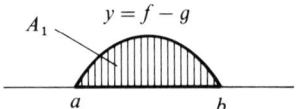

Cavalieri's principle says that $A_1 = A_2$. Prove that this is true by showing, in order,
(a) $A_1 = \int_a^b (f - g)$.
(b) $A_2 = \int_a^b f - \int_a^b g$.
(c) $A_1 = A_2$. [Use (a), (b), and the rule for sums and constant multiples.]

7.3. Is the integral of a product equal to the product of the integrals?

8. The Fundamental Theorem   177

7.4. Use Eudoxus' method to calculate
  (a) $\int_0^1 (x^2 - 4x + 5)\,dx$.
  (b) $\int_0^4 (x^2 - 4x + 5)\,dx$.

7.5. It is true that $\int_1^4 (x^2 - 4x)\,dx = 6 - \int_1^4 5\,dx$. Why is this true? *Hint*. In this section we have shown that $6 = \int_1^4 (x^2 - 4x + 5)\,dx$. Utilize the appropriate properties of the integral.

7.6. Find these integrals quickly, by using the values of related integrals already calculated, together with appropriate properties of the integral.
  (a) $\int_1^4 (-x^2 + 4x - 5)\,dx$. *Answer*: -6.
  (b) $\int_1^4 (3x^2 - 12x + 15)\,dx$.
  (c) $\int_1^4 (\pi x^2 - 4\pi x + 5\pi)\,dx$.
  (d) $\int_1^4 (x^2 - 4x)\,dx$. *Answer*: -9.

7.7. Attempt to calculate $\int_0^1 (1/x^2)\,dx$ by Eudoxus' method.
  (a) Show that $S_n = \sum_{k=1}^{n} n/k^2$.
  (b) What is S_1? S_2? S_3? *Partial answer*: $S_2 = 2.5$.
  (c) Show that S_n always exceeds n. *Hint*. Show $S_n = n(1 + \cdots)$.
  (d) Does Limit S_n exist? *Answer*: In view of part (c), S_n cannot tend to a limit, since it grows arbitrarily large as n increases.
  (e) Does $\int_0^1 (1/x^2)\,dx$ exist? *Hint*. By definition, the integral is equal to Limit S_n. Use part (d).
  (f) Does your answer to part (e) contradict the existence theorem for integrals? Why not?

§8. The Fundamental Theorem

The fundamental theorem shows the connection between the two branches of calculus, *differential* and *integral*. The connection is really between Fermat's method and Eudoxus' method, of course. To prepare the way for the fundamental theorem, let us review Fermat's method, using the notation of Leibniz.

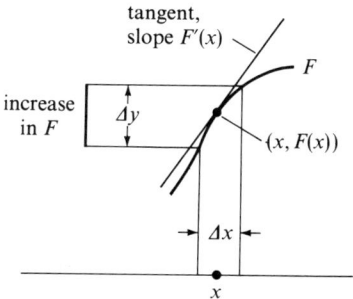

Consider a function F and a point x, and let Δx be length of a small interval that contains x. Then, by Fermat's method, it follows that

$$F'(x) = \underset{\Delta x \to 0}{\text{Limit}} \frac{\Delta y}{\Delta x}.$$

This means, roughly speaking, that the number $\Delta y/\Delta x$ is very close to $F'(x)$ when Δx is very close to (but not equal to) zero. In symbols,

$$F'(x) \approx \frac{\Delta y}{\Delta x}, \quad \text{provided } \Delta x \approx 0,$$

where "$\approx$" stands for "approximate equality". It will serve our purpose to rewrite this as

$$F'(x) - \frac{\Delta y}{\Delta x} \approx 0, \quad \text{provided } \Delta x \approx 0. \tag{28}$$

Here, x is fixed, and the expression $F'(x) - (\Delta y/\Delta x)$ varies in terms of Δx. Let the letter o stand for this expression:

$$F'(x) - \frac{\Delta y}{\Delta x} = o. \tag{29}$$

By (28), we know something about the variable o:

$$o \approx 0, \quad \text{provided } \Delta x \approx 0. \tag{30}$$

From (29), upon multiplying through by Δx,

$$F'(x)\Delta x - \Delta y = o\,\Delta x,$$
$$F'(x)\Delta x = \Delta y + o\,\Delta x. \tag{31}$$

Equation (31), in connection with the information given in (30), is the key to the proof of the fundamental theorem. Note that (31) is, so to speak, what one gets by beginning with Fermat's method and "undoing it". Roughly speaking, (31) says that when the derivative is multiplied by Δx, a small change in x, you get a close approximation to Δy, the corresponding change in y. Remember that o is not 0, but rather a variable tending to 0 as $\Delta x \to 0$.

The Fundamental Theorem of Calculus. *If f is a continuous function with domain $a \leq x \leq b$, then*

$$\int_a^b f(x)\,dx = F(b) - F(a),$$

where F is any antiderivative of f.

PROOF. (Didn't we prove this already when we proved the area principle? Answer: No. The integral of f is not always an area. The fundamental theorem asserts that the antiderivative method works even when the function f is not always positive.)

8. The Fundamental Theorem

Since we know that F is an antiderivative of f, equation (31) says

$$f(x)\Delta x = \Delta y + o\,\Delta x,$$

where Δy is the change in F corresponding to the change Δx in x. Applying this to the k-th subinterval in Eudoxus' method, we have

$$f(x_k)\Delta x = \text{change in } F \text{ on } k\text{-th subinterval} + o_k\,\Delta x.$$

Hence the approximating sum to the integral $\int_a^b f$ can be expressed as

$$\sum_{k=1}^{n} f(x_k)\Delta x = \sum_{k=1}^{n} \text{change in } F \text{ on } k\text{-th subinterval} + \sum_{k=1}^{n} o_k\,\Delta x. \quad (32)$$

It is obvious that

$$\sum_{k=1}^{n} \substack{\text{change in } F \text{ on}\\ k\text{-th subinterval}} = \substack{\text{change in } F \text{ on the}\\ \text{entire interval } a \leq x \leq b}$$
$$= F(b) - F(a).$$

Therefore, from (32),

$$\sum_{k=1}^{n} f(x_k)\Delta x = F(b) - F(a) + \sum_{k=1}^{n} o_k\,\Delta x,$$

and, taking the limit as $\Delta x \to 0$, we get

$$\int_a^b f(x)\,dx = F(b) - F(a) + \int_a^b 0\,dx \quad [\text{by (30)}]$$
$$= F(b) - F(a). \qquad \square$$

The careful reader may feel that the last step in the proof given above does not justify adequately the fact that

$$\underset{\Delta x \to 0}{\text{Limit}} \sum_{k=1}^{n} o_k\,\Delta x = \int_a^b 0\,dx. \quad (33)$$

The careful reader is right. Although (33) is surely made plausible by (30), it has not been justified rigorously in the above proof. Rigorous proof of (33) is better deferred to a course in analysis.

EXERCISES

8.1. (a) Evaluate the integral $\int_{-1}^{1} x\,dx$ by the antiderivative method.
(b) Evaluate $\int_{-1}^{1} x\,dx$ by Eudoxus' method. *Answer:* 0.
(c) Do your answers to (a) and (b) agree, as the fundamental theorem asserts?
(d) Can the integral in question be regarded as an *area*?

8.2. Prove that if the continuous curve f crosses the x-axis, then the integral $\int_a^b f$ gives the *algebraic* sum of the areas between the curve f and the axis, counting area above as positive and below as negative. *Hint.* In the picture below, you want to

show that
$$\int_a^b f = A_1 - A_2 + A_3.$$

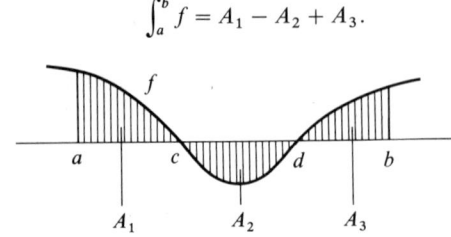

Prove this by giving a reason for each of the following equalities:

$$\int_a^b f = \int_a^c f + \int_c^d f + \int_d^b f$$
$$= \int_a^c (f - 0) - \int_c^d (0 - f) + \int_d^b (f - 0) = A_1 - A_2 + A_3.$$

8.3. Evaluate each of the following integrals by using the fundamental theorem.
   (a) $\int_0^2 (1 - x^3)\,dx$.
   (b) $\int_0^2 3\,dx$.
   (c) $\int_{-1}^4 3x^2\,dx$.
   (d) $\int_{-1}^1 (\pi - \pi x^2)\,dx$.
   (e) $\int_1^{10}(1/x^2)\,dx$.
   (f) $\int_{-10}^{-1}(1/x^2)\,dx$.
   Answers: (d) $4\pi/3$. (e) $\tfrac{9}{10}$. (f) $\tfrac{9}{10}$.

8.4. First express each of the following areas as an integral. Then evaluate the integral, using the fundamental theorem.

(a)
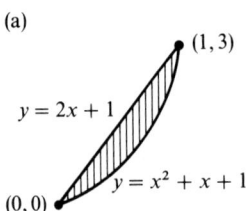

(b)

(1, 2) $y = 1 + 1/x$
(1, 1)   (3, $\tfrac{4}{3}$)
       $y = 1/x$ (3, $\tfrac{1}{3}$)

(c)
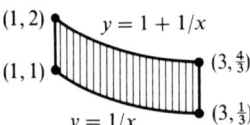

Answer: (a) $\int_0^1 (x - x^2)\,dx = [(x^2/2) - (x^3/3)]_0^1 = \tfrac{1}{6}$ square units.

8.5. (*For careful readers*) What is wrong with the following "calculation"? $\int_{-1}^{1} (1/x^2)\,dx = -1/x\big|_{-1}^{1} = -2$.

8.6. Consider the integral $\int_0^7 \pi \frac{25}{49} x^2\,dx$.
   (a) Evaluate this integral, using the fundamental theorem.
   (b) Draw a picture of an area that is represented by this integral. (On the following pages, we shall see that this same integral also represents a *volume*.)

§9. Integrals and Volumes

Integrals, defined by Eudoxus' method, arise naturally in many contexts having nothing to do with area. Yet the fundamental theorem can still be used to evaluate the integral, provided an appropriate antiderivative can be found. This is why the fundamental theorem is of much more significance than the area principle. Many illustrations of this may be seen in Chapter 8.

One illustration is readily at hand. Let us consider the problem of finding volumes of *solids of revolution*. The only thing we need to know at the outset is the formula for the volume of a cylinder. It is given by the product of the area of the circular base and the height:

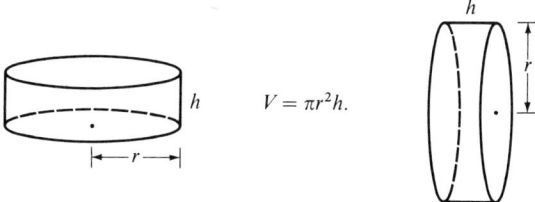

$V = \pi r^2 h.$

Working carefully through an example will enable us to see a shortcut way of working many similar examples. The key is to try to express the volume desired as an integral. The integral can then be evaluated by the fundamental theorem.

EXAMPLE 11. Determine the volume of a cone if its height is 7 feet and if the radius of its base is 5 feet.

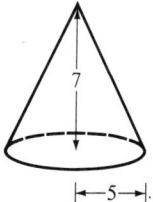

What is meant by the *volume* of a solid figure? This question is easily answered by means of the notion of a *limit*. We can get the volume by approximating it ever more closely and then obtaining it exactly as the limit of our approximations. The desired volume, we shall see, will turn out to be the limit of a sum, just as in Eudoxus' method. That is, the desired volume will turn out to be an integral.

Let us carry out this procedure. If we turn our given cone on its side, we see that it could be regarded as the solid figure obtained by revolving the indicated area 360 degrees about the horizontal axis. Such a solid figure is called a *solid of revolution*. The volume of any solid of revolution is easy to obtain by the method described below.

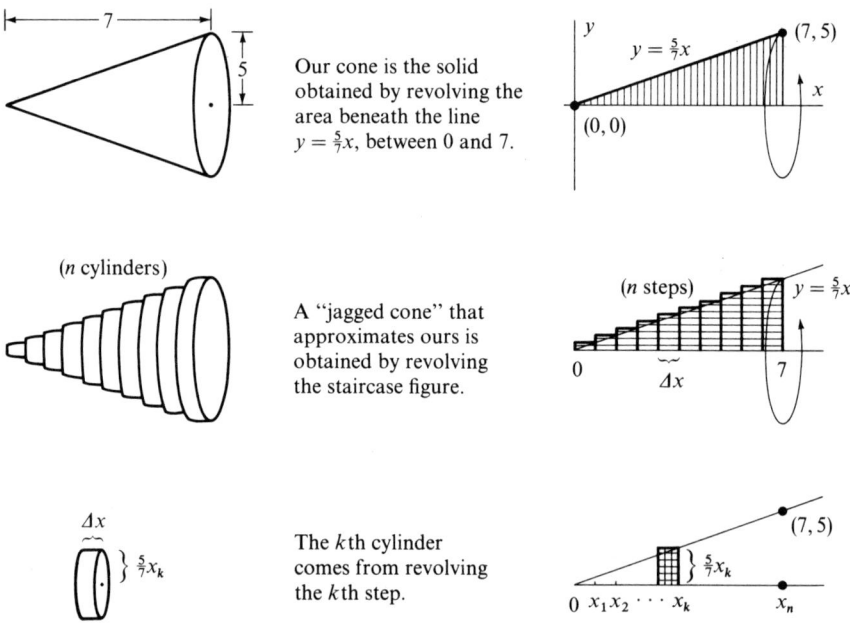

Our cone is the solid obtained by revolving the area beneath the line $y = \frac{5}{7}x$, between 0 and 7.

A "jagged cone" that approximates ours is obtained by revolving the staircase figure.

The kth cylinder comes from revolving the kth step.

From the formula for the volume of a cylinder, the volume of the k-th cylinder is clearly given by

$$\pi \left(\frac{5}{7}x_k\right)^2 \Delta x = \pi \frac{25}{49} x_k^2 \, \Delta x.$$

The jagged cone is made up of n cylinders. Its volume is the sum of the volumes of these cylinders:

$$\text{Volume of jagged cone} = \sum_{k=1}^{n} \text{volume of } k\text{-th cylinder} = \sum_{k=1}^{n} \pi \frac{25}{49} x_k^2 \, \Delta x. \tag{34}$$

9. Integrals and Volumes

As $\Delta x \to 0$, the jagged cone approximates our given cone ever more closely. Therefore,

$$\text{Volume of given cone} = \underset{\Delta x \to 0}{\text{Limit}} \text{ (volume of jagged cone)}$$

$$= \underset{\Delta x \to 0}{\text{Limit}} \sum_{k=1}^{n} \pi \frac{25}{49} x_k^2 \, \Delta x \quad [\text{by (34)}].$$

This says that the volume of our given cone is equal to the limit of a sum, i.e., to an *integral*. What integral is it? The domain is surely $0 \le x \le 7$, because the points x_k subdivide that domain. Clearly, then,

$$\underset{\Delta x \to 0}{\text{Limit}} \sum_{k=1}^{n} \pi \frac{25}{49} x_k^2 \, \Delta x = \int_0^7 \pi \frac{25}{49} x^2 \, dx$$

$$= \pi \frac{25}{49} \frac{x^3}{3} \bigg|_0^7 \approx 183.26.$$

The volume of a cone, with $h = 7$ feet and $r = 5$ feet, is then given by

$$\frac{\pi (25)(7)}{3} \approx 183.26 \text{ cubic feet.} \qquad \square$$

One might conjecture that the volume of a cone of height h and radius r is given by $\pi r^2 h/3$. This seems to be what the answer to Example 11 is trying to tell us. The reader is asked to verify this conjecture in an exercise to follow.

Exercises

9.1. Determine the volume of a cone of height h and radius r. (Just work through each step of Example 11, but with h in place of 7 and with r in place of 5.)

9.2. Compare a cone with a cylinder of the same base and height. Using your answer to exercise 9.1, find the ratio of the volume of the cylinder to the volume of the inscribed cone.

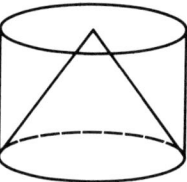

Answer: The ratio is 3:1, first proved by the man himself, Eudoxus of Cnidos, as an application of the method that now bears his name.

9.3. Suppose it is desired to cut a cone parallel to its base, in such a way that the two resulting pieces have the same volume. Where should the cut be made?

9.4. In Example 11, the area beneath the line $y = \frac{5}{7}x$, $0 \le x \le 7$, was revolved about the x-axis, and the volume of the resulting solid of revolution found. Suppose instead we revolve the area beneath the quadratic $y = x^2$, $0 \le x \le 1$.

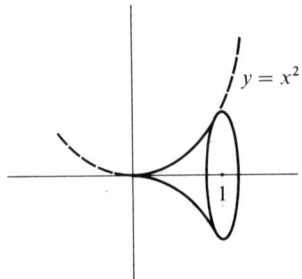

Let V be the volume of the resulting solid, and let the points x_k subdivide the interval $0 \le x \le 1$, as in Eudoxus' method. For each of the equalities that follow, give a reason to justify it.

$$V = \underset{\Delta x \to 0}{\text{Limit}} \sum_{k=1}^{n} \pi x_k^4 \, \Delta x$$

$$= \int_0^1 \pi x^4 \, dx = \pi \int_0^1 x^4 \, dx = \pi \left[\frac{x^5}{5} \right]_0^1 = \frac{\pi}{5} \text{ cubic units.}$$

§10. The Volume of a Solid of Revolution

We found the volume of a cone in Section 9 by regarding the cone as a solid of revolution. Thus its volume could be approximated by a "jagged" solid of revolution, then calculated exactly as an integral. Exactly the same procedure will give us an integral formula for the volume of any solid of revolution.

Consider the solid of revolution obtained by revolving the area beneath a continuous curve f, with domain $a \le x \le b$.

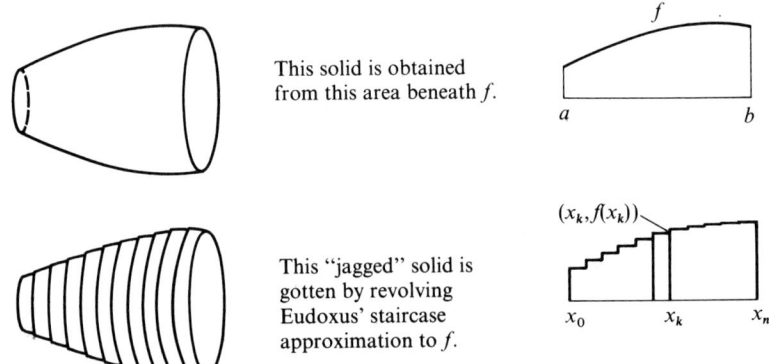

This solid is obtained from this area beneath f.

This "jagged" solid is gotten by revolving Eudoxus' staircase approximation to f.

10. The Volume of a Solid of Revolution

The jagged solid is made up of n cylinders, if the staircase has n steps. The k-th cylinder comes from revolving the k-th step:

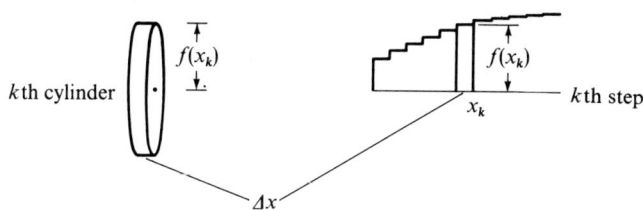

Since the volume of the k-th cylinder is $\pi(f(x_k))^2 \Delta x$, the volume of the jagged solid is

$$\sum_{k=1}^{n} \pi(f(x_k))^2 \Delta x.$$

As $\Delta x \to 0$, the jagged solid's volume tends to

$$\int_a^b \pi(f(x))^2 \, dx. \tag{35}$$

Formula (35) then gives the volume of the solid of revolution obtained by revolving the area beneath the curve f, from $x = a$ to $x = b$, about the x-axis.

EXAMPLE 12. Find the volume of a sphere whose radius is 7 meters.

A sphere can be regarded as the solid of revolution obtained by revolving a semicircle about its diameter.

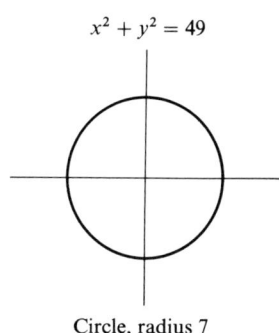

Circle, radius 7

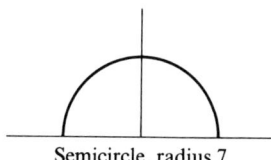

Semicircle, radius 7

The equation

$$y = \sqrt{49 - x^2}, \qquad -7 \le x \le 7,$$

describes a semicircle of radius 7 whose diameter lies on the x-axis. By formula (35), the volume of a sphere of radius 7 meters is given by

$$\int_{-7}^{7} \pi(\sqrt{49-x^2})^2 \, dx = \pi \int_{-7}^{7} (49-x^2) \, dx$$

$$= \pi \left[49x - \frac{x^3}{3} \right]_{-7}^{7}$$

$$= \pi 7^3 \left(\frac{4}{3}\right) \text{ cubic meters.} \qquad \square$$

One might conjecture that the volume of a sphere of radius r is given by $4\pi r^3/3$. That could be what the answer to Example 12, where $r = 7$, is trying to tell us. The reader is asked to verify this conjecture in an exercise to follow.

EXERCISES

10.1. Use formula (35) to find the volumes of the solids of revolution obtained by revolving the areas under each of the following curves.
(a) $f(x) = x^2, 0 \leq x \leq 5$.
(b) $f(x) = x^2, -4 \leq x \leq 4$.
(c) $f(x) = x + 1, 0 \leq x \leq 3$.
(d) $f(x) = \sqrt{1+x^2}, -1 \leq x \leq 2$.
Answers: (a) 625π cubic units. (c) 21π cubic units.

10.2. Consider the integral $\int_1^3 (\pi/x^2) \, dx$. Draw a picture of
(a) a figure in the plane whose area is given by this integral.
(b) a solid of revolution whose volume is given by this integral.

10.3. Determine the volume of a sphere of radius r. (Just work through the steps of Example 12 with r in place of 7.)

10.4. Consider a sphere in comparison with a cylinder in which the sphere is inscribed. Using your answer to exercise 10.3, find the ratio of the volume of the cylinder to the volume of the sphere.

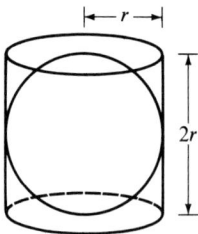

Answer: The ratio is 3:2, as first proved by Archimedes in the third century B.C. (See the appendix on Archimedes.)

§11. Isaac Newton

The Plague, in 1664–1665, had at least one fortunate consequence. Cambridge University was forced to shut down. Newton, having just received his B.A. degree, moved back to the English countryside where he had been born on Christmas Day of 1642. Newton delighted in privacy. The next 2 years of secluded life by an apple grove produced astonishing results. Newton came into possession of ideas that would enable him to create modern physics virtually by himself, with the help of calculus, which he also created at the same time.

Newton not only had ideas of his own. He could seee new features hidden in the ideas of others. Whereas Fermat had developed his method only to find tangent lines to curves, Newton saw in this same method the means of defining a derived function that would measure instantaneous rates of change. Newton used this to study the physics of motion, of which he postulated certain "universal laws". When he put these laws together with his calculus and with the law of gravitation (also discovered on the farm), Newton derived the equations governing the motion of the planets about the sun.

Johannes Kepler (1571–1630) had earlier made the significant discovery that the planets travel in elliptical orbits, but Kepler could not explain why. Newton knew how to explain that this was no more mysterious than the fall of an apple from the tree. When Newton overcame his secretive nature and finally revealed in 1687 the magnitude of his work, the effect was overwhelming. Newton, it was said, had "explained the universe". That was, of course, an overstatement, whose repetition finally prompted the playful couplet of Alexander Pope,

> Nature and Nature's laws lay hid in night;
> God said, "Let Newton be!" and all was light.

Nevertheless, it is generally conceded that Newton's *Mathematical Principles of Natural Philosophy* (1687) remains the greatest single work in the history of science. Perhaps never before or since has so much been uncovered at a single stroke. Aided by 20 years of thought, Newton wrote it, start to finish, in 18 months.

An aura of mystery still surrounds the man:

> ... Newton with his prism and silent face,
> The marble index of a mind forever
> Voyaging through strange seas of thought, alone.

So wrote William Wordsworth near the dawn of the nineteenth century, upon marveling at a statue of Newton celebrating his work in optics.

Newton's highest compliment came from his only rival, who published (in 1684) the first paper on the calculus. There would be bitter years of

[1]

PHILOSOPHIÆ
NATURALIS
Principia
MATHEMATICA

Definitiones.

Def. I.

Quantitas Materiæ est mensura ejusdem orta ex illius Densitate & Magnitudine conjunctim.

AEr duplo densior in duplo spatio quadruplus est. Idem intellige de Nive et Pulveribus per compressionem vel liquefactionem condensatis. Et par est ratio corporum omnium, quæ per causas quascunq; diversimode condensantur. Medii interea, si quod fuerit, interstitia partium libere pervadentis, hic nullam rationem habeo. Hanc autem quantitatem sub nomine corporis vel Massæ in sequentibus passim intelligo. Innotescit ea per corporis cujusq; pondus. Nam ponderi proportionalem esse reperi per experimenta pendulorum accuratissime instituta, uti posthac docebitur.

B   Def.

Figure 2. The first page of Newton's *Principia* (1687)

controversy over who first discovered the calculus. But,

> Taking mathematics from the beginning of the world to the time of Newton, what he has done is much the better half.
>
> Leibniz

Although others owned bits and pieces of the calculus, Newton was the first to have the whole subject at his command. He rose like Archimedes above the age in which he lived, moved by a spirit impervious to time.

> I do not know what I may appear to the world; but to myself I seem to have been only like a boy playing on the seashore, and diverting myself in now and then finding a smoother pebble or a prettier shell than ordinary, whilst the great ocean of truth lay all undiscovered before me.
>
> Newton

The transport of calculus from the seashore to the stars: that was Newton's accomplishment. A dream of old Pythagoras had been realized at last.

Problem Set for Chapter 6

1. A boat travels along a straight course. At time t hours past noon, its speed is $t^2 - 4t + 10$ km/hr. How far does the boat travel between three o'clock and six o'clock?

2. Sketch the quadratic curve $y = t^2 - 4t + 10$, and find the area beneath this curve, between $t = 3$ and $t = 6$.

3. If A is the area indicated in the figure below, find dA/dt. (Your answer should be expressed in terms of t, of course.)

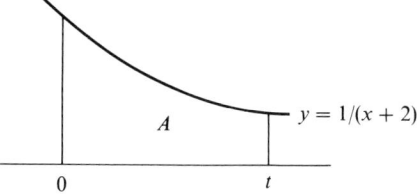

4. Find the indicated area. (In each case, split the area into two pieces by drawing an appropriate vertical line, find the area of each piece separately, and add.)

(a)
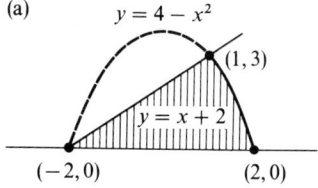

(b)
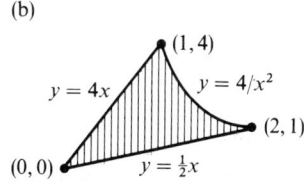

5. Draw a picture of the area represented by the integral $\int_0^2 4\,dt$, and evaluate the integral by finding the area of your picture.

6. Evaluate the integral $\int_0^2 \sqrt{4-t^2}\,dt$, after first drawing a picture of the area it represents.

7. Find the indicated area by using your answers to problems 5 and 6. Why is the general area principle of no use to you here?

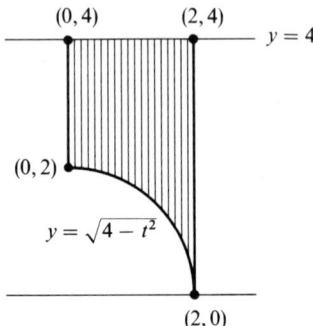

8. Find the indicated area by splitting it into three parts, finding the area of each part, and adding.

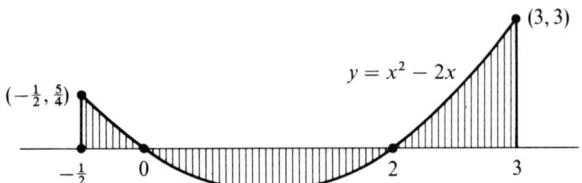

9. Calculate the following integrals directly from their definition by Eudoxus' method.
   (a) $\int_0^2 4x\,dx$.
   (b) $\int_1^4 (2x-1)\,dx$.
   (c) $\int_a^b x\,dx$.

10. Consider the integral $\int_{-1/2}^3 (x^2 - 2x)\,dx$.
    (a) Illustrate Eudoxus' method in calculating this integral.
    (b) Illustrate the fundamental theorem in calculating this integral.
    (c) Explain why it is to be expected that your answers to parts (a) and (b) agree with each other, yet disagree with your answer to problem 8.

11. (a) By using the appropriate rules for derivatives, and simplifying your answer, show that
$$\text{if } F(x) = \frac{3x^2}{2\sqrt{1+x^3}}, \quad \text{then } F'(x) = \frac{3x(4+x^3)}{4(1+x^3)^{3/2}}.$$

   (b) Evaluate the integral
$$\int_0^2 \frac{3x(4+x^3)}{4(1+x^3)^{3/2}}\,dx.$$

12. Find the indicated areas, by any means.

(a)
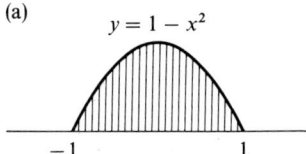
$y = 1 - x^2$

(b)
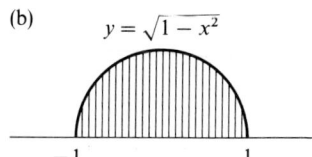
$y = \sqrt{1 - x^2}$

13. Find the volumes of the solids of revolution obtained by revolving the areas of problem 12 about the x-axis.

14. (a) Find an equation of the line joining (1, 2) and (4, 5).
    (b) Find the volume of the frustrum of a cone obtained by revolving the indicated area about the x-axis.

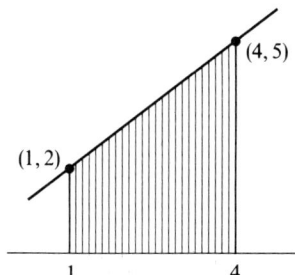

15. Find the volume of the indicated flower pot, shaped like the frustrum of a cone.

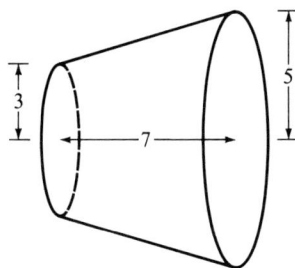

Hint. The volume is that obtained from revolving the area beneath the line joining (0, 3) and (7, 5).

16. Find a formula (in terms of r_1, r_2, and h) for the volume of a frustrum of a cone with the indicated dimensions.

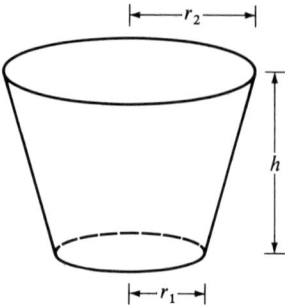

17. A grapefruit half, shaped like a hemisphere of radius 3 inches, is sliced in two, as indicated.

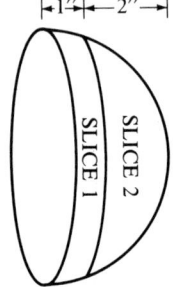

    (a) Find the volume of each slice.
    (b) Which slice has greater volume?
    (c) Where should the slice be made in order to divide the grapefruit into pieces of equal volume? (Your answer may be expressed as the solution to a certain cubic.)

18. Consider the integral $\int_0^5 \pi x^2 \, dx$.
    (a) Draw a picture of a figure whose area is given by this integral.
    (b) Draw a picture of a solid of revolution whose volume is given by this integral.

19. The equation $(x^2/a^2) + (y^2/b^2) = 1$ has an *ellipse* as its graph. Let A be the area inside the ellipse. Justify each of the following equalities in the calculation of A.

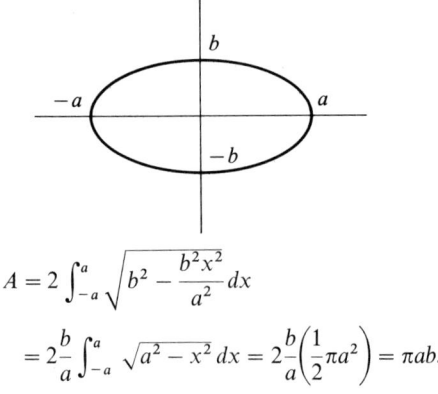

$$A = 2\int_{-a}^{a} \sqrt{b^2 - \frac{b^2 x^2}{a^2}}\, dx$$

$$= 2\frac{b}{a}\int_{-a}^{a} \sqrt{a^2 - x^2}\, dx = 2\frac{b}{a}\left(\frac{1}{2}\pi a^2\right) = \pi ab.$$

20. If the ellipse of problem 19 is revolved about the x-axis, an *ellipsoid of revolution* (watermelon) results. Find the volume inside it.

21. The two right triangles below might be regarded as being made up of "identical vertical segments", if their segments are made to correspond as indicated. But it is clear that the two triangles do not have the same area. Does this violate Cavalieri's principle? Why not?

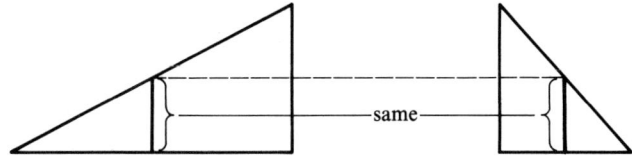

22. In the problem set at the end of Chapter 2, work problem 12 by applying Cavalieri's principle. *Hint.* Go to problem 12, Chapter 2. Turn your head sideways as far as you can. Don't strain your neck.

23. Find the coordinates of the point P lying on the curve $y = 4 - x^2$ that maximizes the area of the indicated triangle PQR. (P must lie between R and Q.)

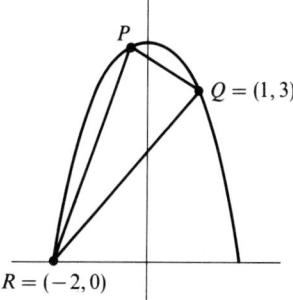

24. Match each of the curves below with its derivative.

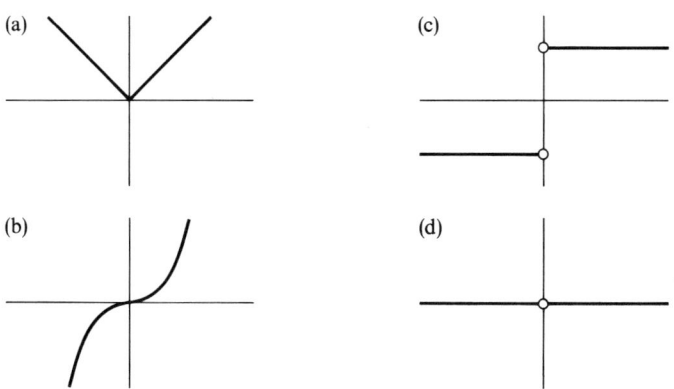

[The curve in (d) coincides with the horizontal axis, but has a hole in it.]

7 A Circle of Ideas

As we have seen, near the end of Chapter 5, the fundamental theorem of calculus veils her face with a circle. In this chapter we shall see that *all of trigonometry hides behind a circle*, as well. According to Plato, Dante, and many others, the circle is the most perfect figure. It is surely rich in ideas.

Trigonometry has quite ancient origins, dating back over 3000 years to the Egyptians and Babylonians. Literally, it means "three-angle measurement", signifying its application to triangles used in land measurement, navigation, astronomy, and the like. Only in comparatively recent times has it been well recognized that trigonometry might be described more properly as the study of circles, rather than triangles.

In this chapter we shall develop the main ideas of trigonometry and apply the methods of differential and integral calculus to *trigonometric* functions (sine, cosine, tangent, etc.). We thus take a giant step forward. Up to now we have restricted our attention almost entirely to *algebraic* functions.

No prior acquaintance with trigonometry is necessary in order to read this chapter.

§1. Trigonometric Functions; Degrees versus Radians

Let us start at the beginning, with a right triangle.

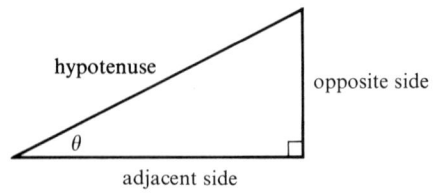

1. Trigonometric Functions: Degrees versus Radians

Consider the angle θ, with "adjacent side" and "opposite side", as indicated. The *sine*, *cosine*, and *tangent* of θ are defined as the following ratios:

$$\text{sine } \theta = \frac{\text{opposite side}}{\text{hypotenuse}}, \quad \text{cosine } \theta = \frac{\text{adjacent side}}{\text{hypotenuse}}, \quad \text{tangent } \theta = \frac{\text{opposite side}}{\text{adjacent side}}.$$

The cases when θ is either 30, 45, or 60 degrees will quickly become familiar to the reader. They look like this. (*See exercises 1.1 and 1.2.*)

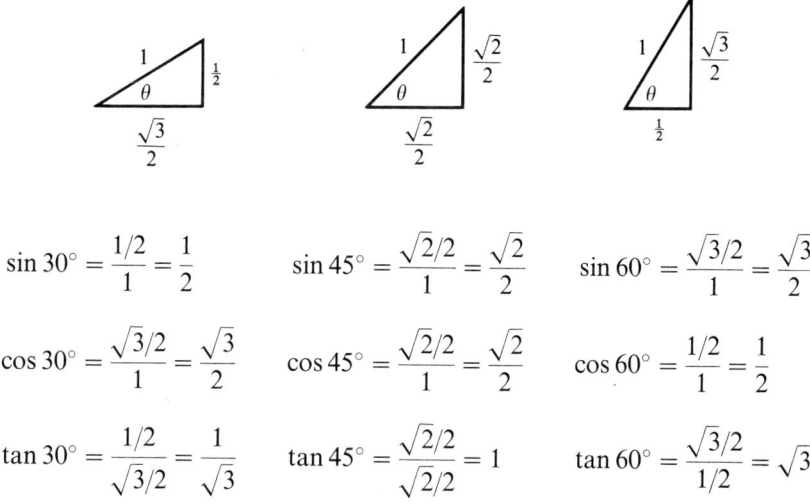

$$\sin 30° = \frac{1/2}{1} = \frac{1}{2} \qquad \sin 45° = \frac{\sqrt{2}/2}{1} = \frac{\sqrt{2}}{2} \qquad \sin 60° = \frac{\sqrt{3}/2}{1} = \frac{\sqrt{3}}{2}$$

$$\cos 30° = \frac{\sqrt{3}/2}{1} = \frac{\sqrt{3}}{2} \qquad \cos 45° = \frac{\sqrt{2}/2}{1} = \frac{\sqrt{2}}{2} \qquad \cos 60° = \frac{1/2}{1} = \frac{1}{2}$$

$$\tan 30° = \frac{1/2}{\sqrt{3}/2} = \frac{1}{\sqrt{3}} \qquad \tan 45° = \frac{\sqrt{2}/2}{\sqrt{2}/2} = 1 \qquad \tan 60° = \frac{\sqrt{3}/2}{1/2} = \sqrt{3}$$

Note that, in each of these cases, the tangent is the quotient of the sine and cosine.

Calculus deals with functions, and we see the beginning of three important trigonometric functions from the work given above.

SIN θ		COS θ		TAN θ	
30	$\frac{1}{2}$	30	$\sqrt{3}/2$	30	$1/\sqrt{3}$
45	$\sqrt{2}/2$	45	$\sqrt{2}/2$	45	1
60	$\sqrt{3}/2$	60	$\frac{1}{2}$	60	$\sqrt{3}$

In the tables given above, the angle θ is measured in *degrees*. When we measure angles by degrees, we are following the practice established by the ancient Babylonians, who decided that a circle should be divided into 360 equal parts:

$$360 \text{ degrees} = 1 \text{ revolution.} \qquad (1)$$

It is worth mentioning that, from (1), we get

$$90 \text{ degrees} = \frac{1}{4} \text{ revolution,}$$

$$60 \text{ degrees} = \frac{1}{6} \text{ revolution,}$$

$$45 \text{ degrees} = \frac{1}{8} \text{ revolution,} \qquad (2)$$

$$30 \text{ degrees} = \frac{1}{12} \text{ revolution,}$$

$$1 \text{ degree} = \frac{1}{360} \text{ revolution.}$$

Rather than measure angles in degrees, we could equally well measure them in fractions of a revolution, as illustrated by the formulas collected in (2). It is remarkable that degree measurement should have lasted so long. Is it not curious that, some 3000 years ago, the Babylonians chose 360 as the numbers of units in a circle? Why not 100, for example, instead? We measure angles today in the same way that Nebuchadnezzar did!

Perhaps the ancient Babylonians thought (and this seems uncertain) that a year is composed of 360 days. If so, then a circle divided into 360 parts is a convenient way of marking the passage of days. The circle, like the Babylonian year, repeats itself after a period of 360 units has been completed. This simple property of a circle, that *circular motion repeats itself after a period of* 1 *revolution*, is of surprising importance. Whenever periodic phenomena appear—as in the rise and fall of a piston, the turning of a wheel, the waves striking the seashore, sound waves hitting the ear—one might look for a mathematical description that involves a circle, appearing in one of its many splendid disguises. We emphasize:

Circular motion is periodic, repeating after 1 revolution. (3)

It is tempting to think that it makes no real difference how angles are measured, whether in degrees or in revolutions, or in whatever other units we may choose to invent. The units in terms of which we measure are purely *conventional*, i.e., chosen for convenience. In calculus a new convention has arisen, and for one reason only. It makes much more convenient the study of the calculus of trigonometric functions. In calculus we measure angles by **radians**, where, by definition,

$$1 \text{ revolution} = 2\pi \text{ radians.} \qquad (4)$$

1. Trigonometric Functions: Degrees versus Radians

A little later we shall see that the advantage of radian measure over degree measure is a simplification in the formulas for the derivatives of the trigonometric functions.

From (1) and (4) it follows that

$$360° = 1 \text{ rev.} = 2\pi \text{ rad,} \tag{5}$$

and the formulas collected in (2) are augmented as follows:

$$90° = \frac{1}{4} \text{ rev.} = \frac{\pi}{2} \text{ rad.}$$

$$60° = \frac{1}{6} \text{ rev.} = \frac{\pi}{3} \text{ rad.}$$

$$45° = \frac{1}{8} \text{ rev.} = \frac{\pi}{4} \text{ rad.} \tag{6}$$

$$30° = \frac{1}{12} \text{ rev.} = \frac{\pi}{6} \text{ rad.}$$

$$1° = \frac{1}{360} \text{ rev.} = \frac{\pi}{180} \text{ rad.}$$

From Section 2 onward, unless otherwise indicated it shall be understood that *angles are measured in radians*. For example,

$$\cos \frac{\pi}{3} = \frac{1}{2},$$

because $\frac{1}{2}$ is the ratio of the adjacent side to the hypotenuse relative to an angle of $\pi/3$ radians (60 degrees). The sine, cosine, and tangent functions, *when defined in terms of radian measure of angles, will be abbreviated by*

$$\sin, \quad \cos, \quad \text{and} \quad \tan,$$

respectively. We shall rarely have need to use the "old-fashioned" functions

$$\text{SIN}, \quad \text{COS}, \quad \text{and} \quad \text{TAN},$$

defined in terms of degree measure.

(Do you see the difference between cos and COS? Whereas $\cos(\pi/3) = \frac{1}{2}$, it is not true that $\text{COS}(\pi/3) = \frac{1}{2}$. Instead, $\text{COS}(\pi/3) \approx 1$, for the cosine of $\frac{\pi}{3}$ *degrees* is approximately equal to 1.)

Our program will be to study the trigonometric functions and see how they arise in many settings related to problems in calculus. We shall find their derivatives, calculate their integrals, and use them in problems of optimization and in problems of related rates.

EXERCISES

1.1. Use the Pythagorean theorem to find the lengths a and b. *Hint.* First convince yourself that $a = b$.

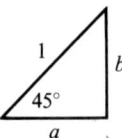

1.2. Use the Pythagorean theorem to find the lengths a and b. *Hint.* First convince yourself that if the triangle is duplicated along the side a, as indicated, then an equilateral triangle results. From this it is obvious what the length b is.

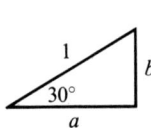 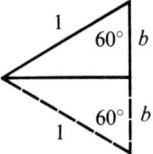

1.3. How many degrees are there in an angle of 1 radian? *Hint.* In equation (5), divide by 2π. *Answer:* 1 radian is approximately 57.3 degrees.

1.4. An arc length of s units is subtended by an angle at the center of a circle of radius 1 unit.

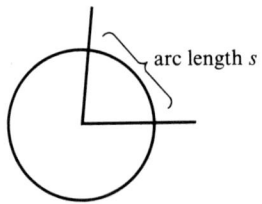

What is the measure of the angle in
(a) revolutions?
(b) degrees?
(c) radians?
Answer: (a) $s/2\pi$ revolutions, since 1 revolution corresponds to an arc length of 2π, the circumference of the unit circle.

1.5. What is the area of the pie-shaped sector in exercise 1.4? *Hint.* The area of the entire unit circle is π square units. What fraction of the unit circle is taken up by the pie-shaped sector?

1.6. An arc length of s units is subtended by an angle at the center of a circle of radius r units. What is the measure of the angle in
(a) revolutions?
(b) degrees?
(c) radians?
Answer: (b) $180s/\pi r$ degrees.

2. Circle Functions

1.7. What is the arc length subtended by an angle of t radians on a circle of radius
  (a) 1 unit?
  (b) r units?

1.8. Convert from degrees to radians:
  (a) 15.
  (b) 12.
  (c) 540.
  (d) 75.
  (e) 6.
  (f) 20.

1.9. Convert from radians to degrees:
  (a) $\pi/12$.
  (b) 3π.
  (c) $\pi/2$.
  (d) $5\pi/6$.
  (e) $\pi/5$.
  (f) $\pi/6$.

1.10. What is the sine of an angle of $\pi/6$ radians? *Answer:* $\frac{1}{2}$.

1.11. Fill in the table below, where t is measured in radians.

t	$\cos t$	$\sin t$
$\pi/6$		
$\pi/4$		
$\pi/3$		

1.12. Find the coordinates of the indicated points lying on the unit circle.

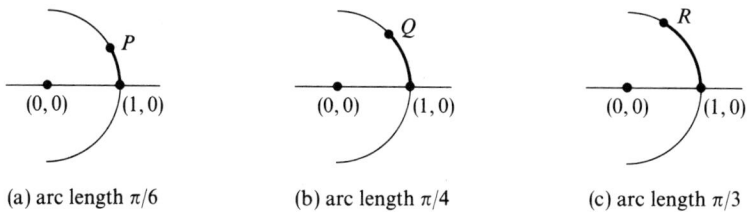

(a) arc length $\pi/6$ (b) arc length $\pi/4$ (c) arc length $\pi/3$

Answer: (b) On the unit circle an arc length of $\pi/4$ units corresponds to an angle of 45 degrees at the center of the circle. By exercise 1.1, $Q = (\sqrt{2}/2, \sqrt{2}/2)$. *Note the similarity in the answers of exercise 1.11 and 1.12.*

§2. Circle Functions

Up to now we have been concerned mainly with angles t lying between 0 and 90 degrees, or 0 and $\pi/2$ radians. *From now on, we shall measure angles in radians,* unless otherwise indicated. One can give a simple description of how to find $\cos t$ and $\sin t$, when t is measured in radians. The description is based upon the *unit circle* (the circle of radius 1 unit, with center at the origin).

Beginning at the point (1,0) on the unit circle, mark off an arc length of t units, traveling *counterclockwise* around the circle. The point on the circle at the end of this arc must (see exercise 1.12) have coordinates given by $(\cos t, \sin t)$.

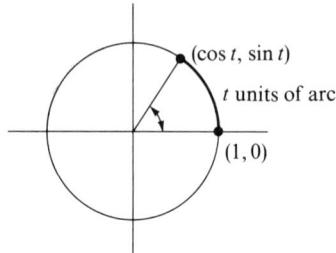

An angle of t radians

The description of the functions sin and cos in terms of a circle agrees with their description, given in Section 1, as ratios of certain sides of a right triangle. The great advantage of describing sin and cos as "circle functions" is that there is no longer any need to restrict the values of t to lie between 0 and $\pi/2$. For any value of t whatever, *positive* or *negative*, it makes sense to speak of $\cos t$ and of $\sin t$, defined as circle functions.* (If t is negative, then you go clockwise instead of counterclockwise in marking off the arc length.)

Let us give some examples of this. The reader will find it handy to be familiar with the sixteen points on the unit circle below. (Actually, if you only know the points P_1, P_2, and P_3, you can easily figure out the coordinates of the rest.)

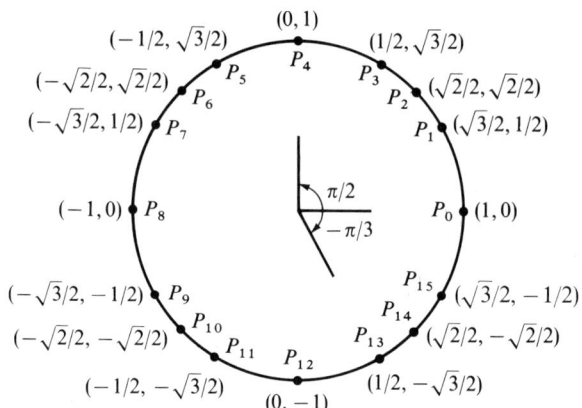

The unit circle, and sixteen points that everyone should know

---

* Some authors say "circular" functions. The phrase "circle function" is supposed to suggest that the function's domain is, in essence, the circle. A function living on a circle is a circle function, just as a man living on earth is an earth man (and not an "earthen" man).

2. Circle Functions

EXAMPLE 1. Find $\sin(\pi/2)$.

We follow the description of the sine as a circle function, and move counterclockwise through an arc length of $\pi/2$ units, beginning at $(1,0)$. We thus move a quarter of the way around the circumference, since $\pi/2$ is one-fourth of 2π, the length of the circumference, and land at P_4. The point P_4 then has coordinates given by $(\cos(\pi/2), \sin(\pi/2))$. Hence,

$$\left(\cos\frac{\pi}{2}, \sin\frac{\pi}{2}\right) = P_4 = (0, 1). \tag{7}$$

From (7) we see that $\sin(\pi/2) = 1$. □

EXAMPLE 2. Find $\cos(\pi/2)$.

It follows from (7) that $\cos(\pi/2) = 0$. □

EXAMPLE 3. Find $\cos(-\pi/3)$.

Beginning at $(1,0)$ on the unit circle and moving *clockwise* through an arc length of $\pi/3$ puts us at the point P_{13}. Therefore,

$$\left(\cos\left(-\frac{\pi}{3}\right), \sin\left(-\frac{\pi}{3}\right)\right) = P_{13} = \left(\frac{1}{2}, -\frac{\sqrt{3}}{2}\right), \tag{8}$$

and it follows that $\cos(-\pi/3) = \frac{1}{2}$. □

EXAMPLE 4. Find $\sin \pi$.

An arc length of π represents half the circumference. Beginning at $(1,0)$ and going halfway around the circle puts us at P_8, so

$$(\cos \pi, \sin \pi) = P_8 = (-1, 0). \tag{9}$$

Therefore, $\sin \pi = 0$. □

EXAMPLE 5. Find $\sin 3\pi$.

Moving from $(1,0)$ through an arc length of 3π units takes us through one and one-half turns of the circle, stopping at P_8. Therefore,

$$(\cos 3\pi, \sin 3\pi) = P_8 = (-1, 0), \tag{10}$$

and $\sin 3\pi = 0$. □

Having defined the sine and cosine functions, we define the tangent function to be their quotient.

Definition. The **tangent function** is denoted by tan, and defined as

$$\tan t = \frac{\sin t}{\cos t}, \quad \text{provided } \cos t \neq 0.$$

Whenever $\cos t = 0$, then $\tan t$ is undefined. Thus, from Example 2, $\tan(\pi/2)$ is undefined. It is obvious that the tangent function is undefined when (and only when) the terminal point of the arc length in question is either P_4 or P_{12}.

From Example 4 and Example 5 we see that

$$\tan \pi = 0 = \tan 3\pi.$$

It is obvious that the tangent function takes the value 0 when (and only when) the terminal point of the arc length in question is either P_0 or P_8. The tangent is positive in the first quadrant (between the points P_0 and P_4), negative in the second quadrant (between P_4 and P_8), positive in the third (between P_8 and P_{12}), and negative in the fourth (between P_{12} and P_0).

EXERCISES

2.1. Find each of the following, using the definition of sin and cos as circle functions.
   (a) $\sin(-\pi/3)$. *Hint.* Use equation (8).
   (b) $\cos \pi$. *Hint.* Use equation (9).
   (c) $\cos(5\pi/3)$. *Hint.* It is obvious (*why?*) that $\cos(5\pi/3) = \cos(-\pi/3)$. See Example 3.
   (d) $\sin(5\pi/3)$.
   (e) $\sin 0$.
   (f) $\cos 0$.

2.2. Use the definition $\tan = \sin/\cos$ to find each of the following.
   (a) $\tan(-\pi/3)$. *Answer:* $\tan(-\pi/3) = (\sin(-\pi/3))/(\cos(-\pi/3)) = (-\sqrt{3}/2)/\frac{1}{2} = -\sqrt{3}$, from equation (8).
   (b) $\tan(5\pi/3)$.
   (c) $\tan 0$.
   (d) $\tan 2\pi$.
   (e) $\tan(3\pi/4)$. *Answer:* -1.
   (f) $\tan(5\pi/4)$.
   (g) $\tan(-\pi/6)$.
   (h) $\tan 1984\pi$.

2.3. Find the tangent of the angle made by the positive x-axis and the line segment
   (a) joining $(0,0)$ and $(2,3)$. *Answer:* $\frac{3}{2}$.
   (b) joining $(0,0)$ and $(4,7)$.
   (c) joining $(0,0)$ and $(2,-1)$. *Answer:* $-\frac{1}{2}$.
   (d) whose equation is $y = bx$.
   (e) joining $(0,0)$ and $(0,3)$. *Answer:* Tangent is undefined.

2.4. Draw an angle whose tangent is
   (a) 3.
   (b) -2.
   (c) 1776.
   (d) -10.

   Partial answer: (b) The angle whose initial side is the positive x-axis and whose terminal side is the line segment joining $(0,0)$ and $(1,-2)$ will do.

3. Trigonometric Identities

2.5. What is the range of the tangent function? *Answer*: The range of tan is unrestricted if the domain is taken to be $-\pi/2 < t < \pi/2$.

2.6. Find the range of the cosine function if the domain is specified by
   (a) $0 \leq t \leq \pi/2$.
   (b) $-\pi/2 < t < \pi/2$.
   (c) $0 < t \leq \pi$.
   (d) $-\pi/6 \leq t < \pi/3$.
   (e) $-\pi \leq t < \pi$.
   (f) $0 < t \leq 2\pi$.
   (g) unrestricted.
   Answers: (d) If $y = \cos t$ and if $-\pi/6 \leq t < \pi/3$, then the range is $\frac{1}{2} < y \leq 1$. In (e), (f), and (g) the range is $-1 \leq y \leq 1$.

2.7. For each of the seven domains specified in exercise 2.6, find the range of the sine function. *Answers*: (d) If $y = \sin t$ and if $-\pi/6 \leq t < \pi/3$, then the range is $-\frac{1}{2} \leq y < \sqrt{3}/2$. In (e), (f), and (g) the range is $-1 \leq y \leq 1$.

2.8. In which of the four quadrants is the sine function positive? In which is the cosine function positive?

2.9. The functions sin and cos are defined for any number whatever. What is meant in the last footnote by the statement that the domain of sin and of cos is "in essence" a *circle*?

§3. Trigonometric Identities

An *identity* is an equation that holds for all values of the variables involved. Thus the equations

$$t + t = 2t, \qquad (t+s)(t-s) = t^2 - s^2, \qquad r(s+t) = rs + rt$$

are familiar *algebraic* identities.

Identities that involve the circle functions are called *trigonometric* identities. By virtue of the fact that $(\cos t, \sin t)$ lies on the unit circle, its coordinates $x = \cos t$ and $y = \sin t$ must satisfy the circle's equation $x^2 + y^2 = 1$. That is,

$$(\cos t)^2 + (\sin t)^2 = 1, \tag{11}$$

for all values of the variable t. This important trigonometric identity occurs frequently. So frequently, in fact, that one tires of forever writing the parentheses in such expressions as $(\cos t)^2$. Instead, by long-standing tradition one writes $\cos^2 t$, thereby freeing parentheses to enclose more exotic phrases elsewhere. Similarly,

$$\sin^2 t \text{ is an abbreviation for } (\sin t)^2. \tag{12}$$

Here is a list of a few trigonometric identities, beginning with (11) rewritten using the convention (12).

$$\cos^2 t + \sin^2 t = 1. \tag{13}$$

$$\cos(-t) = \cos t. \tag{14}$$

$$\sin(-t) = -\sin t. \tag{15}$$

$$\tan(-t) = -\tan t. \tag{16}$$

$$\cos(t + 2\pi) = \cos t. \tag{17}$$

$$\sin(t + 2\pi) = \sin t. \tag{18}$$

$$\tan(t + 2\pi) = \tan t. \tag{19}$$

$$\cos(s + t) = \cos s \cos t - \sin s \sin t. \tag{20}$$

$$\cos(s - t) = \cos s \cos t + \sin s \sin t. \tag{21}$$

$$\cos\left(\frac{\pi}{2} - t\right) = \sin t. \tag{22}$$

$$\sin\left(\frac{\pi}{2} - t\right) = \cos t. \tag{23}$$

$$\sin(s + t) = \sin s \cos t + \cos s \sin t. \tag{24}$$

$$\sin(s - t) = \sin s \cos t - \cos s \sin t. \tag{25}$$

The most important of these are the *cosine sum law* (20) and the *sine sum law* (24), along with their analogues, the *cosine difference law* (21) and the *sine difference law* (25). It takes a little work to prove these, while formulas (13) through (19) are easy to verify.

Formula (13) has already been proved. To see why (14) and (15) are true, all one has to do is study the picture below.

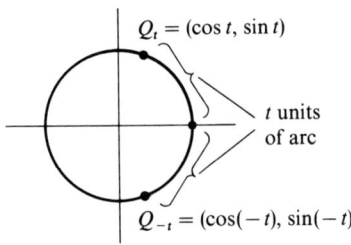

The unit circle. The points Q_t and Q_{-t} have the same first coordinate, hence, $\cos t = \cos(-t)$; and they have opposite second coordinates; showing that $\sin(-t) = -\sin t$

3. Trigonometric Identities

Now that (14) and (15) have been verified, it is easy to prove (16):

$$\tan(-t) = \frac{\sin(-t)}{\cos(-t)} \quad \text{(by definition of tan)}$$

$$= \frac{-\sin t}{\cos t} \quad [\text{by (14) and (15)}]$$

$$= -\tan t. \qquad \square$$

Formulas (17) through (19) can be proved at a single stroke, if one wishes, just by putting statements (3) and (4) together with the definition of the circle functions. But we shall verify statements (17) through (19) by a different means later.

To prove the cosine sum law it will be helpful to know a simple fact about the length of a chord in terms of the size of the central angle associated with it.

Formula for Chordal Lengths. *On the unit circle, the length of a chord subtended by a central angle of θ radians is given by the formula*

$$L = \sqrt{2 - 2\cos\theta}.$$

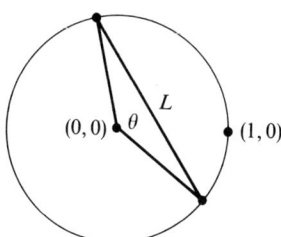

PROOF. It is easier to calculate the length L if first we rotate the whole figure until one side of the central angle lies on the positive x-axis:

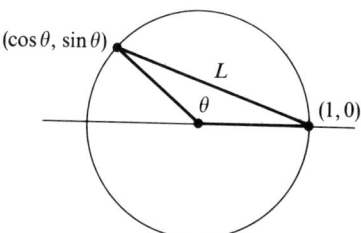

Let us calculate L^2, using the distance formula in the plane:

$$L^2 = (\cos\theta - 1)^2 + (\sin\theta - 0)^2$$
$$= \cos^2\theta - 2\cos\theta + 1 + \sin^2\theta$$
$$= 2 - 2\cos\theta \qquad (26)$$

since $\cos^2 + \sin^2 = 1$ by (13). Taking the positive square root of both sides of (26) yields the desired formula. □

Let us now prove the cosine sum law. For given s and t, let $\theta = s + t$, and consider a chord on the unit circle subtended by θ:

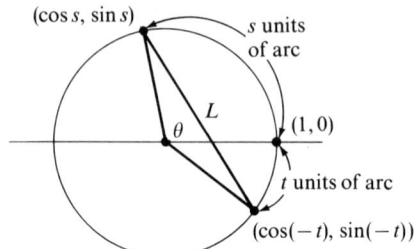

From the distance formula,

$$\begin{aligned}
L^2 &= (\cos s - \cos(-t))^2 + (\sin s - \sin(-t))^2 \\
&= (\cos s - \cos t)^2 + (\sin s + \sin t)^2 \quad \text{[by (14) and (15)]} \\
&= \cos^2 s - 2\cos s \cos t + \cos^2 t + \sin^2 s + 2\sin s \sin t + \sin^2 t \\
&= 2 - 2\cos s \cos t + 2\sin s \sin t,
\end{aligned} \qquad (27)$$

since $\sin^2 + \cos^2 = 1$.

On the other hand, from (26), with $\theta = s + t$,

$$L^2 = 2 - 2\cos(s + t). \qquad (28)$$

Equations (27) and (28) give two different expressions for the same quantity L^2. If we set these two expressions equal to each other the cosine sum law (20) results. □

The sine sum law is proved in the next section.

Exercises

3.1. Put equations (27) and (28) together and derive the cosine sum law.

3.2. In the cosine sum law, let $s = 2\pi$ and simplify the right-hand side of the resulting equation to derive formula (17).

3.3. In the cosine sum law, let $s = \theta$ and $t = \theta$ to prove the *double-angle formula*
$$\cos 2\theta = \cos^2 \theta - \sin^2 \theta.$$

3.4. The cosine sum law holds if we replace s and t by anything we choose. Replace t by $-t$ in (20), then use (14) and (15) to prove (21).

3.5. In (21), let $s = \pi/2$ and derive formula (22).

3.6. Formula (22) holds if we replace t by anything we choose. Replace t by $(\pi/2) - t$ in (22), and thus derive formula (23).

3.7. The formula for chordal length applies to a *unit* circle. Deduce from it the general formula for a circle of radius *r* units.

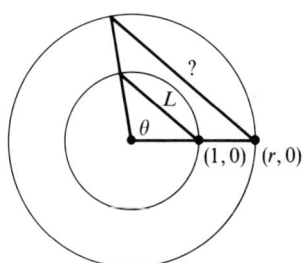

(This is easy. Just guess the indicated length, using the fact that the big circle is just like the small circle, but blown up by a factor of *r*.)

3.8. The formula for chordal lengths holds for any angle θ.

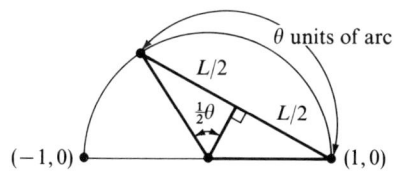

If $0 < \theta < \pi$, there is a simpler formula, given by

$$L = 2\sin\frac{\theta}{2}.$$

Prove this simpler formula from the figure here, and then deduce the *half-angle formula*

$$\sin\frac{\theta}{2} = \sqrt{\frac{1-\cos\theta}{2}}, \quad 0 < \theta < \pi.$$

3.9. The half-angle formula of exercise 3.8 holds if $0 < \theta < \pi$. Plug in each of the following numbers into the half-angle formula, and see whether it still holds. (Remember that $\sqrt{\ }$ denotes the *positive* root.)
 (a) $\theta = 0$.
 (b) $\theta = \pi$.
 (c) $\theta = 2\pi$.
 (d) $\theta = 3\pi$.
 (e) $\theta = -\pi$.
Answer: The formula fails in (d) and (e).

3.10. Pretend the figure of exercise 3.8 is a lake of radius 1 kilometer. A girl at the point $(1,0)$ wishes to get to the point $(-1,0)$ by swimming the length L and then running the rest of the way along the shoreline.
 (a) How far must she run? *Answer*: $\pi - \theta$ kilometers.
 (b) If she can swim 3 km/hr and run 5 km/hr, how long will it take her to reach her destination? *Answer*: $\frac{2}{3}\sin\frac{1}{2}\theta + \frac{1}{5}(\pi - \theta)$ hours.

3.11. Sketch the graphs of the sine and cosine functions. *Answer*: (See Section 5.)

§4. More Identities

Let us first prove the sine sum law (24). We begin with formula (22), replacing t by $s + t$:

$$\sin(s + t) = \cos\left(\frac{\pi}{2} - (s + t)\right)$$

$$= \cos\left(\frac{\pi}{2} - s - t\right)$$

$$= \cos\left(\left(\frac{\pi}{2} - s\right) - t\right)$$

$$= \cos\left(\frac{\pi}{2} - s\right)\cos t + \sin\left(\frac{\pi}{2} - s\right)\sin t \quad \text{[by (21)]}$$

$$= \sin s \cos t + \cos s \sin t \quad \text{[by (22) and (23)]}.$$

This proves (24), and (25) is easily derived from (24). □

From the identities already derived one could develop infinitely many new ones just by making appropriate substitutions. The reader with no previous acquaintance with trigonometry is apt to be bewildered by the multitude of identities. In practice, however, most of the useful identities can be seen as consequences of the sine sum law, the cosine sum law, or the law $\sin^2 + \cos^2 = 1$. Be sure you know these three identities, at least. The double-angle and half-angle formulas, for instance, follow easily from them.

Double- and Half-Angle Formulas. *The following are trigonometric identities.*

(a) $\cos 2\theta = \cos^2 \theta - \sin^2 \theta$.
(b) $\sin 2\theta = 2 \sin \theta \cos \theta$.
(c) $\sin^2 \frac{1}{2}\theta = (1 - \cos \theta)/2$.
(d) $\cos^2 \frac{1}{2}\theta = (1 + \cos \theta)/2$.

PROOF. The double-angle formulas (a) and (b) follow from the cosine and sine sum laws, respectively, by letting $s = t = \theta$. Both (c) and (d) follow from (a), together with the identity $\sin^2 + \cos^2 = 1$, which can be rewritten as

$$\cos^2 \theta = 1 - \sin^2 \theta, \tag{29}$$

or as

$$\sin^2 \theta = 1 - \cos^2 \theta. \tag{30}$$

When (29) is used to make a substitution in (a) we get

$$\cos 2\theta = 1 - 2\sin^2 \theta, \tag{31}$$

4. More Identities

which, when solved for the expression $\sin^2 \theta$, becomes

$$\sin^2 \theta = \frac{1 - \cos 2\theta}{2} \tag{32}$$

from which (c) follows upon replacing θ with $\frac{1}{2}\theta$.
When (30) is used to make a substitution in (a) we get

$$\cos 2\theta = 2\cos^2 \theta - 1. \tag{33}$$

Solving (33) for the expression $\cos^2 \theta$ yields

$$\cos^2 \theta = \frac{1 + \cos 2\theta}{2}, \tag{34}$$

from which (d) follows upon replacing θ with $\frac{1}{2}\theta$. □

The reader will have observed that in the course of proving the double- and half-angle formulas we have derived six additional trigonometric identities (29) through (34) as well. Occasionally these come in handy, particularly (32) and (34), but it is better to be able to derive them quickly than to trust them to memory. They can be deduced in a couple of lines by anyone who knows the sine and cosine sum laws. The product formulas can also be deduced quickly.

Product Formulas. *The following are identities.*

(a) $\cos s \cos t = \frac{1}{2}(\cos(s - t) + \cos(s + t))$.
(b) $\sin s \sin t = \frac{1}{2}(\cos(s - t) - \cos(s + t))$.
(c) $\sin s \cos t = \frac{1}{2}(\sin(s - t) + \sin(s + t))$.

PROOF. The product formulas (a) and (b) each come from putting together the cosine difference law and the cosine sum law:

$$\cos(s - t) = \cos s \cos t + \sin s \sin t.$$
$$\cos(s + t) = \cos s \cos t - \sin s \sin t.$$

Adding together the respective sides of these identities yields

$$\cos(s - t) + \cos(s + t) = 2 \cos s \cos t,$$

from which (a) follows upon dividing by 2.
If we subtract instead of add, we get

$$\cos(s - t) - \cos(s + t) = 2 \sin s \sin t,$$

from which (b) follows upon dividing by 2.

The product formula (c) results from playing off the sine sum law against the sine difference law in the same spirit. The verification of (c) is left to the reader. □

With the aid of the identities at our disposal we can extend our knowledge of the circle far beyond the sixteen points that everybody knows. These sixteen points come, basically, from knowing about right triangles containing angles of π/6, π/4, and π/3 radians (30, 45, and 60 degrees). We can use that knowledge to learn about some other angles.

EXAMPLE 6. Find the cosine of 15°, or π/12 radians, by noting that this angle is the difference of two angles we already know about.

The point is that $15 = 60 - 45$, or $\pi/12 = \pi/3 - \pi/4$. We can thus employ the cosine difference law:

$$\cos\frac{\pi}{12} = \cos\left(\frac{\pi}{3} - \frac{\pi}{4}\right)$$

$$= \cos\frac{\pi}{3}\cos\frac{\pi}{4} + \sin\frac{\pi}{3}\sin\frac{\pi}{4}$$

$$= \frac{1}{2}\left(\frac{\sqrt{2}}{2}\right) + \frac{\sqrt{3}}{2}\left(\frac{\sqrt{2}}{2}\right)$$

$$= \frac{\sqrt{2}}{4}(1 + \sqrt{3})$$

$$= 0.96593\ldots,$$

from a table of square roots. □

EXAMPLE 7. Find the cosine of 15°, or π/12 radians, by noting that the angle is half of an angle we already know about.

From the half-angle formula (d) given above,

$$\cos^2\frac{\pi}{12} = \frac{1 + \cos(\pi/6)}{2}$$

$$= \frac{1 + (\sqrt{3}/2)}{2} = \frac{2 + \sqrt{3}}{4}.$$

Since we know that $\cos \pi/12$ is positive (the cosine being positive in the first quadrant), we take the positive square root of both sides to get

$$\cos\frac{\pi}{12} = \frac{1}{2}\sqrt{2 + \sqrt{3}} = 0.96593\ldots. \quad \square$$

EXAMPLE 8. Find the sine of 75°, or 5π/12 radians.

4. More Identities

Since $75 = 45 + 30$, or $5\pi/12 = (\pi/4) + (\pi/6)$, this is a job for the sine sum law.

$$\sin\frac{5\pi}{12} = \sin\left(\frac{\pi}{4} + \frac{\pi}{6}\right)$$

$$= \sin\frac{\pi}{4}\cos\frac{\pi}{6} + \cos\frac{\pi}{4}\sin\frac{\pi}{6}$$

$$= \frac{\sqrt{2}}{2}\left(\frac{\sqrt{3}}{2}\right) + \frac{\sqrt{2}}{2}\left(\frac{1}{2}\right)$$

$$= \frac{\sqrt{2}}{4}(\sqrt{3} + 1) = 0.96593 \ldots . \qquad \square$$

How could this example have been done more quickly? Hint. Put formula (23) together with Example 6.

Exercises

4.1. Using the sine sum law (24), deduce the sine difference law (25).

4.2. The proof of the product formula (c) was left to the reader, with a hint as to how it might be done. Do it.

4.3. Find the sine of $\pi/12$ radians by using the method of Example 6.

4.4. Find the sine of $\pi/12$ radians by using the method of Example 7.

4.5. Find $\sin(7\pi/12)$ by any method.

4.6. The identity (18) has already been proved. Give a different proof by using the sine sum law. Then use (17) and (18) to deduce identity (19).

4.7. Derive the *tangent sum law*:

$$\tan(s + t) = \frac{\tan s + \tan t}{1 - \tan s \tan t}.$$

Hint. Since $\tan = \sin/\cos$, we know that

$$\tan(s + t) = \frac{\sin(s + t)}{\cos(s + t)}.$$

Use the sine sum law in the numerator and use the cosine sum law in the denominator. Then divide everything in sight by the product $\cos s \cos t$.

4.8. Using the tangent sum law of exercise 4.8, together with the identity (16), derive a *tangent difference law. Answer*: $\tan(s - t) = (\tan s - \tan t)/(1 + \tan s \tan t)$.

4.9. (a) Prove a double-angle formula for the tangent.
(b) Prove a half-angle formula for the tangent.
Answer: (b) From the half-angle formulas (c) and (d) we get, by dividing, the tangent half-angle formula $\tan^2 \frac{1}{2}\theta = (1 - \cos\theta)/(1 + \cos\theta)$.

4.10. In the next section we shall need to use the fact that, in the diagram below, *the arc length P'Q' is less than the length of the line segment P'Q*. Prove this.

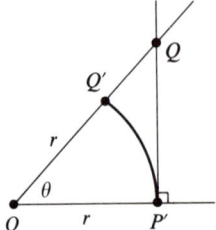

Here, θ is an angle between 0 and $\pi/2$ radians

Suggestion. Let θ be the measure, *in radians*, of the central angle $P'OQ'$, and justify each of the following steps.
(1) Area of sector $P'OQ' <$ area of triangle $P'OQ$.
(2) $(\theta/2\pi)(\pi r^2) < \tfrac{1}{2} r$ (length $P'Q$).
(3) $\theta r <$ length $P'Q$.
(4) Arc length $P'Q' <$ length $P'Q$.

§5. The Derivative of the Sine Function

The sine and cosine functions are pictured below. By (17) and (18), the wavy pattern on an interval of length 2π repeats itself again and again.

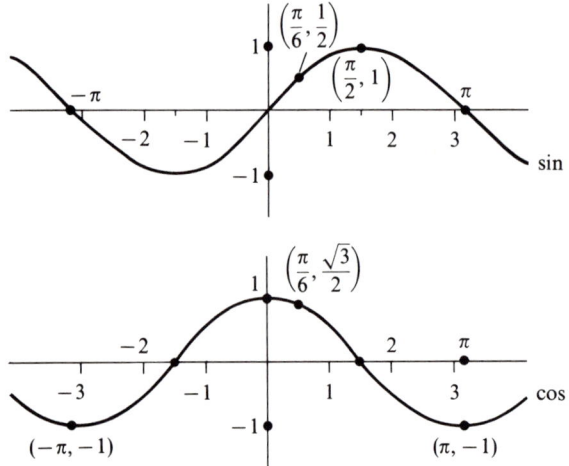

This picture makes it plausible that the derivative of the sine function is the cosine function. Let us make this *guess*:

$$(\sin)' = \cos. \qquad (35)$$

The rest of this section is devoted to the proof of (35). This takes a little more work than might be expected, because the sine function, not being

5. The Derivative of the Sine Function

algebraic, demands careful attention. We must go back to basics whenever we are alone in strange seas. We go back to Fermat's method here.

If $f(x) = \sin x$, then (using the sine sum law) we have

$$f(x + h) - f(x) = \sin(x + h) - \sin x$$
$$= \sin x \cos h + \cos x \sin h - \sin x$$
$$= \sin x (\cos h - 1) + \cos x \sin h.$$

Applying Fermat's method, we divide by nonzero h and take the limit to get

$$f'(x) = \underset{h \to 0}{\text{Limit}} \sin x \left(\frac{\cos h - 1}{h} \right) + \cos x \left(\frac{\sin h}{h} \right). \tag{36}$$

What is this limit? Will it be equal to $\cos x$, in accordance with our guess (35) above? We cannot answer until first we investigate the behavior of the two expressions in parentheses,

$$\frac{\cos h - 1}{h} \quad \text{and} \quad \frac{\sin h}{h},$$

to determine their limits as h tends to zero. In order for our guess to be correct, the limit of the first should be 0 and the limit of the second should be 1.

Let us first investigate the expression $(\sin h)/h$ as $h \to 0$. In the figure below, think of h as being very close to zero.

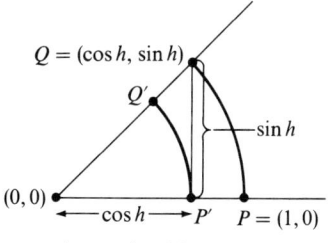

An angle of h radians

It is clear (see exercise 4.10) that the line segment $P'Q$ has a length smaller than that of the arc PQ, but greater than the length of the arc $P'Q'$. Since the coordinates of Q are $(\cos h, \sin h)$, this means that the quantity

$$\sin h \text{ lies between } h \cos h \text{ and } h. \tag{37}$$

Reason: The arc $P'Q'$, being subtended by an angle of h radians on a circle of radius $\cos h$, has length $h \cos h$; while the line segment $P'Q$ has length $\sin h$, and the arc PQ has length h. Using (37) and dividing by nonzero h, we see that when $h \approx 0$,

$$\frac{\sin h}{h} \text{ lies between } \cos h \text{ and } 1. \tag{38}$$

The information in (38) is all we need to find the limit of $(\sin h)/h$ as $h \to 0$, for it is clear that $\cos h \to \cos 0 = 1$ (the cosine function being continuous at the point 0). Thus $(\sin h)/h$, being squeezed between $\cos h$ and 1, must also tend to 1:

$$\underset{h \to 0}{\text{Limit}} \frac{\sin h}{h} = 1. \tag{39}$$

We have just finished the hard part, having found the limit of the second term in parentheses in (36). The limit of the first term there succumbs to the following trick.

$$\frac{\cos h - 1}{h} = \frac{\cos h - 1}{h} \left(\frac{\cos h + 1}{\cos h + 1} \right)$$

$$= \frac{\cos^2 h - 1}{h(\cos h + 1)}$$

$$= \frac{-\sin^2 h}{h(\cos h + 1)}$$

$$= \frac{\sin h}{h} \left(\frac{-\sin h}{\cos h + 1} \right),$$

which, as h approaches 0, tends to

$$1 \left(\frac{-0}{1+1} \right) = 0,$$

by (39) and by the continuity at 0 of the sine and cosine functions. Therefore,

$$\underset{h \to 0}{\text{Limit}} \frac{\cos h - 1}{h} = 0. \tag{40}$$

We can now complete the calculation left hanging in equation (36). From (36), if $f(x) = \sin x$, then

$$f'(x) = (\sin x)(0) + (\cos x)(1) = \cos x,$$

by (39) and (40). Our guess (35) is confirmed. □

The derivation just given may seem at first to rely too much upon the picture that shows h as a positive quantity. However, the statement (37) holds if h is negative as well, as the reader is asked to demonstrate. Thus (38) holds for any small nonzero h.

EXERCISES

5.1. Show that (37) holds if h is a small negative number. [In this case all three quantities mentioned in statement (37) are negative.]

5.2. Find the limits of each of the following quantities, as h tends to zero. *Hint.* (b) Go through the derivation preceding equation (40), but with h^2 in the denominator instead of h.
   (a) $(\tan h)/h$.
   (b) $(\cos h - 1)/h^2$.
   (c) $((\cos h - 1)/h^2)^2$.
   Answers: (a) 1. (b) $-\frac{1}{2}$. (c) $\frac{1}{4}$.

5.3. Find the limits of each of the following, as x tends to zero.
   (a) $(\sin 2x)/x$.
   (b) $(\cos 2x - 1)/x$.
   (c) $(\cos 2x - 1)/x^2$.
   Answer: (a) $(\sin 2x)/x = 2(\sin 2x)/2x = 2(\sin h)/h$ if we let $h = 2x$. Since $h \to 0$ as $x \to 0$, we have
   $$\underset{x \to 0}{\text{Limit}} \frac{\sin 2x}{x} = \underset{h \to 0}{\text{Limit}} 2\frac{\sin h}{h} = 2(1) = 2.$$

5.4. Use the chain rule to find the derivatives of the functions given by each of the following rules.
   (a) $\sin 2x$.
   (b) $\sin \pi x$.
   (c) $\sin((\pi/2) - t)$.
   (d) $\sin(1 + t^2)$.
   (e) $\sin(\sqrt{x})$.
   (f) $\sin 3\theta$.
   (g) $\sin^2 \theta$.
   (h) $\sin(\theta^2)$.
   Hint. (c). See Example 10 in the next section. Answers: (a) $2 \cos 2x$. (d) $2t \cos(1 + t^2)$. (g) $2 \sin \theta \cos \theta$. (h) $2\theta \cos(\theta^2)$.

5.5. Find the derivative of the cosine function in two ways:
   (a) using the identity $\cos t = \sin((\pi/2) - t)$, together with your answer to exercise 5.4(c).
   (b) beginning with the identity $\sin^2 + \cos^2 = 1$, using the rule for squares to write
   $$2(\sin)(\sin)' + 2(\cos)(\cos)' = 0,$$
   and continuing from here.

5.6. The *secant* is defined as the reciprocal of the cosine, i.e., $\sec t = 1/\cos t$, whenever $\cos t \neq 0$. Prove the important identity
   $$\sec^2 t = 1 + \tan^2 t.$$
   Hint. In identity (13), divide everything by $\cos^2 t$.

§6. More Derivatives of Trigonometric Functions

To find the derivative of the cosine function, let $f(x) = \cos x$ and use the cosine sum law to simplify the difference quotient:

$$\frac{f(x+h) - f(x)}{h} = \frac{\cos(x+h) - \cos x}{h}$$

$$= \frac{\cos x \cos h - \sin x \sin h - \cos x}{h}$$

$$= \cos x \left(\frac{\cos h - 1}{h}\right) - \sin x \left(\frac{\sin h}{h}\right).$$

Taking the limit of the difference quotient as h tends to 0, we get
$$f'(x) = (\cos x)(0) - (\sin x)(1) \quad [\text{by (39) and (40)}]$$
$$= -\sin x.$$
Thus, *the derivative of the cosine function is the negative of the sine function.* □

The derivative of the tangent function can be found by using the quotient rule. Since $\tan = \sin/\cos$,
$$(\tan)' = \left(\frac{\sin}{\cos}\right)'$$
$$= \frac{(\cos)(\sin)' - (\sin)(\cos)'}{(\cos)^2}$$
$$= \frac{(\cos)(\cos) - (\sin)(-\sin)}{(\cos)^2}$$
$$= \frac{1}{(\cos)^2} \quad [\text{by (13)}].$$

The derivative of the tangent function is thus the square of the reciprocal of the cosine function. In trigonometry the reciprocal of the cosine is called the **secant**, i.e.,
$$\sec x = \frac{1}{\cos x}.$$
Thus the derivative of the tangent can be expressed more briefly as
$$(\tan)' = \left(\frac{1}{\cos}\right)^2 = \sec^2.$$
The derivative of the tangent is the square of the secant.

There are two more trigonometric functions which come into play less frequently, the **cosecant** and the **cotangent**, defined as the reciprocals of the sine and the tangent functions, respectively:
$$\csc x = \frac{1}{\sin x}, \quad \cot x = \frac{1}{\tan x}.$$
The derivatives of the six trigonometric functions are given below.

$$(\sin)' = \cos \qquad (\csc)' = -\csc \cot$$
$$(\cos)' = -\sin \qquad (\sec)' = \sec \tan$$
$$(\tan)' = \sec^2 \qquad (\cot)' = -\csc^2$$

The formulas in the first column have been proved above. The reader is asked to verify the formulas in the second column.

Now that we know the derivatives of the trigonometric functions we can apply the techniques of the calculus to trigonometry.

6. More Derivatives of Trigonometric Functions

EXAMPLE 9. The point $(\pi/6, 1/2)$ is on the curve $f(x) = \sin x$. Write an equation of the tangent line at this point.

The slope of the tangent line is

$$(\sin)'\frac{\pi}{6} = \cos\frac{\pi}{6} = \frac{\sqrt{3}}{2}.$$

x	$f(x)$	$f'(x)$
$\pi/6$	$\frac{1}{2}$	$\sqrt{3}/2$
x	$\sin x$	$\cos x$

An equation of the line through $(\pi/6, 1/2)$ with slope $\sqrt{3}/2$ is given by

$$y - \frac{1}{2} = \frac{\sqrt{3}}{2}\left(x - \frac{\pi}{6}\right). \qquad \square$$

EXAMPLE 10. Apply the chain rule to find dy/dt, where $y = \sin((\pi/2) - t)$.

Here we have the chain $y = \sin x$ and $x = (\pi/2) - t$. By the chain rule,

$$\frac{dy}{dt} = \frac{dy}{dx}\frac{dx}{dt} = (\cos x)(-1).$$

Hence,

$$\frac{dy}{dt} = -\cos x$$

$$= -\cos\left(\frac{\pi}{2} - t\right)$$

$$= -\sin t, \quad \text{by (22)}. \qquad \square$$

EXAMPLE 11. A particle moves along the y-axis in such a way that its position at time t seconds is given by

$$y = 5\cos t.$$

Find its instantaneous velocity and acceleration when $t = \pi/3$.

The instantaneous velocity is given by

$$v = \frac{dy}{dt} = -5\sin t,$$

and the acceleration a is given by

$$a = \frac{dv}{dt} = -5\cos t.$$

Thus, when $t = \pi/3$, the velocity is $-5\sqrt{3}/2$ units per second (the particle is thus moving downward), and the acceleration is $-5(\frac{1}{2})$ units/sec per second (showing that the particle is accelerating downward). $\qquad \square$

The motion described is the projection upon the y-axis of the curve $y = 5\cos t$.

 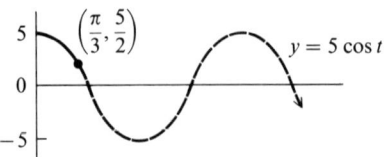

EXAMPLE 12. Find the indicated area.

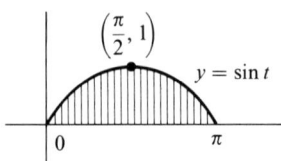

The area is given by the integral

$$\int_0^\pi \sin t \, dt.$$

By the fundamental theorem of calculus, this is equal to

$$[-\cos t]_0^\pi,$$

since $(-\cos)' = \sin$. The area is then

$$-\cos \pi - (-\cos 0) = -(-1) - (-1) = 2$$

square units. □

EXAMPLE 13. Find the value of θ at which the function given by

$$f(\theta) = 3600 \cos \theta (1 + \sin \theta), \qquad 0 \le \theta \le \frac{\pi}{2},$$

assumes its maximum value.

The maximum value of f (see the end of Chapter 4) must occur either at an endpoint of the domain or at a critical point in the interior of the domain. Let us check the values of f at the endpoints first:

$$f(0) = 3600(1)(1) = 3600. \tag{41}$$

$$f\left(\frac{\pi}{2}\right) = 3600(0)(2) = 0. \tag{42}$$

To find a critical point we take the derivative, using the product rule, and simplify using trigonometric identities:

$$\begin{aligned} f'(\theta) &= 3600 \cos \theta (\cos \theta) - 3600 \sin \theta (1 + \sin \theta) \\ &= 3600(\cos^2 \theta - \sin \theta - \sin^2 \theta) \\ &= 3600(1 - \sin^2 \theta - \sin \theta - \sin^2 \theta) \\ &= 3600(1 - \sin \theta - 2\sin^2 \theta). \end{aligned}$$

6. More Derivatives of Trigonometric Functions

When is the derivative f' equal to zero? Only when
$$2\sin^2\theta + \sin\theta - 1 = 0.$$
This is a quadratic equation, *not in θ, but in the expression* $\sin\theta$. The quadratic formula says this expression is then given by
$$\sin\theta = \frac{-1 \pm \sqrt{9}}{4} = \begin{cases} -1 \\ \frac{1}{2} \end{cases}.$$
In the domain $0 \leq \theta \leq \pi/2$, it is never true that $\sin\theta = -1$. Hence the only critical point in the domain occurs when
$$\sin\theta = \frac{1}{2},$$
$$\theta = \frac{\pi}{6}.$$
The value of f at the critical point $\pi/6$ is
$$f\left(\frac{\pi}{6}\right) = 3600\left(\frac{\sqrt{3}}{2}\right)\left(\frac{3}{2}\right) = 4676.5 \ldots . \qquad (43)$$
The maximum of f must occur at an endpoint or at a critical point. Inspection of (41), (42), and (43) shows that the maximum occurs when $\theta = \pi/6$. □

Exercises

6.1. Verify the formulas given in this section for the derivatives of the secant, cosecant, and cotangent functions.

6.2. The point $(\pi/6, \sqrt{3}/2)$ lies on the curve $y = \cos x$. Write an equation of the tangent line at that point. *Answer*: $y - (\sqrt{3}/2) = -\frac{1}{2}(x - (\pi/6))$.

6.3. In the situation described in Example 11, find the velocity and acceleration when
 (a) $t = \pi/2$.
 (b) $t = \pi$.
 (c) $t = 7\pi/4$.

6.4. Find the area beneath the curve $y = \cos t$, between
 (a) $t = 0$ and $t = \pi/3$.
 (b) $t = -\pi/4$ and $t = \pi/6$.

6.5. Find the derivatives of the following.
 (a) $\cos(1 + t^3)$.   (b) $\cos^2(1 + t^3)$.
 (c) $t^2 \sin t$.   (d) $\sqrt{2 + \cos(t^2)}$.
 (e) $(\sin x)^3/3$.   (f) $(\cos x)/x$.
 (g) $\cos\sqrt{2 + x}$.   (h) $\sin(\cos x)$.
 Answers: (b) $-6t^2 \cos(1 + t^3)\sin(1 + t^3)$.
 (e) $\sin^2 x \cos x$. (g) $(-\sin\sqrt{2 + x})/2\sqrt{2 + x}$.
 (h) $-\sin x \cos(\cos x)$.

6.6. The equation $2\sin^2\theta + \sin\theta - 1 = 0$ was solved in Example 13 with the help of the quadratic formula. Solve this equation by factoring the left-hand side instead.

6.7. Find an *antiderivative* of each of the following expressions. Just use the trial-and-error method until you hit upon what you are after.
(a) $\sin 2x$.
(b) $\sin \pi x$.
(c) $\cos 3x$.
(d) $\cos \pi x$.

6.8. Guess a general formula for an antiderivative of (a) $\sin kx$ and (b) $\cos kx$, where k is a constant. *Answers*: (a) $(-\cos kx)/k$. (b) $(\sin kx)/k$.

6.9. Use the result of exercise 6.8 together with the product formulas obtained in Section 4 to find antiderivatives of each of the following products.
(a) $\cos 2x \cos 5x$.
(b) $\cos^2 x$.
(c) $\sin 4x \sin x$.
(d) $\sin^2 x$.
(e) $\sin 3x \cos x$.
(f) $\sin(-2x)\cos(-4x)$.
Answers: (d) $(2x - \sin 2x)/4$. (e) $((-\cos 2x)/4) - ((\cos 4x)/8)$.

6.10. If the area pictured in Example 12 is revolved about the horizontal axis, a football-shaped solid is generated. Find the volume of this solid of revolution. *Hint*. You will need to use your answer to exercise 6.9(d) to evaluate the integral in the formula for the volume. *Answer*: $\pi^2/2$ cubic units.

§7. Optimization and Inverse Trigonometric Functions

One cannot study the calculus of trigonometric functions very long without beginning to feel the need for the *inverses* of these functions. (If a function is regarded as a pair of columns of numbers, then its *inverse* is obtained by interchanging the columns.) Let us look at a couple of optimization problems here and see if we don't feel that need.

EXAMPLE 14. A farmer has three identical pieces of rigid wooden fencing, each 60 feet in length. These, together with a long straight fence already standing, will enclose a garden plot.

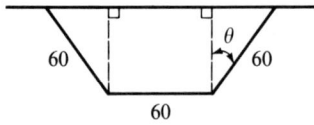

What should the indicated angle θ be in order that the area enclosed be as large as possible?

7. Optimization and Inverse Trigonometric Functions

Let A be the enclosed area, in square feet, so that A is a function of θ. What is the domain of this function? It is reasonable to consider values of θ that lie in the interval from 0 to a right angle:

$$0 \leq \theta < \frac{\pi}{2}.$$

The area A is the sum of the areas of a rectangle and two identical triangles.

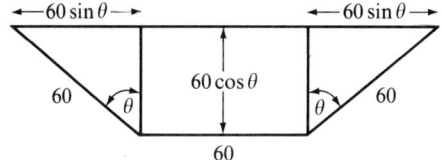

$$A = 60(60 \cos \theta) + 2\left(\frac{1}{2}\right)(60 \cos \theta)(60 \sin \theta)$$

$$= 3600(\cos \theta)(1 + \sin \theta).$$

This gives a trigonometric rule expressing A in terms of θ. From our work in Example 13, we see that when θ is $\pi/6$ radians (30 degrees), the area is 4676.5 ... square feet, which is the maximum possible area. □

There are a couple of remarks worth making about Example 14. First, the endpoint $\pi/2$ is missing from the domain of the function under consideration, yet is included in the domain of the function of Example 13, whose maximum occurs when

$$\sin \theta = \frac{1}{2}. \tag{44}$$

Obviously no harm is done by including or excluding an endpoint when the maximum turns out to occur at an interior point of the domain.

In Example 14, if the question had been to find the value of θ *minimizing* the area, then the answer should be that *there is no such θ* (unless one is willing to allow the three pieces of fencing to be nailed on top of the fence already standing). That is, the function of Example 14 does not attain a minimum value, while the function of Example 13 attains its minimum when $\theta = \pi/2$.

A more important consideration to be dealt with, however, is that frequently one must solve such equations as (44) in optimization problems involving trigonometric functions. Look again at equation (44). To find θ, one must find the arc length whose sine is $\frac{1}{2}$, since measurement in radians is essentially measurement by length of subtended arc. That is, from (44) we get

$$\theta = \text{the arc length whose sine is } \frac{1}{2}, \tag{45}$$

which we can express more briefly by writing

$$\theta = \arcsin\frac{1}{2}. \tag{46}$$

Equation (46), where *arcsin* is read "arc sine", is simply an abbreviation for the longer statement (45). It turns out that minor chaos can ensue if we are not absolutely clear about the meaning of *arcsin*.

Convention. *The expression arcsin shall be understood to designate an arc length lying between* $-\pi/2$ *and* $\pi/2$ *(inclusive).*

Thus $\arcsin\frac{1}{2}$ is NOT equal to $5\pi/6$, to $-11\pi/6$, or to $13\pi/6$, even though these are all arcs with sine of $\frac{1}{2}$. There is only one number lying between $-\pi/2$ and $\pi/2$ whose sine is $\frac{1}{2}$, and that is $\pi/6$. Therefore,

$$\arcsin\frac{1}{2} = \frac{\pi}{6},$$

and from equation (46) we get

$$\theta = \frac{\pi}{6}.$$

The point just made is easy to forget. Let us illustrate the point with a picture.

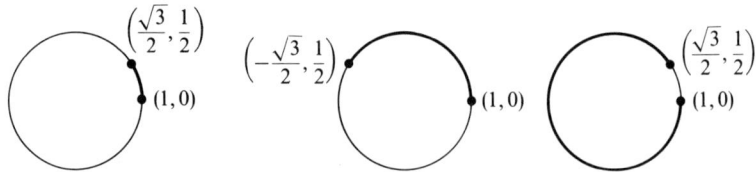

There are (infinitely) many arc lengths having a sine of $\frac{1}{2}$: $\pi/6$, $5\pi/6$, $-11\pi/6$, $13\pi/6$, etc. Only one of these, however, lies between $-\pi/2$ and $\pi/2$, and that is the one designated as $\arcsin\frac{1}{2}$.

The reader may surely wonder why we dwell so long on such a simple point about the meaning of *arcsin*. The reason is that if this point is understood, then

$$\arcsin$$

is a *function*: if $-1 \leq x \leq 1$, then arcsin defines a rule that produces *exactly one* number in $-\pi/2 \leq y \leq \pi/2$ that corresponds to x. It is an important, useful function that enables us to write down a convenient expression for the solution of such an equation as (44). It behooves us to study this function and to study the analogous functions *arccos* and *arctan* (see the exercises

7. Optimization and Inverse Trigonometric Functions

below) from the point of view of calculus. After we become more familiar with these inverse functions, we must inquire about their derivatives.

Let us look at another optimization problem first.

EXAMPLE 15. Suppose, in Example 14, one of the pieces of fencing is 100 feet instead of 60 feet. What angle θ will maximize the area of the trapezoid? It is intended that we should maximize the area A of

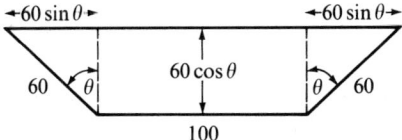

Here we have

$$A = 6000 \cos \theta + 3600 \cos \theta \sin \theta,$$

where the domain is $0 \leq \theta < \pi/2$. We could proceed just as in Example 13, but let us take a different approach that is more flamboyant, making use of the second derivative and of trigonometric identities.

By a double-angle formula, $2 \cos \theta \sin \theta = \sin 2\theta$. The expression for A can be rewritten as

$$A = 6000 \cos \theta + 1800 \sin 2\theta,$$

from which we can take two derivatives quickly:

$$\frac{dA}{d\theta} = -6000 \sin \theta + 3600 \cos 2\theta,$$

$$\frac{d^2 A}{d\theta^2} = -6000 \cos \theta - 7200 \sin 2\theta.$$

The second derivative $d^2A/d\theta^2$ is obviously negative throughout the domain $0 \leq \theta < \pi/2$. Therefore (why?), we shall find the maximum area A if we find a critical point. A critical point satisfies the equation

$$0 = \frac{dA}{d\theta} = -6000 \sin \theta + 3600(1 - 2\sin^2 \theta)$$

by the identity (31). This equation simplifies to

$$6 \sin^2 \theta + 5 \sin \theta - 3 = 0.$$

By the quadratic formula,

$$\sin \theta = \frac{-5 \pm \sqrt{97}}{12} = \begin{cases} -1.237\ldots \\ 0.404\ldots \end{cases}$$

Since the sine function never takes a value less than -1, the critical point θ must satisfy
$$\sin\theta = 0.404\ldots, \qquad 0 \le \theta < \frac{\pi}{2}.$$
Therefore,
$$\theta = \arcsin 0.404\ldots$$
yields the maximum area A. □

EXERCISES

7.1. In Example 15 we found that the optimal value of θ occurred when $\sin\theta = 0.404\ldots$.
  (a) Find $\cos\theta$. *Hint.* $\sin^2 + \cos^2 = 1$.
  (b) Draw a picture of the trapezoid that maximizes the area in Example 15, giving all dimensions to several decimal places.
  (c) What is the maximal area in Example 15?

7.2. A farmer has three pieces of rigid wooden fencing, of lengths 30 meters, 30 meters, and 45 meters, respectively. These, together with a long straight fence already standing, will enclose a trapezoidal area. Draw a picture of the trapezoid, giving its dimensions, that maximizes the enclosed area.

7.3. Find the arcsin of each of the following.
  (a) $\sqrt{3}/2$.
  (b) $-\sqrt{3}/2$.
  (c) -1.
  (d) 1.
  (e) $-\frac{1}{2}$.
  (f) 0.
  (g) $\sin(5\pi/6)$.
  (h) $\sin(\pi/3)$.
  (i) $\sin(3\pi/4)$.
  Answers: (b) $-\pi/3$. (g) $\arcsin(\sin(5\pi/6)) = \arcsin\frac{1}{2} = \pi/6$.

7.4. Find the range of each of the following functions, with indicated domains.
  (a) $f(\theta) = \cos\theta + 2\sin\theta\cos\theta$, $0 \le \theta \le \pi/2$.
  (b) $f(\theta) = 4\cos\theta + 3\sin\theta\cos\theta$, $0 \le \theta \le \pi/2$.
  (c) $f(\theta) = 4\cos\theta + 3\sin\theta$, $0 \le \theta \le \pi/2$.
  Answer: (c) $3 \le f(\theta) \le 5$.

7.5. The **arctangent function** is defined as follows: If x is any number whatever, $\arctan x$ is the arc length y in the range $-\pi/2 < y < \pi/2$ whose tangent is x. Find the arctangent of each of the following.
  (a) 0.
  (b) 1.
  (c) $\sqrt{3}$.
  (d) $-\sqrt{3}$.
  (e) -1.
  (f) $\tan 2\pi$.
  (g) $\tan(3\pi/4)$.
  (h) $\tan(-\pi/4)$.
  Answers: (b) $\pi/4$. (f) 0. (g) $-\pi/4$.

7.6. Evaluate each of the following.
  (a) $\arctan t\,|^{1}_{-1}$.
  (b) $\arcsin x\,|^{1/2}_{0}$.
  (c) $\arcsin\theta\,|^{1}_{-1}$.
  Answers: (a) $\pi/2$. (b) $\pi/6$. (c) π.

7.7. Consider the function given by

$$f(\theta) = \frac{2}{3}\sin\frac{\theta}{2} - \frac{1}{5}\theta + \frac{\pi}{5}, \qquad 0 \le \theta \le \pi.$$

(*This function arose in exercise 3.10.*)
(a) Find $f'(\theta)$ and find $f''(\theta)$.
(b) Show that $f''(\theta) < 0$ if $0 < \theta < \pi$.
(c) Explain why it follows from part (b) that the function f cannot attain its minimum value at any point θ satisfying $0 < \theta < \pi$.
(d) What should the girl in exercise 3.10 do, in order to reach her destination as quickly as possible?
Partial answer: (a) $f'(\theta) = \frac{1}{3}\cos\frac{1}{2}\theta - \frac{1}{5}$.

7.8. The **arccosine function** is defined as follows: If x is a number in the domain $-1 \le x \le 1$, then arccos x is the number y in the range $0 \le y \le \pi$ whose cosine is x.
(a) Why do we not specify $-\pi/2 \le y \le \pi/2$, as we did in the case of the arctangent and the arcsine?
(b) Find the arccosine of each of the following.
   (i) 0.   (ii) 1.
   (iii) $\sqrt{2}/2$.   (iv) $-\sqrt{2}/2$.
   (v) $\cos 2\pi$.   (vi) $\sin(\pi/6)$.
Answers: (iv) $3\pi/4$. (vi) $\pi/3$.

7.9. Consider again the function f of exercise 7.7.
(a) Show that $f'(\theta) = 0$ when $\theta = 2\arccos\frac{3}{5}$.
(b) Show that if $\cos\frac{1}{2}\theta = \frac{3}{5}$, and if $0 \le \theta \le \pi$, then $\sin\frac{1}{2}\theta = \frac{4}{5}$. *Hint.* $\sin^2 + \cos^2 = 1$.
(c) Show that the largest number in the range of the function f is given by

$$\frac{8}{15} - \frac{2}{5}\arccos\frac{3}{5} + \frac{\pi}{5} = 0.79 \ldots.$$

(d) Sketch the graph of the curve f.

§8. Inverse Functions and Their Derivatives

One way to look at a function is to regard it as a pair of columns of numbers, the first column having no number repeated. Suppose the columns are interchanged, so that the second column becomes the first. Then we have the **inverse relationship**. *If* the inverse relationship is a function, then it is called the **inverse function**. The inverse function is denoted f^{-1} if the original function is called f.

The notation f^{-1}, read "f inverse", has become traditional despite the fact that it is not wholly satisfactory. It might be better to call the inverse function $f^{\leftarrow}$, for f^{-1} is easily confused with $1/f$, which is something entirely different. Thus $\sin^{-1}$ does NOT mean $1/\sin$, despite the fact that $\sin^2$ means $(\sin)^2$. As we shall see below, $\sin^{-1}$ means arcsin.

As an example, or rather a "nonexample", consider the function f given by

$$f(x) = \sin x, \quad 0 \leq x.$$

In this case there is no inverse function, because the inverse relationship is not a function:

sin x	x
0	0
0	π
1	$\pi/2$
1	$3\pi/2$
(etc.)	

The first column has a number (in fact, many numbers) repeated.

In contrast, the function f given by

$$f(x) = \sin x, \quad -\frac{\pi}{2} \leq x \leq \frac{\pi}{2},$$

has an inverse:

-1	$-\pi/2$
$-\sqrt{2}/2$	$-\pi/4$
0	0
$\sqrt{2}/2$	$\pi/4$
1	$\pi/2$

The inverse relationship here is a function. There is only one value between $-\pi/2$ and $\pi/2$ having a given value for its sine.

The inverse relationship here should be recognized immediately as being nothing other than the arcsine function. That is, if

$$f(x) = \sin x, \quad -\frac{\pi}{2} \leq x \leq \frac{\pi}{2},$$

then

$$f^{-1}(x) = \arcsin x, \quad -1 \leq x \leq 1.$$

An inverse function is obtained by interchanging the columns of the original function. Therefore the range of the original function becomes the domain of the inverse function. Likewise the domain of the original function becomes the range of its inverse. The curve f^{-1} has the same shape as f, but with the axes interchanged.

8. Inverse Functions and Their Derivatives

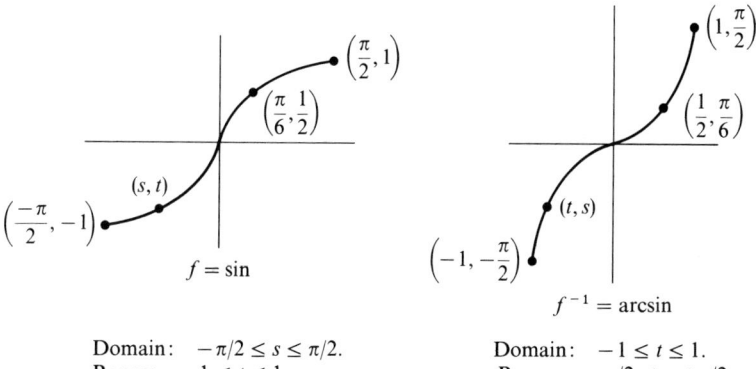

Domain: $-\pi/2 \leq s \leq \pi/2$.
Range: $-1 \leq t \leq 1$.

Domain: $-1 \leq t \leq 1$.
Range: $-\pi/2 \leq s \leq \pi/2$.

It is easy to find the derivative of f^{-1} if one already knows the derivative of f. After all, if one can construct tangent lines to the curve f, one ought to be able to do the same for the curve f^{-1}. However, there is a slight subtilty that arises, owing to the interchange of the axes mentioned above. Before we find the derivative of the arcsine function it may be well to discuss thoroughly a simpler example.

EXAMPLE 16. Consider the function given by

$$f(s) = s^2, \quad 0 < s.$$

This function has an inverse function (why?). Since the point (π, π^2) is on the curve f, it follows that (π^2, π) is on the curve f^{-1}. Find the slope of the tangent line to the curve f^{-1} at the point (π^2, π).

Since f is the squaring function, its derivative is the doubling function. Therefore we know the information in this table:

f	slope of tangent
π	π^2	2π

and we are asked to fill in properly the question mark in this table:

f^{-1}	slope of tangent
π^2	π	?

This, it turns out, is easy. *The answer is $1/2\pi$, as may be seen in several ways.*

First Solution. Let us get an explicit formula for f^{-1} and then find its derivative. For the squaring function f with the domain of all positive numbers, *the inverse function is the square root function*. To see this (if it is not already apparent) we reason as follows. Since $f(s) = s^2$ the points (s, s^2) make up the curve f. Hence, the curve f^{-1} is made up of points of the form (s^2, s), showing that

$$f^{-1}(s^2) = s.$$

If we let t replace s^2 (so that $\sqrt{t}$ must replace s), this becomes

$$f^{-1}(t) = \sqrt{t}.$$

By the square root rule,

$$(f^{-1})'(t) = \frac{1}{2\sqrt{t}}.$$

The slope of the tangent line to the curve f^{-1} at the point (π^2, π) is then given by

$$(f^{-1})'(\pi^2) = \frac{1}{2\sqrt{\pi^2}} = \frac{1}{2\pi}.$$

Second Solution. Let us look at this from a geometric point of view, mixed with a little algebra. We know that the slope of the tangent line to the curve f at (π, π^2) is given by 2π. Therefore, an equation of the tangent line, in x-y coordinates, is

$$y - \pi^2 = 2\pi(x - \pi). \tag{47}$$

Recall how the curve f^{-1} is related to the curve f by the interchange of coordinates: (x, y) is on f if and only if (y, x) is on f^{-1}.

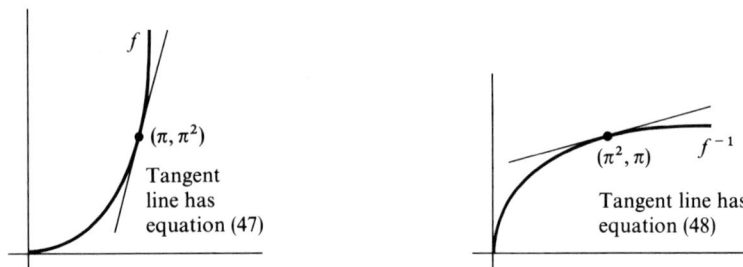

Tangent line has equation (47)

Tangent line has equation (48)

Therefore, if we interchange x and y in an equation of a tangent line to f, such as (47), should we not expect to get an equation of a tangent line to f^{-1}? That is,

$$x - \pi^2 = 2\pi(y - \pi),$$

or

$$y - \pi = \frac{1}{2\pi}(x - \pi^2) \tag{48}$$

will be a line tangent to f^{-1} at (π^2, π). Obviously, the line with equation (48) has slope $1/2\pi$.

8. Inverse Functions and Their Derivatives

Third Solution. A third way to do Example 16 is to use the chain rule to find the derivative of f^{-1}. We have the chain

$$t = f(s) \quad \text{and} \quad s = f^{-1}(t). \tag{49}$$

By the chain rule,

$$\frac{dt}{ds}\frac{ds}{dt} = 1,$$

$$f'(s)(f^{-1})'(t) = 1,$$

$$(f^{-1})'(t) = \frac{1}{f'(s)} \quad \text{if } f'(s) \neq 0. \tag{50}$$

Note that equation (50) gives $(f^{-1})'(t)$, not in terms of t, but in terms of s, where s and t are related by the chain (49). Thus, in our case, equation (50) becomes

$$(f^{-1})'(\pi^2) = \frac{1}{f'(\pi)} = \frac{1}{2\pi}. \qquad \square$$

In the third solution to Example 16 we have essentially proved a general theorem on the calculus of inverse functions.

Rule for Inverse Functions. *If a differentiable function f has an inverse function f^{-1}, then its derivative is given by formula (50), where s and t are related by the chain (49).*

Let us not take time to give a formal proof of this rule, which is somewhat cumbersome. What it says is that if we have the information in the table here:

$$\begin{array}{c|c|c} \curvearrowright f \searrow & & \\ \hline s & t & m \end{array}$$

then (assuming f^{-1} exists and assuming $m \neq 0$) we are entitled to the information here:

$$\begin{array}{c|c|c} \curvearrowright f^{-1} \searrow & & \\ \hline t & s & 1/m \end{array}$$

EXAMPLE 17. Consider the function given by $f(s) = s^2, 0 < s$. Find a formula for the derivative of f^{-1}.

If $t = f(s) = s^2$, then $f'(s) = 2s$ and (50) becomes

$$(f^{-1})'(t) = \frac{1}{2s}, \quad \text{where } s = \sqrt{t}.$$

That is,

$$(f^{-1})'(t) = \frac{1}{2\sqrt{t}}.$$

This is of course the well-known formula for the derivative of the square root function. □

Before finding the derivative of the arcsine function, let us do one more example in the same spirit, to find the derivative of the *n-th root function*.

EXAMPLE 18. Consider the function given by $f(s) = s^n$, $0 < s$, where n is a positive integer. The inverse f^{-1} is called the *n-th root function*. Find its derivative.

If $t = f(s) = s^n$, then $f'(s) = ns^{n-1}$ and (50) becomes

$$(f^{-1})'(t) = \frac{1}{ns^{n-1}}, \quad \text{where } t = s^n.$$

To get this derivative in terms of t we write

$$(f^{-1})'(t) = \frac{1}{n}\left(\frac{s}{s^n}\right) = \frac{1}{n}\left(\frac{t^{1/n}}{t}\right) = \frac{1}{n}t^{(1/n)-1}.$$

Thus the derivative of the n-th root function $t^{1/n}$ is given by the rule $(1/n)t^{(1/n)-1}$. Note that when $n = 2$, this reduces to the result obtained in Example 17. □

EXAMPLE 19. Consider the function given by $f(s) = \sin s$, $-\pi/2 \leq s \leq \pi/2$. The inverse f^{-1} is called the *arcsine function*. Find its derivative.

If $t = f(s) = \sin s$, then $f'(s) = \cos s$ and (50) becomes

$$(f^{-1})'(t) = \frac{1}{\cos s}, \quad \text{where } s = \arcsin t.$$

Thus the answer could be expressed by

$$(f^{-1})'(t) = \frac{1}{\cos(\arcsin t)}. \tag{51}$$

It turns out, however, that the expression (51) can be simplified to an algebraic rule, simply by pondering the question, *What is* $\cos(\arcsin t)$?
It is, of course,

$$\cos s, \quad \text{where } s = \arcsin t.$$

That is, $\cos(\arcsin t)$ is simply

$$\cos s, \quad \text{where } \sin s = t, \quad -\frac{\pi}{2} \leq s \leq \frac{\pi}{2}.$$

Now things become familiar, because

$$\cos^2 s + \sin^2 s = 1,$$
$$\cos^2 s + t^2 = 1,$$
$$\cos s = \pm\sqrt{1 - t^2}.$$

8. Inverse Functions and Their Derivatives

But (*important!*) the arc s must lie between $-\pi/2$ and $\pi/2$, where the cosine is never negative. Therefore, the negative root should be discarded and we have proved that

$$\cos(\arcsin t) = \cos s = \sqrt{1 - t^2}.$$

Thus equation (51) can be simplified to

$$(\arcsin)'(t) = \frac{1}{\sqrt{1 - t^2}}. \qquad \square \quad (52)$$

Something has happened here that is intriguing. Who would have thought that the derivative of a *trigonometric* function such as arcsin would turn out to be an *algebraic* rule? Or, to look at things in reverse, who would have thought that an antiderivative of an algebraic rule like $1/\sqrt{1-t^2}$ would *not* be algebraic?

Exercises

8.1. The graph of a function cannot be cut more than once by a vertical line. Suppose a *horizontal* line cuts the graph of a function f in two or more places. Does f have an inverse function f^{-1}?

8.2. Consider the cosine function. Does it have an inverse if its domain is taken as
 (a) $-\pi/2 \le s \le \pi/2$?
 (b) $0 \le s \le \pi$?
 Apply the test of exercise 8.1.

8.3. Write an equation of the tangent line to the curve $y = \arcsin x$ at the point $(1/2, \pi/6)$.

8.4. Given the information that the point $(1, 2)$ is on a certain curve f, that $f'(1) = 3$, and that the function f has an inverse function, find the slope of the tangent line to the curve f^{-1} at $(2, 1)$. *Answer*: $\frac{1}{3}$, by equation (50).

8.5. The point $(\pi/4, 1)$ is on the graph of $f(s) = \tan s$. What is the slope of the tangent line to the curve $f^{-1}(t) = \arctan t$ at the point $(1, \pi/4)$? *Hint*. This is done just like exercise 8.4 if you can find $f'(\pi/4)$.

8.6. The point $(\pi/3, 1/2)$ is on the graph of $f(s) = \cos s$. Find the slope of the tangent line to the curve $f^{-1}(t) = \arccos t$ at the point $(1/2, \pi/3)$.

8.7. Use the chain rule to find the derivatives of the following.
 (a) $\arcsin 5x$. (b) $\arcsin((x-3)/3)$.
 (c) $\arcsin \sqrt{x}$. (d) $\arcsin \sqrt{(6-x)/x}$.
 (e) $\arcsin(\sin x)$. (f) $\arcsin(\cos x)$.
 (g) $\sin(\arcsin x)$. (h) $\cos(\arcsin x)$.
 Answers: (a) $5/\sqrt{1-25x^2}$. (e) $(\cos x)/\sqrt{1-\sin^2 x} = (\cos x)/\sqrt{\cos^2 x}\,(=1, -1,$ or undefined, depending upon x). The answer to part (b) is, if simplified, seen to be the negative of twice the answer to part (d). (This fact is useful in problem 18 at the end of the chapter.)

8.8. Evaluate the following integrals by applying the fundamental theorem of calculus.
  (a) $\int_0^{1/2} (1/\sqrt{1-t^2})\,dt$.
  (b) $\int_{-1/2}^{\sqrt{3}/2} (1/\sqrt{1-t^2})\,dt$.
  Answer: (b) $\pi/2$.

8.9. If $f(x) = \arcsin(1/x)$, show that
  (a) $f'(x) = -1/x\sqrt{x^2-1}$ if $1 < x$.
  (b) $f'(x) = 1/x\sqrt{x^2-1}$ if $x < -1$.
  (c) $f'(x)$ is undefined if $-1 \le x \le 1$.

8.10. The expression $t^{m/n}$, for $0 < t$, where m is an integer and n a positive integer, is defined by
$$t^{m/n} = (t^{1/n})^m.$$
Regarding this as being given by the chain of relations composed of the n-th root followed by the m-th power, find dy/dt, where $y = t^{m/n}$. Answer: $dy/dt = (m/n)t^{(m/n)-1}$.

§9. Implicit Differentiation

The derivatives of the three most important inverse trigonometric functions are summarized here.

$$(\arcsin)'(t) = \frac{1}{\sqrt{1-t^2}}, \quad -1 < t < 1.$$

$$(\arccos)'(t) = \frac{-1}{\sqrt{1-t^2}}, \quad -1 < t < 1.$$

$$(\arctan)'(t) = \frac{1}{1+t^2}, \quad \text{for all } t.$$

The first of these has already been proved and the second, similar to it, is left to the reader as an exercise. Let us prove the third.
If $t = f(s) = \tan s$, then $f'(s) = \sec^2 s$ and (50) becomes

$$(f^{-1})'(t) = \frac{1}{\sec^2 s}, \quad \text{where } t = \tan s.$$

Therefore (see exercise 5.6),

$$(\arctan)'(t) = \frac{1}{\sec^2 s} = \frac{1}{1+\tan^2 s} = \frac{1}{1+t^2}. \quad \square$$

An alternate approach to the theory of inverse functions is by the method of *implicit differentiation*. Sometimes it shortens one's work. Let us do a

9. Implicit Differentiation

problem in related rates first by the usual method, and then by making use of this new technique.

EXAMPLE 20. A plane is flying at a constant altitude of 3000 feet and at a constant speed of 600 feet per second. An observer on the ground measures the angle θ as indicated in the diagram. Find the rate of change of θ, 2 seconds after the plane has passed directly over the observer.

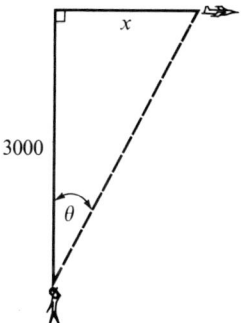

This is a related rates problem. We know the rate of change of x:

$$\frac{dx}{dt} = 600,$$

and we know a relation between x and θ:

$$\tan \theta = \frac{x}{3000}, \qquad (53)$$

or

$$\theta = \arctan \frac{x}{3000}.$$

This may be regarded as a chain given by $\theta = \arctan u$, where $u = x/3000$, so that

$$\frac{d\theta}{dx} = \frac{1}{1+u^2}\frac{du}{dx} = \frac{1}{1+(x/3000)^2}\frac{1}{3000}.$$

Two seconds after the plane passes directly overhead we have $x = 1200$ (why?). Plugging this value of x into the above expression yields

$$\left.\frac{d\theta}{dx}\right|_{x=1200} = \frac{1}{3480}.$$

However, it is not $d\theta/dx$ that is required in this example, but rather $d\theta/dt$. By the chain rule, when $x = 1200$, we have

$$\frac{d\theta}{dt} = \frac{d\theta}{dx}\frac{dx}{dt} = \frac{1}{3480}600 = \frac{5}{29} \text{ radians per second.}$$

Note that the answer comes out in radians per second not in degrees per second. *Why?*

Alternate solution. Look carefully at equation (53). We know that θ is a function of t even though the variable t does not appear in (53). The rule giving θ in terms of t is hidden, or *implicit*, in equation (53), because x depends upon t. In this equation, let us take the derivative, not with respect to x, but with respect to t. This is called *implicit differentiation*. We get

$$\sec^2 \theta \frac{d\theta}{dt} = \frac{1}{3000} \frac{dx}{dt} = \frac{1}{5}, \qquad (54)$$

since $dx/dt = 600$.

When $x = 1200$, the picture looks like this:

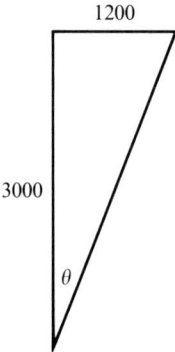

showing that $\tan \theta = \frac{2}{5}$, so

$$\sec^2 \theta = 1 + \tan^2 \theta = 1 + \left(\frac{2}{5}\right)^2 = \frac{29}{25}.$$

Thus, when $x = 1200$, equation (54) becomes

$$\frac{29}{25} \frac{d\theta}{dt} = \frac{1}{5},$$

so

$$\frac{d\theta}{dt} = \frac{5}{29} \text{ rad/sec.} \qquad \square$$

Note that in the alternate solution given above we did not need to know the formula for differentiating the arctangent function. Implicit differentiation made this unnecessary.

EXAMPLE 21. A car parked too close to the movie screen in a drive-in theater begins to back slowly away from the screen in order to obtain a better view. If the screen is 50 feet high, with its base 30 feet above eye level, where should the car stop in order to maximize the angle of vision?

9. Implicit Differentiation

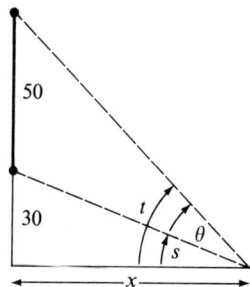

We must find the value of x that maximizes θ, the angle of vision. Let us get an explicit formula expressing the dependence of θ upon x. Clearly, from the diagram,

$$\theta = t - s$$
$$= \arctan \frac{80}{x} - \arctan \frac{30}{x}, \qquad 0 < x.$$

Therefore,

$$\frac{d\theta}{dx} = \frac{1}{1 + (80/x)^2} \cdot \frac{-80}{x^2} - \frac{1}{1 + (30/x)^2} \cdot \frac{-30}{x^2}.$$

Simplifying and setting this equal to zero yields

$$\frac{80}{x^2 + (80)^2} = \frac{30}{x^2 + (30)^2}$$

To solve this for a critical point x, we cross-multiply and collect terms to get

$$(80 - 30)x^2 = (80)^2(30) - (30)^2(80) = 80(30)(80 - 30),$$
$$x^2 = 80(30),$$
$$x = \sqrt{2400}$$
$$\approx 48.99 \text{ ft.}$$

It should be evident that the angle of vision θ is *maximal* when $x \approx 49$ feet. This is the only critical point in the domain. The maximum must occur here, because near the "ends" of the domain (when x is near zero or when x is very large) the angle of vision is tending to zero.

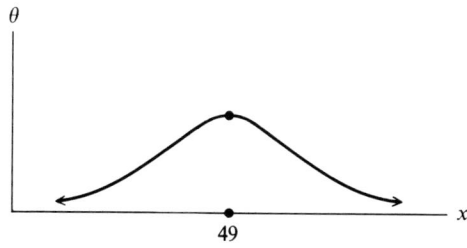

Example 21 can also be done by implicit differentiation, beginning with the equation

$$\tan\theta = \tan(t - s) = \frac{\tan t - \tan s}{1 + \tan t \tan s}, \qquad (55)$$

and then getting the right-hand side of this equation in terms of x. The reader is asked to carry this out in an exercise below. □

EXERCISES

9.1. By following carefully the line of reasoning developed in Example 19, but making necessary modifications, show that $(\arccos)'(t) = -1/\sqrt{1-t^2}$.

9.2. Consider the function given by $y = \arcsin t + \arccos t$, with domain $-1 < t < 1$.
(a) Find y'.
(b) Deduce from part (a) that $y(t) = C$ for some constant C.
(c) Find C.
(d) Deduce the trigonometric identity

$$\arccos t = \frac{\pi}{2} - \arcsin t, \qquad -1 \le t \le 1.$$

(e) The identity in part (d) was proved, not by using trigonometry, but by using calculus. Can you prove the identity without using calculus?

9.3. Consider the function given by $y = \arctan(\tan x)$. Note that it is undefined when $x = \pm\pi/2, \pm 3\pi/2$, etc.
(a) Show that $dy/dx = 1$, except when $x = \pm\pi/2, \pm 3\pi/2$, etc.
(b) Use the fundamental principle of integral calculus to conclude that, for some constants C and D,

$$\arctan(\tan x) = x + C, \qquad -\frac{\pi}{2} < x < \frac{\pi}{2}.$$

$$\arctan(\tan x) = x + D, \qquad \frac{\pi}{2} < x < \frac{3\pi}{2}.$$

(c) Find C and D. *Hint.* Plug in $x = 0$ and $x = \pi$.
(d) Deduce the trigonometric identities

$$\arctan(\tan x) = x, \qquad -\frac{\pi}{2} < x < \frac{\pi}{2}.$$

$$\arctan(\tan x) = x - \pi, \qquad \frac{\pi}{2} < x < \frac{3\pi}{2}.$$

(e) Simplify the expression $\arctan(\tan x)$, where $-3\pi/2 < x < -\pi/2$. (*See problem 28 at the end of the chapter.*)

9.4. A plane is flying at a constant altitude of 5000 meters at a speed of 800 m/sec. An observer on the ground measures the angle θ between the plane and a vertical line, as in Example 20. Three seconds after the plane passes directly overhead, find the rate of change of θ in
(a) radians per second.
(b) degrees per second.

9.5. An observer 5000 meters from the launching pad watches a rocket in its vertical takeoff. When the altitude of the rocket is 1000 meters, its speed is 500 m/sec. Find the rate of change of the angle of inclination of the rocket, as observed at this instant.

9.6. A tall statue has its base on a pedestal 10 feet above eye level. If the statue itself is 30 feet tall, where should an observer stand in order that he may enjoy the widest possible angle of vision? *Answer*: Twenty feet from the pedestal.

9.7. Beginning with equation (55) and reading the suggestion that follows that equation, give an alternate solution to Example 21 using implicit differentiation.

9.8. Evaluate each of the following integrals by using the fundamental theorem of calculus.
   (a) $\int_0^{\sqrt{3}} (1/(1+t^2))\,dt$.
   (b) $\int_{-1}^{1} (1/(1+t^2))\,dt$.
   Answer: (b) $[\arctan t]_{-1}^{1} = (\pi/4) - (-\pi/4) = \pi/2$.

9.9. Use implicit differentiation to find the slope of the tangent line to the given curve at the indicated point P.
   (a) $x^2 + y^2 = 25$, $P = (3, -4)$.
   (b) $x^2 y^3 + y^2 + 2y = 3x + 1$, $P = (1, 1)$.
   (c) $y \sin x + \arctan y = 0$, $P = (0, 0)$.
   Hint. (b) $x^2(3y^2 y') + 2xy^3 + 2yy' + 2y' = 3$. When $x = y = 1$, this becomes $3y' + 2 + 2y' + 2y' = 3$. Solve for y'.

§10. Summary

The sine and cosine functions are defined in terms of a circle. As t increases, the point $(\cos t, \sin t)$ periodically traces out the unit circle in a counterclockwise manner. The other trigonometric functions are defined in terms of the sine and cosine, giving rise to numerous trigonometric identities.

We adopt the convention of measuring angles by radians because this convention simplifies the rules for differentiating the trigonometric functions. The inverse trigonometric functions, surprisingly, have derivatives that are given by algebraic rules.

The circle is rich in ideas.

Problem Set for Chapter 7

1. In the third century B.C., Eratosthenes of Cyrene measured a famous angle (see Chapter 2, Section 4) and found it to be approximately one-fiftieth of a revolution. What is the measure of this angle in degrees? in radians?

2. Apply the sine, cosine, and tangent sum laws to simplify each of the following.
   (a) $\sin(t + \pi)$.
   (b) $\cos(t + \pi)$.
   (c) $\tan(t + \pi)$.

3. Find the limit, as h tends to zero, of each of the following.
   (a) $(\sin 2h)/h$.
   (b) $(\sin h^2)/h$.
   (c) $(\sin h^2)/(\sin^2 h)$.
   (d) $(1 - \cos h)/3h^2$.

4. Write an equation of the tangent line to the curve $y = \tan x$ at the point $(\pi/4, 1)$.

5. Find the derivative of each of the following. Do not simplify your answers.
   (a) $x \sin x$.
   (b) $\sin x \cos x$.
   (c) $\sin(x \cos x)$.
   (d) $\sin \pi x$.
   (e) $\sin \frac{1}{2}x$.
   (f) $\sec^2 x$.
   (g) $\sec x \tan x$.
   (h) $\sqrt{5 + \tan x}$.
   (i) $\tan^2 x$.
   (j) $\tan^2 \pi x$.
   (k) $\cos^2(3x - 4)$.
   (l) $\cos(\cos x)$.
   (m) $(\sin x)/x$.
   (n) $(\cos 3x)/(x^2 + 1)$.

6. Find both coordinates of an inflection point on the curve
   (a) $y = \sin x$.
   (b) $y = \cos x$.
   (c) $y = \tan x$.

7. Fill in the table, and use the information to sketch the curve $y = \tan x$, $-\pi/2 < x < \pi/2$.

x	y	y'	y''
$-\pi/3$			
$-\pi/6$			
0			
$\pi/6$			
$\pi/3$			

8. Use the rule found in problem 2(c), together with the curve sketched in 7, to sketch the curve $y = \tan x$ on the domain $-\pi \le x \le \pi$ (with the points $-\pi/2$ and $\pi/2$ deleted from the domain of course).

9. A garden plot is to be in the shape of a trapezoid, as in Example 15. One side of the plot will be bordered by a long fence already standing, while the other three sides are to be 80 feet (opposite the long fence), 50 feet, and 50 feet. Make a sketch, giving dimensions, indicating how the plot should be laid out in order to maximize the enclosed area.

10. Work problem 9, assuming the three sides to be 50 feet (opposite the long fence), 80 feet, and 80 feet.

11. A person standing on the side of a circular lake of radius 1 kilometer wants to get to the point directly opposite him on the other side. His motorboat can travel 4 km/hr and he can jog 6 km/hr. Prove that by no combination of boat travel and jogging can he get to his destination is less than half an hour.

12. A particle moves along the y-axis in such a way that its position is given by

$$y = 4\cos t + 3\sin t, \qquad 0 \le t.$$

   (a) Find its position, velocity, and acceleration when $t = \pi/3$.
   (b) Find its position, velocity, and acceleration when $t = 3\pi/2$.
   (c) Show that the particle's velocity is zero whenever $\tan t = \frac{3}{4}$.
   (d) Specify the range of positions y assumed by the particle.

13. Find the area, between $t = 0$ and $t = \pi/3$, beneath each of the following curves.
    (a) $y = \cos t$.
    (b) $y = 5\sin t$.
    (c) $y = \sec^2 t$.

14. By carrying out each of the following steps, prove by Fermat's method that $(\tan)' = \sec^2$.
    (a) Find the limit, as $h \to 0$, of $(\tan h)/h$.
    (b) Apply the tangent sum law to the expression $\tan(x + h)$.
    (c) Using the result of part (b) and combining terms with a common denominator, show that

    $$\tan(x + h) - \tan x = \frac{(1 + \tan^2 x)\tan h}{1 - \tan x \tan h}.$$

    (d) In the equation of part (c), divide both sides by nonzero h. Then take the limit, using the result of part (a).

15. Consider each of the following statements and indicate whether it is true or false.
    (a) $\sin 0 = 0$.
    (b) $\operatorname{SIN} 0 = 0$.
    (c) $\sin 30 = \frac{1}{2}$.
    (d) $\operatorname{SIN} 30 = \frac{1}{2}$.
    (e) $\sin(\pi/6) = \frac{1}{2}$.
    (f) $\arcsin(\pi/6) = 30$.
    (g) $\arcsin \frac{1}{2} = 30$.
    (h) $\arcsin \frac{1}{2} = 5\pi/6$.
    (i) $\arcsin(\sin x) = x$ for all x.
    (j) $\arcsin(\sin(5\pi/6)) = \pi/6$.
    (k) $\arctan(\cos 0) = \pi/4$.
    (l) $\cos(\arctan 0) = \pi/4$.
    (m) $\operatorname{arccos}(-1) = -\pi$.
    (n) $\arccos(-1) = \pi$.
    (o) $\tan(\arctan 6) = 6$.
    (p) $\arctan(\tan 6) = 6$.
    (q) $\tan(-x) = -\tan x$.
    (r) $\sin(\arcsin x) = x$ for all x.

16. By writing $\cos 3x = \cos(2x + x)$ and applying the cosine sum law, derive an identity expressing a "triple-angle formula" for the cosine. Can you express your formula in terms of cos alone, without using sin?

17. Take the derivative of both sides of the triple-angle formula obtained in problem 16, then solve for the expression sin 3x, thus obtaining by calculus a triple-angle formula for the sine function.

18. Consider the function given by
$$f(x) = \arcsin\left(\frac{x-3}{3}\right) + 2\arcsin\sqrt{\frac{6-x}{6}},$$
defined on the domain $0 \le x \le 6$.
   (a) Show that $\pi/2$ is in the range of f.
   (b) Show that $\pi/2$ is the largest number in the range of f.
   (c) Show that $\pi/2$ is the smallest number in the range of f.
   (d) Deduce a trigonometric identity that holds on the domain $0 \le x \le 6$.

19. Find the indicated areas.

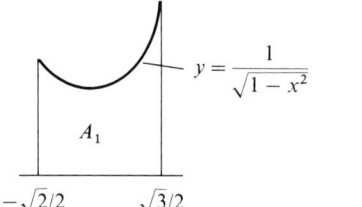
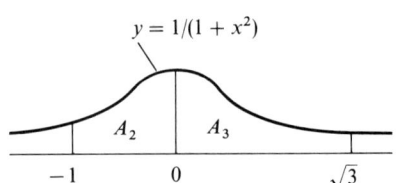

20. An observer 5000 feet from a launching pad watches a rocket in its vertical takeoff and measures its angle θ of inclination as it rises.

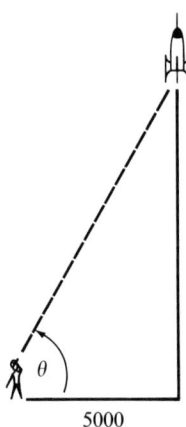

When θ is 60 degrees, the observer finds that $d\theta/dt$ is 3 degrees per second.
   (a) When $\theta = \pi/3$ radians, what is $d\theta/dt$ in radians per second?
   (b) What is the altitude of the rocket when $\theta = \pi/3$ radians?
   (c) What is the speed of the rocket when $\theta = \pi/3$ radians?

21. A field-goal kicker has a problem. The ball must be kicked from a point on the indicated line.

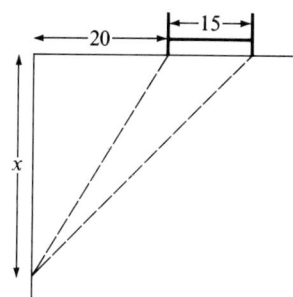

Which point x should be chosen in order to open up the widest angle between the goal posts?

22. The base of a statue is 12 feet above eye level, while the statue itself is 25 feet tall. Where should an observer stand in order to have the best view of the statue?

23. Evaluate each of the following integrals. To find antiderivatives, use the technique developed in exercises 6.8 and 6.9.
    (a) $\int_0^{3\pi} \cos t \, dt$.
    (b) $\int_0^{\pi} \sin 5t \, dt$.
    (c) $\int_{-\pi}^{\pi} \sin^2 t \, dt$.
    (d) $\int_0^{\pi} \cos^2 t \, dt$.
    (e) $\int_0^{\pi/2} \sin 2x \cos 3x \, dx$.
    (f) $\int_0^{\pi} \sin 2x \sin 5x \, dx$.
    (g) $\int_0^{\pi/2} \cos 3x \cos x \, dx$.
    (h) $\int_0^{\pi/4} \sin 2x \sin 5x \, dx$.

24. Evaluate the integral $\int_0^{\pi} L(\theta) \, d\theta$, where $L(\theta) = \sqrt{2 - 2\cos\theta}$. *Hint.* To find an antiderivative of L, first express L in simpler terms, using the result of exercise 3.8.

25. The inverse of each of the functions in the following list is also in the list. Match each function with its inverse.
    (a) The *squaring function*, sending x to x^2, $0 < x$.
    (b) The *reciprocal function*, sending x to $1/x$, $x \neq 0$.
    (c) The *identity function*, sending x to x.
    (d) The *doubling function*, sending x to $2x$.
    (e) The *square root function*, sending x to $\sqrt{x}$, $0 < x$.
    (f) The *halving function*, sending x to $\frac{1}{2}x$.

26. The *identity function*, sending x to x, plays a special role in the theory of inverse functions.
    (a) Show that the composition of a function with its inverse is the identity function, i.e., show
        (i) $f^{-1}(f(x)) = x$ if x is in the domain of f,
        (ii) $f(f^{-1}(x)) = x$ if x is in the domain of f^{-1}.
    (b) The graph of the identity function is simply the straight line of slope 1 through the origin. Show that the graph of f^{-1}, the graph of the identity function, and the graph of f enjoy the following camaraderie: the curves f and f^{-1} are each "reflections" of the other, through the identity function's graph.

27. Given the following table of information about the function f, fill in the question marks properly in the table for f^{-1}.

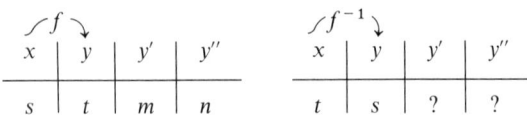

28. Each of the following expressions is of the form $f^{-1}(f(x))$:
    (i) arctan(tan x).
    (ii) arccos(cos x).
    (iii) arcsin(sin x).
    (a) Match each of them with its graph below.

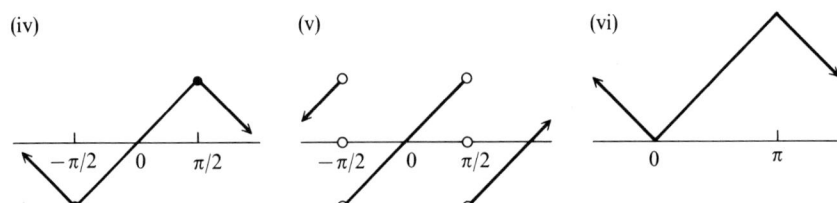

(iv)   (v)   (vi)

(b) As we can see from these graphs, it is not necessarily true that $f^{-1}(f(x)) = x$ for all x. Does this contradict the result of problem 26(a)?
(c) Sketch the curves giving the derivatives of the functions pictured in part (a).

29. The **absolute value function** is sometimes useful in making certain complicated expressions appear in a simpler form. The **absolute value** of a number x is denoted by $|x|$ and defined as

$$|x| = \sqrt{x^2} = \begin{cases} x & \text{if } 0 < x. \\ 0 & \text{if } x = 0. \\ -x & \text{if } x < 0. \end{cases}$$

(a) If $f(x) = |x|$, find $f'(x)$. *Hint.* The derivative does not exist at the point $x = 0$, but if $x \neq 0$, you can find the derivative by applying the square root rule to $\sqrt{x^2}$.
(b) Show that
$$\arccos(\cos x) = |x| \quad \text{if } -\pi \leq x \leq \pi.$$

(c) Show that if $f(x) = \arcsin(\sin x)$, then
$$f'(x) = \frac{\cos x}{|\cos x|}.$$
Hint. See exercise 8.7(e).
(d) Show that $f'(x) = (\sin x)/|\sin x|$ if $f(x) = \arccos(\cos x)$. Specify the values of x for which $f'(x)$ does not exist.

30. What is wrong with the following "argument"? If $f(x) = \arctan(\tan x)$, then by the chain rule

$$f'(x) = \frac{1}{1 + \tan^2 x} \sec^2 x = \frac{1}{\sec^2 x} \sec^2 x = 1.$$

Since $f'(x) = 1$ for all x, it follows that $f(x) = x$. Therefore, $\arctan(\tan x) = x$ for all x.

31. (a) Find dy/dx if $y = \sin(\pi x/180)$.
    (b) Show that $\text{SIN } x = \sin(\pi x/180)$ and $\text{COS } x = \cos(\pi x/180)$, where SIN and COS denote the sine and cosine functions, respectively, relative to degree measure of angles.
    (c) Using parts (a) and (b), prove that $(\text{SIN})'(x) = (\pi/180) \text{ COS } x$.
    (d) Find $(\text{COS})'$, in terms of SIN.

32. (a) Justify each of the following steps.

$$\sin t + \sqrt{3} \cos t = 2\left(\frac{1}{2} \sin t + \frac{\sqrt{3}}{2} \cos t\right)$$

$$= 2\left(\cos \frac{\pi}{3} \sin t + \sin \frac{\pi}{3} \cos t\right)$$

$$= 2 \sin\left(t + \frac{\pi}{3}\right).$$

   (b) Explain how it is obvious from part (a), without using calculus, that the range of the function given by $\sin t + \sqrt{3} \cos t$ is $-2 \le y \le 2$ if the domain is unrestricted.
   (c) Prove, without using calculus, that the range of the function given by $a \sin t + b \cos t$ is

$$-\sqrt{a^2 + b^2} \le y \le \sqrt{a^2 + b^2}$$

   if the domain contains an interval of length 2π.
   (d) Do problem 12, part (d), without using calculus.

33. Use implicit differentiation to find the slope of the tangent line to each of the following curves at the indicated point P.
    (a) $x^2 + y^2 = 169$, $P = (5, 12)$.
    (b) $x^2 + y^2 = 169$, $P = (5, -12)$.
    (c) $y^3 - 4xy + 3x^2 = 0$, $P = (1, 1)$.
    (d) $\sin 4y + \arctan x + \cos(x + y) = 1$, $P = (0, 0)$.

34. Find the "general solution" to the fence problem illustrated in problems 9 and 10. Show that if the fences have lengths a, b, and a, then the angle θ should be chosen as

$$\theta = \arcsin\left(\frac{-b + \sqrt{b^2 + 8a^2}}{4a}\right),$$

in order to maximize the enclosed area.

35. Show that, regardless of the values of a and b in problem 34, the optimal angle θ found in problem 34 can never exceed $\pi/4$ radians. *Hint*. Let $x = b/a$ and regard θ as a function of x.

36. Find the general solution of the angle problem illustrated in problems 21 and 22.

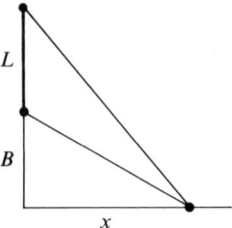

Show that if L is the length of the object in which you are interested, whose base is a distance B from a line on which you are constrained to move, then the optimal value of x is $\sqrt{B(L + B)}$.

37. (*For circle lovers*) Investigate (i.e., prove or disprove) the following conjecture. Given the general situation as found in problems 9 and 10, what you should do, to optimize the area, is to make your fence as nearly like a circle as you can. That is, if you have fencing of lengths a, b, and a, then lay out your fence so that you can put a circle through the four corners, with its diameter along the fence already standing:

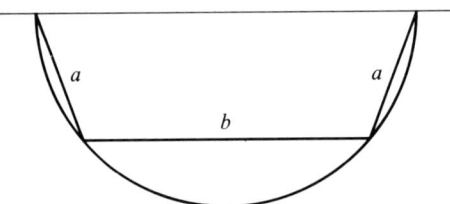

38. (*For circle lovers*) Investigate the following conjecture. Given the general situation as found in problems 21 and 22, what you should do, to optimize the indicated angle, is to find the position on the line where a circle tangent to the line cuts the top and bottom points of the object in which you are interested.

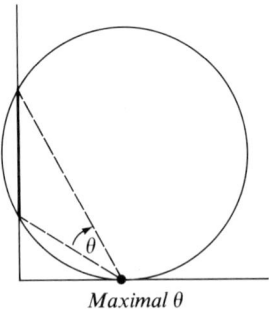

Maximal θ

Can you prove this using only Euclidean geometry?

House of Integrals 8

Reculer pour mieux sauter! ("Draw back to leap!") say the French, and we shall bear in mind their good advice. Before we can leap into a higher realm, we must step backwards, in two senses. We must consider more carefully some fundamental principles that we have already met but still do not really know. And we must think a little more about pulling back from a given function to an antiderivative. Does every continuous function, for example, have an antiderivative? We shall see.

The word *transcendental* has been used in mathematics for well over a century. It means what it says, referring to something that transcends, or rises above, whatever realm has already been attained. A *transcendental function* is one whose rule of correspondence transcends the realm of ordinary algebra. For example, the function F given by

$$F(x) = \arctan x$$

is transcendental. The rule for calculating $\arctan x$ cannot be expressed in terms of algebraic operations (sums, quotients, powers, roots, etc.) applied to the variable x.

In contrast, the function f given by

$$f(x) = \frac{1}{1+x^2}$$

is not transcendental. The function f is an *algebraic function* because the rule for f is stated in terms of quotients, sums, and squares—all familiar algebraic operations. Almost all the functions we encountered before Chapter 7 were algebraic, as is, for example, the function given by

$$g(x) = \frac{1}{x}.$$

It is curious that, as in the case of F and f given above, *an antiderivative F of an algebraic function f need not be algebraic*. Does this mean that techniques for finding antiderivatives will be more involved than the techniques we already know for finding derivatives? In one sense the answer to this question is *yes*, but in another sense, as we shall see, an antiderivative of any continuous function can be found with almost ridiculous ease.

Before we can see clearly how simple it is to write down, for example, an antiderivative of the function g given above, we must learn to feel more at home with the integral. Calculus consists of two branches, differentiation and integration, and so far we have placed more emphasis upon the former. It should be remembered, however, that integration is almost exactly 2000 years older than differentiation, Eudoxus living 2000 years before Fermat. Let us try to do justice to the integral, to meet it in friendship, to enter its house.

§1. Fundamentals: Existence and Uniqueness

In varying degrees of depth we have discussed the following fundamental principles:

F1. *On a connected domain, two antiderivatives of the same function differ by an additive constant.*

F2. *If f is a continuous function with domain $a \leq x \leq b$, then the integral of f exists.* [*That is, in Eudoxus' method when the limit is taken, the limit will exist, producing the number to be denoted $\int_a^b f(x)\,dx$.*]

F3. *The integral $\int_a^b f(x)\,dx$ can be evaluated by calculating $F(b) - F(a)$, where F is any antiderivative of f.*

These principles might be called, respectively, the **uniqueness** theorem, the **existence** theorem, and the **evaluation** theorem. The uniqueness theorem **F1** was made plausible in Chapter 5, and the evaluation theorem **F3** was proved in Chapter 6. The existence theorem **F2** was stated in Chapter 6 without proof. In this chapter we shall see some reasons for believing the existence theorem, although we shall not see its complete proof.

The existence theorem **F2** tends to be overshadowed by the evaluation theorem **F3**, despite the fact that **F3** is of no use whatever unless an antiderivative is known. For many purposes the existence theorem is of paramount importance. If we forget it, we can find ourselves in pitfalls.

Pitfall. It is obvious, from Eudoxus' method, that if a function is always positive, then its integral is too:

$$\int_a^b f(x)\,dx \geq 0 \quad \text{if } f(x) \geq 0. \tag{1}$$

1. Fundamentals: Existence and Uniqueness

This must apply to the function given by $f(x) = 1/x^2$, since squares are never negative. By **F3**, since the derivative of $-1/x$ is $1/x^2$, we have

$$\int_{-1}^{1} f(x)\,dx = \frac{-1}{x}\bigg|_{-1}^{1} = -1 - \left(\frac{-1}{-1}\right) = -2. \tag{2}$$

Statements (1) and (2) contradict each other. □

How did we let ourselves get into this pitfall? Our trouble comes from being so enchanted with the evaluation theorem **F3** that we tend to forget the existence theorem **F2**. The function $1/x^2$ is not continuous throughout the domain $-1 \leq x \leq 1$, since it is not even defined at the point $x = 0$. If we attempted to apply Eudoxus' method to $1/x^2$ on this domain, we should find that the approximating sums S_n get arbitrarily large and do not tend to a limit. (Compare exercise 7.7 of Chapter 6.) The integral $\int_{-1}^{1}(1/x^2)\,dx$ does not exist, and of course the evaluation theorem **F3** cannot be used to evaluate a nonexisting number!

One reason for learning the existence theorem, then, is simply to avoid looking silly. An even better reason for making friends with **F2** will be discussed in the next section.

EXERCISES

1.1. Among the following rules, specify the ones you can recognize to be algebraic functions.
   (a) $\sin x$.
   (b) $1/\sqrt{1-x^2}$.
   (c) $\arcsin x$.
   (d) The area of a circle of radius x.
   (e) The area beneath the squaring function, between 0 and x.
   (f) The area beneath the curve $y = 1/(1+t^2)$, between 0 and x.
   (g) The area beneath the curve $y = 1/t$, between 1 and x.
   Answers: (b), (d), and (e) are algebraic.

1.2. Here are some conjectures. For each of them, tell whether it is true or false, and why.
   (a) If a transcendental function has a derivative, its derivative must also be a transcendental function.
   (b) If an algebraic function has a derivative, its derivative must also be an algebraic function.
   (c) An antiderivative of a transcendental function must be transcendental.
   (d) An antiderivative of an algebraic function must be algebraic.

1.3. Every one of the following "calculations" is ridiculous. Explain why, in each case.
   (a) $\int_0^2 (1/\sqrt{1-x^2})\,dx = \arcsin 2$.
   (b) $\int_0^1 (1/(1+x^2))\,dx = \arctan 1 - \arctan 0 = 45$.
   (c) $\int_0^2 (1/(1-x)^2)\,dx = 1/(1-x)\big|_0^2 = -2$.
   (d) $\int_0^\pi \sec^2 x\,dx = \tan x\big|_0^\pi = \tan \pi - \tan 0 = 0$.

1.4. Of the integrals that follow, some exist and some do not. Which of these can be proved to exist by appealing to the existence theorem **F2**? (*You are not asked to evaluate the integrals.*)

(a) $\int_0^2 (1/(1+x))\,dx$.

(b) $\int_0^2 (1/(1-x))\,dx$.

(c) $\int_{-1}^1 |x|\,dx$.

(d) $\int_{-10}^{10} (1/(x^2-x+1))\,dx$.

(e) $\int_{-1}^1 \arcsin x\,dx$.

(f) $\int_{-10}^{10} (1/(x^2-x-1))\,dx$.

(g) $\int_0^2 \tan t\,dt$.

(h) $\int_0^5 \sin^3 t \cos^2 t\,dt$.

(i) $\int_0^1 \sqrt{1-2t}\,dt$.

(j) $\int_{-5}^3 t^2 \sin t \arctan(\cos t)\,dt$.

Hint. A function is continuous at a point, if it is differentiable at the point. However, it can be continuous without being differentiable. *Answers*: (a), (c), (d), (e), (h), and (j) exist by **F2**.

1.5. The importance of **F1** lies in the fact that it makes a statement about uniqueness. It says that *the* antiderivative of a given function is *not* unique, that is, there is more than one antiderivative. Use **F1** to prove this more positively expressed variant:

Uniqueness Theorem (*An Alternate Statement of F1*). *Assume that F and G are antiderivatives of the same function, defined on a connected domain. If, for some point c it is true that $F(c) = G(c)$, then $F(x) = G(x)$ for all x in the domain.*

§2. What is a Number?

Sometimes it is contended that the most novel development arising out of calculus is a change in mankind's conception of *number*. There is no doubt that we think of numbers today in a way different from those ancient Greeks for whom "number" meant a positive quantity that could be measured by a ratio of integers. On the other hand some Greeks, such as Eudoxus and Archimedes, would easily recognize the modern conception of number as having been borrowed largely from their own work. This is another subject better pursued in depth by a course in analysis, but we must touch upon it lightly here.

In the beginning there were only positive whole numbers, 1, 2, 3, etc., the playthings of the Pythagoreans, who first saw the value of numbers to philosophy. To increase their power the Pythagoreans "enlarged" their conception of number to include positive rationals, that is, numbers of the form m/n, where m and n are positive integers. Their discovery of irrational points on the "number line", however, showed that their conception of number was still too narrow to accomplish what they wished. They could not (and they proved they could not!) even measure with their numbers the hypotenuse of a right triangle whose legs were of unit length.

It is clear from the experience of the Pythagoreans that if *numbers* are to measure arbitrary quantities, then *there must be a* number *corresponding to each length of line segment*. This is what a number "really" is. The **real number system** must contain numbers capable of measuring any quantity, whether positive, negative, or zero.

2. What is a Number?

How can a quantity be measured, particularly if it is an irrational quantity? The modern answer is, of course, by using the notion of *limit*, by approximating it ever more closely by quantities we already know, and by thinking of the number as the limit of our approximations.

Today, for example, we think of $\sqrt{2}$ as being a perfectly proper *real number*, whereas Pythagoras had to think of it as a certain *length*, and not a "number" at all. When we write

$$\sqrt{2} = 1.414\ldots \quad \textit{(note the three dots!)},$$

we do *not* mean $\sqrt{2} = 1.414$ (which, of course, is false, because the square of 1.414 is not equal to 2), but that a sequence beginning

$$1, \quad 1.4, \quad 1.41, \quad 1.414, \ldots$$

is getting progressively closer to what we *do* mean. The real number $\sqrt{2}$ is defined by specifying a sequence

$$1, \quad \frac{14}{10}, \quad \frac{141}{100}, \quad \frac{1414}{1000}, \ldots$$

of positive rationals whose squares

$$1, \quad 1.96, \quad 1.9881, \quad 1.999396, \ldots$$

are tending to 2.

As the example of $\sqrt{2}$ illustrates, to define a real number, it is only necessary to specify how it may be approximated to any desired degree of accuracy. That is, *to define a real number, it is only necessary to specify a way to approximate it as a limit.*

The idea is that a real number is supposed to represent a point on the number line. While the *point* itself we conceive of as a "sharply defined" object, it may well be impossible to hit the point with a rational number. The best we can do is to specify how the point may be approximated ever more closely.

The number π is defined as the area of a circle of radius 1 unit. The sequence 3, 3.1, 3.14, 3.141, 3.1415, 3.14159, ... is the beginning of a sequence of rationals tending to π as a limit: $\pi = 3.14159\ldots$.

What has all this got to do with integrals? A whole lot, indeed! An *integral* is defined—by Eudoxus' method—as a limit, with the sequence of numbers tending to that limit being the approximating sums calculated in the method. There is a moral in this. When we see an integral, such as

$$\int_1^2 \frac{1}{t} dt, \tag{3}$$

we should think of it as representing a number no less sharply defined than numbers like $\sqrt{2}$ or π. We must avoid the temptation to think that with no antiderivative of $1/t$, the expression (3) is worthless because it cannot be "evaluated" by the fundamental theorem **F3**. *This is the wrong attitude to take*, and anyone who takes this attitude cannot make friends with integrals and cannot, probably, make much further progress in this book.

On the contrary, the expression (3) is a perfectly wonderful number, perfectly well defined. It is defined by Eudoxus' method as the number that is the limit of approximating sums of the form

$$\sum_{k=1}^{n} f(t_k) \Delta t, \qquad n = 1, 2, 3, \ldots . \tag{4}$$

By taking n larger and larger, we specify a way of approximating the real number (3) to any desired number of decimal places. The only question is whether the limit of the approximating sums (4) exists, as n increases without bound. And here is where the importance of the fundamental existence theorem **F2** is seen. Because $f(t) = 1/t$ is differentiable, therefore *continuous*, on the interval $1 \leq t \leq 2$, the existence theorem assures us that the approximations in (4) do indeed tend to a limit.

The integral (3), with the help of **F2**, is seen to have built into itself a way to calculate the terms of its decimal expansion, to as many places as desired. And this can be done by pure arithmetic, without any knowledge of antiderivatives or of derivatives. One need only be a humble computer to plug a large value of n into (4) and to find that

$$\int_1^2 \frac{1}{t} dt = 0.693147 \ldots . \tag{5}$$

By the same token, a computer knowing nothing about calculus, but only about addition and multiplication, can discover that

$$\int_1^3 \frac{1}{t} dt = 1.09861 \ldots . \tag{6}$$

Integrals, if they exist, express numbers just as well as any other means of expressing numbers. The expression (3) is every bit as "definite" as the expression π.

Actually, integrals of the reciprocal function $1/t$ are of more than incidental interest, as we shall see in Chapter 9. Here are a few more results, given without proof. By plugging into the expression (4) a large value of n one can discover, for example, that

$$\int_1^4 \frac{1}{t} dt = 1.38629 \ldots, \tag{7}$$

$$\int_1^6 \frac{1}{t} dt = 1.79176 \ldots, \tag{8}$$

$$\int_1^{10} \frac{1}{t} dt = 2.30259 \ldots, \tag{9}$$

$$\int_1^{12} \frac{1}{t} dt = 2.48491 \ldots . \tag{10}$$

2. What is a Number?

The integrals (5) through (10), and all integrals of the form

$$\int_1^x \frac{1}{t} dt, \tag{11}$$

are of particular interest, as we shall see. It is important to realize that an expression such as (11), despite its complicated symbolism, simply stands for a certain real number whenever x is given a particular value. The number is perfectly definite, if $x > 1$, since it can be calculated to arbitrarily many decimal places by Eudoxus' method.

Exercises

2.1. The Greeks used only positive numbers. When was the number *zero* introduced into our number system? (Use an encyclopedia to find out.)

2.2. Criticize this statement: "Zero is nothing." *Answer*: Zero is a number, naming a certain point on the number line. *Nothing* is not a number. Zero is therefore not nothing.

2.3. What motivation do we have to enlarge our number system to include negative numbers? *Hint*. Think about the problem of naming points on the "left half" of the number line, or of measuring the value of overdrawn bank accounts, or of reading freezing temperatures on the Celsius scale.

2.4. What is the motivation for including in the real number system numbers that are not rational?

2.5. With our more sophisticated notion of *number* it can sometimes be quite a problem to tell when numbers are equal.
(a) Archimedes knew that $\sqrt{3}$ is irrational, and doubtless suspected that π is too. Attempting to find very close rational approximations, he found that $\sqrt{3}$ lies between $265/153$ and $1351/780$, while π lies between $3\frac{10}{71}$ and $3\frac{1}{7}$. Anybody can show that $\sqrt{2}$ lies between $\frac{1414}{1000}$ and $\frac{1415}{1000}$. Using the information here, and *without using the decimal system* (Archimedes did not), give a convincing argument to prove that $\sqrt{2} + \sqrt{3} \neq \pi$.
(b) It is true that

$$\frac{\sqrt{2}}{4}(1 + \sqrt{3}) = 0.9659\ldots,$$

and it is also true that

$$\frac{1}{2}\sqrt{2 + \sqrt{3}} = 0.9659\ldots.$$

On the basis of this information alone, can one infer that $(\sqrt{2}/4)(1 + \sqrt{3}) = \frac{1}{2}\sqrt{2 + \sqrt{3}}$? *Answer*: No. All we know from this is that the numbers are "equal" to four decimal places.
(c) Prove that the two numbers in part (b) that agree to four decimal places are in fact equal. *Hint*. There are a couple of ways of doing this. One way is to use the results of Examples 6 and 7 of Chapter 7.

2.6. When was the decimal system introduced into mathematics? (Use an encyclopedia find out.)

2.7. Let $L(x)$ be defined as the expression (11).
   (a) What is $L(2)$? *Answer*: By (5), $L(2) = 0.693147\ldots$.
   (b) What is $L(3)$?
   (c) What is $L(2) + L(3)$?
   (d) What is $L(6)$?
   (e) What is $L(3) + L(4)$?
   (f) What is $L(12)$?
   (g) Do we have enough information to conclude that $L(2) + L(3) = L(6)$ or that $L(3) + L(4) = L(12)$ or that $L(2) + L(2) = L(4)$? *Answer*: No. But we also do not know enough to prove they are *not* equal.

2.8. It can be shown by Eudoxus' method that

$$\int_1^{1024} \frac{1}{t}\, dt = 6.93147\ldots.$$

   (a) Explain how one might have *guessed* this. *Hint*. $1024 = 2^{10}$.
   (b) Make a *guess* (no proof called for) as to the value of each of the following integrals:

$$\int_1^{30} \frac{1}{t}\, dt, \qquad \int_1^{24} \frac{1}{t}\, dt, \qquad \int_1^{9} \left|\frac{1}{t}\right| dt.$$

Answer: (b) $3.4012\ldots, 3.1781\ldots, 2.1972\ldots.$

§3. Approximating $\int_a^b f$

In the preceding section it was emphasized that the number $\int_a^b f$ has built into itself a means by which it can be approximated to any desired degree of accuracy. As a practical matter, it is of interest to know how close we come by specifying a particular approximation.

EXAMPLE 1. The integral $\int_1^3 (1/t)\, dt$ exists by **F2**. Calculate the approximating sum

$$S_n = \sum_{k=1}^{n} f(t_k)\, \Delta t,$$

for (a) $n = 4$, (b) $n = 8$, and (c) $n = 16$.

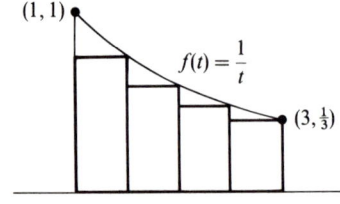

3. Approximating $\int_a^b f$

Note that for (a) we have $\Delta t = (3-1)/4 = \frac{1}{2}$, and therefore

$$S_4 = \sum_{k=1}^{4} \frac{1}{1+(k/2)}\left(\frac{1}{2}\right)$$

$$= \sum_{k=1}^{4} \frac{1}{2+k}$$

$$= \frac{1}{3} + \frac{1}{4} + \frac{1}{5} + \frac{1}{6} = 0.95.$$

For (b) we have $\Delta t = \frac{1}{4}$, and, when simplified,

$$S_8 = \sum_{k=1}^{8} \frac{1}{4+k} = \frac{1}{5} + \cdots + \frac{1}{12} = 1.020\ldots.$$

For (c) we have $\Delta t = \frac{1}{8}$, and, when simplified,

$$S_{16} = \sum_{k=1}^{16} \frac{1}{8+k} = \frac{1}{9} + \cdots + \frac{1}{24} = 1.058\ldots. \qquad \square$$

In Example 1, the approximating sums S_4, S_8, S_{16} are getting closer to the "right answer", which, by (6), is $1.09861\ldots$. *If we do not know the right answer already, how can we tell how close to it we come by calculating an approximating sum S_n?* This question motivates the following theorem, which tells us the maximum possible error we can make.

Theorem on Approximating Sums. *Suppose the curve f is falling on the domain $a \le t \le b$. Then the approximating sum S_n to the integral $\int_a^b f$ differs from the integral by no more than the quantity*

$$(f(a) - f(b))\Delta t. \tag{12}$$

That is [*since $\Delta t = (b-a)/n$*], *the number $\int_a^b f$ must satisfy the inequality*

$$S_n \le \int_a^b f \le S_n + \frac{(f(a)-f(b))(b-a)}{n}. \tag{13}$$

PROOF. A formal proof of this theorem could be given, but the reader will find it easier to understand this theorem by simply studying the picture below.

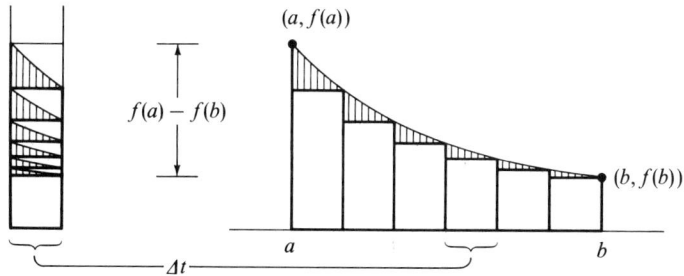

S_n is the area beneath the staircase with n steps

From the picture it is obvious that the shaded area is the difference between S_n and $\int_a^b f$. If each component of the shaded area is pushed horizontally over to the vertical axis on the left, it becomes obvious that the entire shaded area cannot exceed the area of the rectangle whose base is Δt and whose height is $f(a) - f(b)$. The error in the approximation S_n is then

$$\int_a^b f - S_n = \text{shaded area}$$
$$\leq \text{rectangle of base } \Delta t, \text{ and}$$
$$\text{height } f(a) - f(b)$$
$$= (f(a) - f(b))\Delta t.$$

This inequality shows that the quantity (12) exceeds the error in the approximation S_n. The inequality (13) is apparent when one adds the quantity S_n to each side of the inequality just above. □

The theorem just proved applies only to *falling* curves. (An analogous result holds for *rising* curves, but this is left to the reader as an exercise. If the curve is neither rising nor falling, then the considerations made above do not apply.)

Since the curve given by $f(t) = 1/t$ is falling on the domain $1 \leq t \leq 3$, we may apply our theorem to it. Using the results of Example 1, we see that the cases $n = 4$, $n = 8$, and $n = 16$ make the inequality (13) become, respectively,

$$0.95 \leq \int_1^3 \frac{1}{t} dt \leq 0.95 + 0.333,$$

$$1.02 \ldots \leq \int_1^3 \frac{1}{t} dt \leq 1.02 + 0.167,$$

$$1.058 \ldots \leq \int_1^3 \frac{1}{t} dt \leq 1.058 + 0.083.$$

Note that successive doublings of n (4, 8, 16) produce successive halvings (0.333, 0.167, 0.083) of the error estimation. We can thus get as close to the integral as we please, but we have to stop sometime of course. The last inequality above shows the integral between 1.058 and 1.141. A good *guess* would be the average of these, which is 1.099:

$$\int_1^3 \frac{1}{t} dt \approx 1.099.$$

This average will be slightly larger than the integral here, because the curve $y = 1/t$ is concave up. (*What does concavity have to do with this?*) Nevertheless, 1.099 is a satisfactory approximation to $1.09861 \ldots$, the number obtained by calculating S_n for very large n.

What we have just done makes it easy to give a proof of an existence theorem for falling curves.

Existence Theorem (Special Case). *Let f be a falling curve with domain $a \leq t \leq b$. Then $\int_a^b f$ exists.*

4. Definition of $\int_a^b f$ if $b \le a$

PROOF. The inequality (13) shows how to specify the number $\int_a^b f$ to any desired degree of accuracy. This is all that is required to assert that a real number exists. (Reread Section 2, where real numbers are discussed.) □

EXERCISES

3.1. The integral $\int_1^2 (1/t)\, dt$ exists by **F2**. Calculate the approximating sum for
   (a) $n = 4$.
   (b) $n = 8$.

3.2. What does inequality (13) become, for the integral $\int_1^2 (1/t)\, dt$ when
   (a) $n = 4$?
   (b) $n = 8$?

3.3. The integral $\int_0^1 (1/(1+t^2))\, dt$ exists by **F2**.
   (a) Calculate its approximating sum S_4.
   (b) What does inequality (13) become when $n = 4$?
   (c) Use your answer to part (b) to make a guess at an approximation to the number π.
   Answers: (b) $0.72029 \le \pi/4 \le 0.72029 + 0.1250$. (c) Taking the average of these bounds as an estimate for $\pi/4$, we might guess $\pi \approx 3.131$, which we expect (why?) to be low.

3.4. Prove that if the curve f is *rising* on the interval $a \le t \le b$, then

$$S_n - \frac{(f(b) - f(a))(b-a)}{n} \le \int_a^b f \le S_n.$$

3.5. Prove that if f is a rising curve, then $\int_a^b f$ exists.

3.6. The existence theorem **F2** asserts that if f is continuous on $a \le t \le b$, then $\int_a^b f$ exists. We have not given a proof of this, but we have established the existence of the integral for a rising curve and for a falling curve. Can you make an argument showing that the integral exists for a continuous curve that is sometimes rising and sometimes falling?

3.7. Give an example of a discontinuous function whose integral exists. *Hint*. Continuity was never used in the proofs of this section. Consult problem 3 at the end of this chapter.

§4. Definition of $\int_a^b f$ if $b \le a$.

Up to now, whenever we saw the expression

$$\int_a^b f, \qquad (14)$$

it was either explicitly stated, or else understood, that $a < b$. What should expression (14) be understood to mean if $b \le a$? We make the following definition. The motivation for making this definition will become clear on the pages that follow.

Definition. *The expression* (14) *is defined as*

(a) *the number calculated by Eudoxus' method if* $a < b$;
(b) *zero if* $a = b$;
(c) *the negative of* $\int_b^a f$ *if* $b < a$.

For example, $\int_1^1 (1/t)\,dt = 0$ by this definition, and

$$\int_1^0 \frac{1}{1+t^2}\,dt = -\int_0^1 \frac{1}{1+t^2}\,dt$$

$$= -[\arctan t]_0^1 = -\frac{\pi}{4}.$$

This definition may seem peculiar, but there is a pragmatic reason for making it. With this definition, the evaluation theorem **F3** is true without the restriction that $a < b$.

The Fundamental Theorem (Extended). *Let* f *be a continuous function defined on some interval containing the points* a *and* b. *Then, regardless of whether* $a < b$, *it is true that*

$$\int_a^b f = F\Big|_a^b \qquad \text{if } F' = f.$$

PROOF. We already know equality holds between $\int_a^b f$ and $F|_a^b$ when $a < b$. We must prove equality for the other two cases. If $a = b$, then the integral by definition is equal to 0, but so is $F|_a^a = F(a) - F(a)$. Thus equality holds if $a = b$.

The remaining case is $b < a$. By the definition just given, in this case we have

$$\int_a^b f = -\int_b^a f$$

$$= -[F]_b^a \quad \text{(by the original } \mathbf{F3}\text{)}$$

$$= -(F(a) - F(b))$$

$$= F\Big|_a^b.$$

Equality therefore holds if $b < a$ as well. □

Now that the expression (14) has a wider significance, it sometimes requires a little thought to determine for what values a and b the expression (14) makes sense.

EXAMPLE 2. Specify the values of x for which the following integrals exist.

(a) $\int_{1/2}^x \sin$.
(b) $\int_{1/4}^x \arcsin$.
(c) $\int_0^x \arctan$.

4. Definition of $\int_a^b f$ if $b \le a$

The integral (a) exists for all x, since the sine function is everywhere continuous. Another way of seeing this is to note that by the fundamental theorem, as just extended,

$$\int_{1/2}^{x} \sin = [-\cos]_{1/2}^{x} = -\cos x + \cos \frac{1}{2} \tag{15}$$

for all x.

The integral (b) does not exist for all x, since arcsin is continuous only on the domain $-1 \le x \le 1$. By **F2** the integral (b) exists on this domain. If we have no antiderivative of arcsin, we cannot give a simplified expression for (b), in contrast to the simplification of the integral (a) in equation (15).

By **F2** the integral (c) exists for all x, since arctan is continuous everywhere. □

In Example 2 each of the functions in question has a *name* (sin, arcsin, arctan). If the function has no name, but is specified by an algebraic rule, a dummy variable must be used. (Compare exercise 3.3 of Chapter 5.)

EXAMPLE 3. Specify the values of x for which the following integrals exist.

(a) $\int_1^x (1/t)\,dt$.
(b) $\int_{-1}^x (1/t)\,dt$.
(c) $\int_{1/2}^x \sqrt{4 - t^2}\,dt$.

Here, t is a dummy variable. The answer would be unaffected if t were replaced by any other variable (except in this case x, since x is already being used in another context as the upper limit of the integral). Since the algebraic rule $1/t$ has a discontinuity at the point $t = 0$ and nowhere else, the integral (a) exists on the domain $0 < x$, by **F2**.

The integral (b) exists on the domain $x < 0$. (*Why?*)

The square root function is continuous wherever it is defined, and it is defined only when the expression beneath the radical sign is not negative. The expression $4 - t^2$ is nonnegative when $-2 \le t \le 2$. Therefore the integral (c) exists by **F2** on the domain $-2 \le x \le 2$. [Note that the answer to (c) is *not* $-2 \le t \le 2$, because t is a dummy variable. The problem has to do with x, not t.] □

EXERCISES

4.1. Use the fundamental theorem, as extended in this section, to calculate each of the following integrals.
(a) $\int_2^0 t^2\,dt$.
(b) $\int_2^x t^2\,dt$.
(c) $\int_3^3 \sin^5 t\,dt$.
(d) $\int_3^1 \cos t\,dt$.
(e) $\int_2^0 s^2\,ds$.
(f) $\int_2^x s^2\,ds$.
(g) $\int_3^3 \sin^5 s\,ds$.
(h) $\int_3^1 \cos s\,ds$.

Answers: (a) $-\frac{8}{3}$. (b) $(x^3/3) - (8/3)$. (e) $-8/3$. (f) $(x^3/3) - (8/3)$.

4.2. Consider each of the following statements and tell whether it is true or false.
(a) $\int_2^x t^2 \, dt = \int_2^x s^2 \, ds = \int_2^x L^2 \, dL$.
(b) The derivative of $\int_{1/2}^x \sin t \, dt$ is $\sin x$. (See (15) for another way to express $\int_{1/2}^x \sin t \, dt$.)
(c) $\int_2^x t^2 \, dt$ is an antiderivative of the squaring function.
(d) The derivative of $\int_2^x t^2 \, dt$ is given by t^2.
Answers: (a), (b), and (c) are true. In (d) the derivative is x^2.

4.3. Specify the natural domains of the following. That is, specify in each case the values of x for which the expression makes sense.
(a) $L(x) = \int_1^x (1/t) \, dt$.
(b) $F(x) = \int_0^x (1/(t-1)(t+3)) \, dt$.
(c) $G(x) = \int_0^x \sqrt{(t+1)(t+3)} \, dt$.
Answers: (a) $0 < x$. (b) $-3 < x < 1$. (c) $-1 \leq x$.

4.4. Verify each of the following, using the extended definition of the integral.
(a) $(1/(b-a)) \int_a^b f = (1/(a-b)) \int_b^a f$ if $a < b$.
(b) $(1/(b-a)) \int_a^b f = (1/(a-b)) \int_b^a f$ if $a > b$.
(*This simple result will be surprisingly handy to know on several occasions in the pages ahead.*)

§5. Integrals as Averages

The remaining sections in this chapter deal with the uses of the integral. We first found integrals by considering how to measure areas, but integrals are of value to many people having no interest whatever in studying areas. That is why we have been emphasizing that the integral is a *number*. That number need not be an area. It might instead represent some distance traveled, some quantity of work performed, the magnitude of some force, or the possibility of occurrence of some momentous event. It might be a "moment of inertia", a "quadrature", or an "average". The house of integrals has many sides.

Perhaps the easiest way to enter this house is by considering the notion of an *average*. The average of several quantities is calculated by finding their sum and then dividing by the number of quantities. The average of two quantities has a simple geometric interpretation:

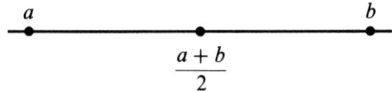

$$\frac{a+b}{2}$$

The average of a and b is halfway in between

What about the average of more than two quantities? Does it have a geometric interpretation? Consider the average of the first five squares 1, 4, 9, 16, and 25. The average is 11, since $(1 + 4 + 9 + 16 + 25)/5 = 11$. In a rather vague sense, the number 11 is the "center" of the numbers 1, 4, 9, 16, and 25.

5. Integrals as Averages

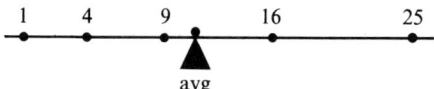

We can say a little more precisely what is meant by *center*. Think about the line above as representing a seesaw whose only weight is concentrated at the points 1, 4, 9, 16, and 25. If equal weights are placed at these points, then a fulcrum placed at the average value, 11, will make the seesaw balance. It will not tilt either way, with the fulcrum at its center. (Note that the center of a seesaw is not necessarily exactly in the middle. If, in the figure above, the fulcrum was put in the middle, at the point 13, the left side of the seesaw would go down and the right side up.)

There is a lot to be learned by playing on seesaws, as Archimedes was first to note. (See the appendix in this book on Archimedes.) Let us pause to play a moment.

Consider the n quantities $1, 2, 3, \ldots, n$. What is their average?

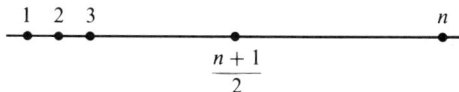

If we think of the seesaw we can guess right away their average, without doing any involved calculation. If equal weights are put at the points $1, 2, 3, \ldots, n$, where should the fulcrum be placed in order to make the seesaw balance? By the symmetry of the situation, it should obviously be placed in the middle, that is, halfway between 1 and n.

$$\frac{n+1}{2} = \text{avg of } 1, 2, 3, \ldots, n.$$

This is a case when we are able to guess the average without using the definition, which would entail adding the n quantities and then dividing by n:

$$\text{avg of } 1, 2, 3, \ldots, n = \frac{1}{n} \cdot \sum_{k=1}^{n} k.$$

Let us combine the preceding two equations to obtain

$$\frac{1}{n} \cdot \sum_{k=1}^{n} k = \frac{n+1}{2}.$$

Multiplying through by n yields

$$\sum_{k=1}^{n} k = n\left(\frac{n+1}{2}\right),$$

which is the familiar formula for the sum of the first n positive integers. We have just derived this old result in a new way, using the notion of an *average*.

What has just been done reveals a use for averages that is sometimes handy to remember. If you can somehow *guess* the average of a number of quantities, then, as above, you can quite easily calculate their sum.

An idea should begin to grow here, in the mind of a student of calculus. Sums are "like" integrals, since integrals are limits of sums. It might be handy to invent, in analogy with the foregoing, the notion of the average of a *function*, and see how it is related to the integral of the function.

To take a concrete example, let us consider what ought to be meant by the average value of the sine function, on the domain $0 \leq x \leq \pi$.

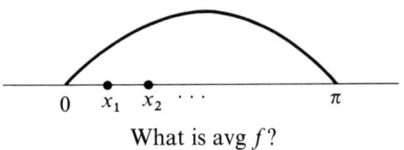

What is avg f?

If we partition the domain into n equal parts, as indicated, then the average value taken by the function f at these n points is, of course,

$$\frac{1}{n} \sum_{k=1}^{n} f(x_k). \tag{16}$$

This expression is the average value of f on the (discrete) set of points $x_1, x_2, \ldots, x_n$. It is natural to define the average of f on the (continuous) domain $0 \leq x \leq \pi$ as the limit of (16), as n increases without bound:

$$\operatorname{avg} f = \operatorname{Limit} \frac{1}{n} \sum_{k=1}^{n} f(x_k). \tag{17}$$

Equation (17) gives us a natural definition of avg f, as a limit of a certain sum. How is this related to the integral $\int_0^\pi f(x)\,dx$, which is also the limit of a sum? Applying Eudoxus' method to the interval $0 \leq x \leq \pi$, we get $\Delta x = \pi/n$, so that

$$\int_0^\pi f(x)\,dx = \operatorname{Limit} \sum_{k=1}^{n} f(x_k)\,\Delta x$$

$$= \operatorname{Limit} \sum_{k=1}^{n} f(x_k) \frac{\pi}{n}$$

$$= \pi \cdot \operatorname{Limit} \frac{1}{n} \sum_{k=1}^{n} f(x_k)$$

$$= \pi \cdot \operatorname{avg} f \quad [\text{by (17)}].$$

Since $\int_0^\pi f(x)\,dx = \pi \cdot \operatorname{avg} f$, it follows that

$$\operatorname{avg} f = \frac{1}{\pi} \int_0^\pi f(x)\,dx.$$

5. Integrals as Averages

Hence, for $f(x) = \sin x$, $0 \le x \le \pi$, we have

$$\operatorname{avg} f = \frac{1}{\pi} \int_0^\pi \sin x \, dx = \frac{1}{\pi}(-\cos x)\Big|_0^\pi = \frac{2}{\pi}.$$

The average value of the sine function, on the domain $0 \le x \le \pi$, is then $2/\pi$.

From considering the preceding example we can see the relation we were seeking to find, between the integral of f and the average of f, over a certain interval. The average of f is the integral of f, divided by the length of the interval. We are thus led naturally to this definition.

Definition. Let f be a continuous function on the interval $a \le x \le b$. The **average value** of f on this interval is denoted by $\operatorname{avg} f$ and defined by the equation

$$\operatorname{avg} f = \frac{1}{b-a} \int_a^b f(x) \, dx.$$

(Continuity of f is assumed, in order that the existence of the integral of f can be guaranteed by **F2**.)

EXAMPLE 4. The average value of the sine function on the interval from 0 to $\pi/3$ is given by

$$\frac{1}{\pi/3} \int_0^{\pi/3} \sin x \, dx = \frac{3}{\pi}[-\cos x]_0^{\pi/3} = \frac{3}{2\pi}. \qquad \square$$

EXAMPLE 5. The average value of the squaring function on the interval from 1 to 5 is given by

$$\frac{1}{5-1} \int_1^5 x^2 \, dx = \frac{1}{4}\frac{x^3}{3}\Big|_1^5 = \frac{31}{3} = 10.33\ldots. \qquad \square$$

(Why isn't the answer equal to 11, which, as we saw above, is the average of the first five squares 1, 4, 9, 16, and 25?)

EXAMPLE 6. If the speed of a motorcycle at time t is given by $v(t) = 40 + 3t^2 - 4t$ miles per hour, then the motorcycle's average speed, between $t = 1$ and $t = 3$, is given by

$$\frac{1}{3-1} \int_1^3 (40 + 3t^2 - 4t) \, dt = \frac{1}{2}[40t + t^3 - 2t^2]_1^3$$

$$= 45 \text{ miles per hour.} \qquad \square$$

EXAMPLE 7. If the height of a rock at time t is given in feet by $h(t) = -16t^2 + 96t$, then the rock's average height, between $t = 0$ and $t = 3$, is given by

$$\frac{1}{3-0} \int_0^3 (-16t^2 + 96t) \, dt = \frac{1}{3}\left[-16\frac{t^3}{3} + 48t^2\right]_0^3 = 96 \text{ feet.} \qquad \square$$

At the beginning of this section we saw how the notion of an average of a collection of quantities can sometimes be an aid to a quick calculation of their sum. Similarly, the notion of the average of a function can sometimes simplify the calculation of its integral. From the definition of avg f, the integral of f is given by

$$\int_a^b f(x)\,dx = (\operatorname{avg} f)(b - a). \tag{18}$$

For certain functions, over certain intervals, it is possible to guess their average value, and thus, by (18), obtain their integral with virtually no work at all. It is particularly easy for the function given by x^n, where n is odd, and where the interval is symmetric about 0 (i.e., has 0 as its midpoint). The average of such a function is obviously 0, a fact that can be seen quickly in several ways. One way is to observe that if $f(x) = x^n$, where n is odd, then

$$f(-x) = -f(x), \tag{19}$$

because $f(-x) = (-x)^n = ((-1)(x))^n = (-1)^n x^n = -x^n = -f(x)$. (*Where did we use the assumption that n is odd?*) From (19) it follows easily that avg f is 0 on an interval symmetric about 0:

$$\begin{array}{c} \overset{-x}{\bullet} \qquad \overset{x}{\bullet} \\ 0 \end{array}$$

$f(x)$ is the negative of $f(-x)$

In approximating avg f by sums like (16), we get cancellation between the value of f at a point x_k and the value of f at $-x_k$, from which it follows that avg f is 0.

Thus we know immediately that such integrals as $\int_{-1}^{1} x^3\,dx$, $\int_{-5}^{5} x^7\,dx$, $\int_{-a}^{a} x^5\,dx$ are all equal to 0 by (18), since in each case the average value of the integrand is 0. We are also allowed to perform such operations as

$$\int_{-3}^{5} x^3\,dx = \int_{-3}^{3} x^3\,dx + \int_{3}^{5} x^3\,dx = \int_{3}^{5} x^3\,dx,$$

$$\int_{-a}^{b} f(x)\,dx = \int_{-a}^{a} f(x)\,dx + \int_{a}^{b} f(x)\,dx = \int_{a}^{b} f(x)\,dx,$$

provided (19) holds.

It is convenient to have a name for functions f for which $f(-x) = -f(x)$. They are called **odd functions**. As we have just seen, avg $f = 0$ on a domain symmetric about 0 if f is an odd function. Besides x, x^3, x^5, etc., odd functions include sin [because $\sin(-x) = -\sin x$], tan [because $\tan(-x) = -\tan x$], and many others. This is sometimes very useful to know. For example, even though we may not as yet know antiderivatives of $\tan x$ or $x^3 \cos x$ or $\cos^2 2x \sin 3x$, we can nevertheless assert that the integrals

$$\int_{-1}^{1} \tan x\,dx, \qquad \int_{-4}^{4} x^3 \cos x\,dx, \qquad \int_{-3}^{3} \cos^2 2x \sin 3x\,dx$$

are all equal to 0, because they are integrals of odd functions over intervals symmetric about 0.

5. Integrals as Averages

Exercises

5.1. (a) How many integers are there between the integer 7 and the integer 10 (inclusive)?
  (b) How many integers are there between the integer m and the integer n (inclusive), assuming that n is greater than m?
  (c) Without adding up the integers from 7 to 10, guess their average. Then use your answer to 5.1(a) to deduce their sum.
  (d) Guess the average of the integers from m to n, inclusive. Then use your answer to 5.1(b) to deduce their sum.
  (e) Check your answer to 5.1(d) by writing
  $$\sum_{k=m}^{n} k = \sum_{k=1}^{n} k - \sum_{k=1}^{m-1} k,$$
  and applying the well-known rule for evaluating the two summations on the right-hand side.

5.2. In each of the following, find the average value of the function on the indicated interval.
  (a) $\sin x$, $0 \le x \le \pi/4$.
  (b) x^2, $0 \le x \le 3$.
  (c) x, $0 \le x \le 4$.
  (d) x^3, $-1 \le x \le 1$. Answer: 0, since x^3 is odd function.
  (e) x^3, $-1 \le x \le 2$.
  (f) $\cos x$, $\pi/2 \le x \le \pi$. Answer: $-2/\pi$.

5.3. A rock is thrown upwards from ground level at an initial speed of 64 feet per second. Find its average height over the first 2 seconds of its flight.

5.4. Which of the following are odd functions?
  (a) $f(x) = 2$.
  (b) $f(x) = 2x$.
  (c) $f(x) = 2\cos x$.
  (d) $f(x) = 2x \cos x$.
  (e) $f(x) = \sin(x^2)$.
  (f) $f(x) = \sin(x^3)$.
  (g) $f(x) = 2x + 2$.
  (h) $f(x) = \sin 2x \sin 3x$.

5.5. A function f that satisfies the equation $f(-x) = f(x)$ for all x is called an **even function**. Which of the functions in exercise 5.4 are even functions?

5.6. Give an example of a function that is neither odd nor even. *Hint.* There is one in exercise 5.4.

5.7. Assume that f is an even function that is continuous.
  (a) Show that avg f on $-a \le x \le 0$ is equal to avg f on $0 \le x \le a$.
  (b) Show, using 5.7(a), that $\int_{-a}^{0} f(x)\,dx = \int_{0}^{a} f(x)\,dx$.
  (c) Show, using 5.7(b), that $\int_{-a}^{a} f(x)\,dx = 2\int_{0}^{a} f(x)\,dx$.

5.8. Without evaluating any of the following integrals, give reasons for justifying the following equalities. Use the facts you have learned about odd and even functions.
  (a) $\int_{-1}^{1}(x^5 + x^4 + x^3 + x^2 + x + 1)\,dx = 2\int_{0}^{1}(x^4 + x^2 + 1)\,dx$.
  (b) $\int_{-r}^{r}(1 - x^2)\,dx = 2\int_{0}^{r}(1 - x^2)\,dx$.
  (c) $\int_{-2}^{4}(\sin x + x)\,dx = \int_{2}^{4}(\sin x + x)\,dx$.
  (d) $\int_{-1}^{1}(x^2 \tan x \cos x)\,dx = 0$.

5.9. Consider each of the following conjectures. If it is true, give a reason why; if false, specify a concrete example where the conjecture fails.
  (a) The sum of two odd functions is itself an odd function.
  (b) The product of two odd functions is itself an even function.
  (c) The product of even functions is an even function.
  (d) The sum of even functions is even.
  (e) If f is not an even function, then f must be an odd function.
  (f) The graph of any even function is "symmetric about the vertical axis".
  (g) If an odd function is continuous at 0, then it takes the value 0 at 0.
  (h) The graph of any odd function is "symmetric with respect to the origin".
  (i) If f is any function whatever, and a function g is defined by $g(x) = f(x) + f(-x)$, then g is an even function.
  (j) If f is any function whatever, and a function h is defined by $h(x) = f(x) - f(-x)$, then h is an odd function.
  (k) Any function whatever can be expressed as the sum of an even function and an odd function.

5.10. In Chapter 5 we defined average speed as the quotient of the distance traveled by the time it took. In this section the average speed was defined as the integral of the speed, divided by the length of the time interval in question. Do these definitions agree?

5.11. On the interval $a \le x \le b$, find the average value of
  (a) $f(x) = C$, where C is a constant.
  (b) $f(x) = x$.
  (c) $f(x) = x^2$.
  (d) $f(x) = x^3$.
  Hints. (c) $b^3 - a^3 = (b-a)(b^2 + ab + a^2)$. (d) $b^4 - a^4 = (b-a)(b^3 + b^2 a + ba^2 + a^3)$

5.12. Frequently we need to speak of the average of a function over more than one interval. To indicate that the interval in question is from a to b, we write
$$\operatorname*{avg}_{a,b} f.$$
  (a) Find $\operatorname{avg}_{0,1}(x^2)$. Answer: $(1/(1-0)) \int_0^1 x^2\, dx = 1/3$.
  (b) Find $\operatorname{avg}_{1,2}(x^2)$. Answer: $(1/(2-1)) \int_1^2 x^2\, dx = 7/3$.
  (c) Find $\operatorname{avg}_{0,2}(x^2)$.
  (d) Is it true that $\operatorname{avg}_{0,2}(x^2) = \operatorname{avg}_{0,1}(x^2) + \operatorname{avg}_{1,2}(x^2)$?

5.13. Let f be a continuous function. Consider each of the following conjectures, and tell whether it is true or false.
  (a) $\operatorname{avg}_{a,b}(cf) = c \cdot \operatorname{avg}_{a,b} f$.
  (b) $\operatorname{avg}_{a,b}(f+g) = \operatorname{avg}_{a,b} f + \operatorname{avg}_{a,b} g$, where g is continuous.
  (c) $\operatorname{avg}_{a,b}(f^2) = (\operatorname{avg}_{a,b} f)^2$.
  (d) $\operatorname{avg}_{a,c} f = \operatorname{avg}_{a,b} f + \operatorname{avg}_{b,c} f$.
  Hint. (d) See 5.12(d).

5.14. Just as we have found it useful to define $\int_a^b f$ when $b \le a$, so it will be useful to define $\operatorname{avg}_{a,b} f$ when $b \le a$. The natural **definition** is as follows:
  (a) $\operatorname{avg}_{a,a} f = f(a)$.
  (b) $\operatorname{avg}_{a,b} f = \operatorname{avg}_{b,a} f$ if $b < a$.
  Use the result of exercise 4.4 to prove that $(1/(b-a)) \int_a^b f$ is equal to $\operatorname{avg}_{a,b} f$, even if $b < a$.

§6. Simpson's Rule

A moral of the preceding section is that if you can somehow guess the average of a function, then you can easily find its integral. For complicated functions we cannot hope to guess correctly. It turns out, however, that for simple functions like quadratics or cubics, it is remarkably simple to find averages.

Let us begin with something familiar. Suppose a teacher is about to assign a grade to a student who has made scores of 70, 90, and 80 on three papers. If the three papers are of equal scope and size, the teacher would probably average the three grades to get a cumulative grade of 80. Suppose, however, that the second paper was the result of a project that was far more ambitious than the first and third assignments, which were of ordinary difficulty. In this case it would be reasonable to count the second grade with more weight than the first and third.

Suppose it was decided to give the second paper four times the weight of the first, and to give the third paper the same weight as the first. Then the teacher would take not an ordinary average of the numbers 70, 90, and 80, but a "one-four-one" *weighted* average:

$$\frac{70(1) + 90(4) + 80(1)}{6} = 85.$$

(We divide by 6, since 6 is the sum of the weights 1, 4, and 1.)

A one-four-one weighted average of three numbers, as first observed by the eighteenth-century English mathematician Simpson, just happens to be exactly what we need in our present endeavor. Simpson found, perhaps simply by the method of trial and error, a simple one-four-one rule for writing down quickly the average of any quadratic or cubic.

Consider the problem of finding the average of the function given by $f(x) = x^2$ on the interval $1 \leq x \leq 3$.

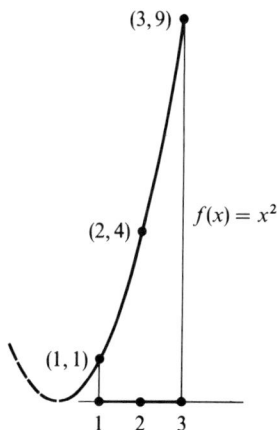

Notice that this function takes the value 1 at the left-hand endpoint, 4 at the midpoint, and 9 at the right-hand endpoint:

x	y
1	1
2	4
3	9

The ordinary average of these values of 1, 4, and 9 is 14/3, which will be a fair approximation to $\text{avg } f$. However, Simpson realized that—at least for quadratics and cubics, and perhaps other functions as well—the value of the function at the *midpoint* of the interval in question has more weight in determining the average of the function than the function's values at the endpoints of the interval. Consider a one-four-one weighted average of the values 1, 4, and 9:

$$\frac{1(1) + 4(4) + 9(1)}{6} = \frac{13}{3}.$$

The average of the function is, by definition,

$$\text{avg } f = \frac{1}{3-1} \int_1^3 x^2 \, dx = \left.\frac{x^3}{6}\right|_1^3 = \frac{13}{3}.$$

Is it just an accident that $\text{avg } f$ is equal to a one-four-one weighted average of the values of f at the left, center, and right of the interval in question?

Let us use the abbreviation

$$\text{smp } f, \quad \text{or} \quad \operatorname*{smp}_{a,b} f,$$

to denote the application of Simpson's one-four-one rule to f on the interval $a \leq x \leq b$:

$$\text{smp } f = \operatorname*{smp}_{a,b} f = \frac{f(a) + 4f((a+b)/2) + f(b)}{6}. \tag{20}$$

We want to compare this with

$$\text{avg } f = \operatorname*{avg}_{a,b} f = \frac{1}{b-a} \int_a^b f(x) \, dx.$$

We have just seen that $\text{smp } f$ is equal to $\text{avg } f$ for $f(x) = x^2$ if $a = 1$ and $b = 3$:

$$\operatorname*{smp}_{1,3} x^2 = \frac{13}{3} = \operatorname*{avg}_{1,3} x^2.$$

EXAMPLE 8. Calculate $\text{smp } f$ for $f(t) = -16t^2 + 96t$ on the interval from 0 to 3, and compare $\text{smp } f$ with $\text{avg } f$.

6. Simpson's Rule

Here, the midpoint is $\frac{3}{2}$, and we have the following table of values.

t	$f(t)$
0	0
$\frac{3}{2}$	108
3	144

Simpson's one-four-one rule gives

$$\operatorname*{smp}_{0,3}(-16t^2+96t) = \frac{0+4(108)+144}{6} = 96.$$

In Example 7 of the preceding section, we found that

$$\operatorname*{avg}_{0,3}(-16t^2+96t) = 96. \qquad \square$$

EXAMPLE 9. Calculate smp f for $f(x) = \sin x$ on the interval from 0 to $\pi/3$, and compare smp f with avg f.

Here we have the table of values as indicated,

x	$f(x)$
0	0
$\pi/6$	$\frac{1}{2}$
$\pi/3$	$\sqrt{3}/2$

and

$$\operatorname*{smp}_{0,\pi/3}(\sin x) = \frac{0+4(\frac{1}{2})+(\sqrt{3}/2)}{6} = \frac{4+\sqrt{3}}{12}, \qquad (21)$$

From Example 4 of the preceding section, we have

$$\operatorname*{avg}_{0,\pi/3}(\sin x) = \frac{3}{2\pi}. \qquad \square \quad (22)$$

EXAMPLE 10. Calculate smp f for $f(x) = x^4$ on the interval from 0 to 4, and compare smp f with avg f.

x	x^4
0	0
2	16
4	256

From the table, we calculate

$$\operatorname*{smp}_{0,4} x^4 = \frac{0+4(16)+256}{6} = 53\frac{1}{3}, \qquad (23)$$

whereas

$$\operatorname*{avg}_{0,4} x^4 = \frac{1}{4}\int_0^4 x^4\,dx = \frac{1}{4}\cdot\frac{x^5}{5}\Big|_0^4 = 51\frac{1}{5}. \qquad \square$$

From these examples we are led to expect close agreement, but not necessarily equality, between smp f and avg f. We do get equality, however, if f is a linear, quadratic, or cubic function.

Proposition on Exactness in Simpson's Rule. *It is true that* smp $f =$ avg f *if f is given by*

(a) $f(x) = C$ (constant function).
(b) $f(x) = x$.
(c) $f(x) = x^2$.
(d) $f(x) = x^3$.

PROOF. We shall prove part (d), leaving the rest to the reader.

x	x^3
a	a^3
$(a+b)/2$	$(a+b)^3/8$
b	b^3

If $f(x) = x^3$, then

$$\operatorname*{smp}_{a,b} f = \frac{a^3 + 4((a+b)/2)^3 + b^3}{6}$$

$$= \frac{1}{6}\left(a^3 + 4\left(\frac{a^3 + 3a^2b + 3ab^2 + b^3}{8}\right) + b^3\right)$$

$$= \frac{1}{6}\left(\frac{12a^3 + 12a^2b + 12ab^2 + 12b^3}{8}\right)$$

$$= \frac{1}{4}(a^3 + a^2b + ab^2 + b^3)$$

$$= \operatorname*{avg}_{a,b} f \quad \text{[by exercise 5.11(d)]}.$$

This proves part (d) of the proposition. The other parts are easier and are left to the reader to verify. $\qquad \square$

It follows from the proposition just proved, as the reader will show in an exercise to follow, that

$$\operatorname{smp} f = \operatorname{avg} f \quad \text{if } f(x) = ax^3 + bx^2 + cx + d.$$

6. Simpson's Rule

Thus, one can calculate exactly any integral of any cubic function by first using Simpson's rule to find the average value of the cubic. This is sometimes a great time-saver. Even for more general functions Simpson's rule is a great aid in guessing a quick *approximation* to an integral, because smp f is generally a good approximation to avg f. Frequently one has neither the time nor the patience to carry out Eudoxus' method to obtain a value of an integral that is correct to many decimal places. Let us introduce some notation to indicate the approximation to an integral that Simpson would make. Let

$$S_a^b f$$

denote this approximation to $\int_a^b f$. That is,

$$S_a^b f = (b-a) \smash{\smash p_{a,b}} f \tag{24}$$

$$\approx (b-a) \operatorname*{avg}_{a,b} f \quad (\text{since smp } f \approx \text{avg } f)$$

$$= \int_a^b f \quad [\text{by (18)}].$$

When one is unable to use the fundamental evaluation theorem **F3** for want of an antiderivative, Simpson's rule to calculate $S_a^b f$ is a welcome relief. Though $S_a^b f$ will not in general be equal to $\int_a^b f$, we can expect to get approximate equality between these two numbers, as shown above. Let us look at a few examples to see how close they can come.

EXAMPLE 11. Estimate the integral $\int_1^2 (1/x)\,dx$ by Simpson's rule.
Here we have

$$\operatorname*{smp}_{1,2}\left(\frac{1}{x}\right) = \frac{1 + 4(\frac{2}{3}) + \frac{1}{2}}{6} = \frac{25}{36} = 0.694\ldots,$$

so that by (24),

$$S_1^2\left(\frac{1}{x}\right) = (2-1)\left(\frac{25}{36}\right) = 0.694\ldots,$$

whereas, from equation (5),

$$\int_1^2 \left(\frac{1}{x}\right) dx = 0.693\ldots. \qquad \square$$

EXAMPLE 12. Estimate the integral $\int_1^3 (1/x)\,dx$ by Simpson's rule.
Here we have

$$\operatorname*{smp}_{1,3}\left(\frac{1}{x}\right) = \frac{1 + 4(\frac{1}{2}) + \frac{1}{3}}{6} = \frac{5}{9},$$

so that
$$S_1^3\left(\frac{1}{x}\right) = (3-1)\left(\frac{5}{9}\right) = 1.11\ldots,$$

whereas, from equation (6),
$$\int_1^3 \frac{1}{x}dx = 1.099\ldots. \qquad \square$$

EXAMPLE 13. Estimate the integral $\int_1^{12}(1/x)\,dx$ by Simpson's rule. Here we have
$$\operatorname*{smp}_{1,12}\left(\frac{1}{x}\right) = \frac{1 + 4(2/13) + (1/12)}{6} = \frac{265}{396},$$
so that
$$S_1^{12}\left(\frac{1}{x}\right) = (12-1)\left(\frac{265}{936}\right) = 3.1\ldots,$$

whereas, from equation (10),
$$\int_1^{12} \frac{1}{x}dx = 2.5\ldots. \qquad \square$$

These examples suggest that Simpson's rule may give a remarkably close estimation of an integral $\int_a^b f$, provided that the length of the interval from a to b is small, but that $S_a^b f$ may be somewhat different from $\int_a^b f$ when the length $b - a$ is large. One would expect, for example, to get a better approximation to $\int_1^{12}(1/x)\,dx$ by repeated applications of Simpson's rule to smaller intervals, say, e.g.,

$$\int_1^{12} \frac{1}{x}dx = \int_1^3 \frac{1}{x}dx + \int_3^6 \frac{1}{x}dx + \int_6^{12} \frac{1}{x}dx$$

$$\approx S_1^3\left(\frac{1}{x}\right) + S_3^6\left(\frac{1}{x}\right) + S_6^{12}\left(\frac{1}{x}\right)$$

$$\approx 1.11 + ? + ? \qquad (25)$$

[The reader is asked to complete the calculation and see if the result is a better approximation to $\int_1^{12}(1/x)\,dx$ than the approximation $S_1^{12}(1/x)$ in Example 13.]

Consider the problem of estimating the number π to several decimal places. Remember that π is defined as the area of the unit circle, so that, by definition, π is equal to

$$2\int_{-1}^{1} \sqrt{1-x^2}\,dx. \qquad (26)$$

6. Simpson's Rule

By Simpson's rule, one then has it that π is approximately equal to

$$2 \cdot S^1_{-1}(\sqrt{1-x^2}) = 2(1-(-1))\left(\frac{2}{3}\right) = 2.67\ldots,$$

which is not a very good approximation, partly because the interval is fairly large, being of length 2, but mainly because $\sqrt{1-x^2}$ is quite unlike any cubic equation, since it has vertical tangents at -1 and at 1. (Roughly speaking, one expects more error with Simpson's estimation, the more unlike a cubic is the curve in question.) We would expect to improve upon our estimation of (26) by noting that the integrand is an even function, so that (26) is equal to

$$4\int_0^1 \sqrt{1-x^2}\, dx \approx 4 \cdot S^1_0(\sqrt{1-x^2})$$

$$= 4(1-0)\left(\frac{1+4(\sqrt{3}/2)+0}{6}\right)$$

$$= \frac{2+4\sqrt{3}}{3} = 2.97\ldots.$$

This is a better approximation to π, because the interval is half as long as before, but the approximation is still not very satisfactory.

Let us be more cunning. Since the sine function is so nice, and since the interval 0 to $\pi/3$ is not too long, we have grounds for supposing that the values in (21) and (22) are quite close. That is,

$$\frac{4+\sqrt{3}}{12} \approx \frac{3}{2\pi}.$$

From this is follows that

$$\pi \approx \frac{18}{4+\sqrt{3}}$$

$$= \frac{72-18\sqrt{3}}{13} \quad \text{(multiplying top and bottom by } 4-\sqrt{3}\text{)}$$

$$\approx \frac{72-18(1.732)}{13}$$

$$= 3.140\ldots.$$

By luck we have hit upon a very good approximation to π, though it is still not so good as that of Archimedes. The wily Syracusan, by delicate estimates with a 96-sided polygon approximating the unit circle, was able to show that π lies between $3\frac{10}{71}$ and $3\frac{1}{7}$. In decimal notation (Archimedes did not use decimal notation, of course), this means that π is greater than 3.1408 and less than 3.1428. We now know that π is irrational, and we have estimates of π to a million decimal places. For over a thousand years, however, Archimedes' estimate was about the best we knew.

Exercises

6.1. Apply Simpson's rule, as indicated.
   (a) $smp_{0,1}(1/(1+x^2))$.
   (b) $smp_{0,\pi/3}(\cos x)$.
   (c) $smp_{0,2}(x^4)$.
   (d) $smp_{2,4}(x^4)$.

6.2. With the help of your answers to exercise 6.1, find $S_0^1(1/(1+x^2))$, $S_0^{\pi/3}(\cos x)$, $S_0^2(x^4)$, and $S_2^4(x^4)$. Then calculate, by the fundamental theorem **F3**, the integrals $\int_0^1(1/(1+x^2))$, $\int_0^{\pi/3}(\cos)$, $\int_0^2(x^4)$, and $\int_2^4(x^4)$, and compare them to their approximations by Simpson's rule.

6.3. In the proposition of this section, only part (d) was proved. Give proofs of parts (a), (b), and (c).

6.4. Prove that Simpson's rule shares two properties with the integral, namely,
   (a) $S_a^b(cf) = c \cdot S_a^b(f)$.
   (b) $S_a^b(f+g) = S_a^b(f) + S_a^b(g)$.

6.5. Using the proposition of this section, together with the properties (a) and (b) of exercise 6.4, and *without using* **F3**, prove that $S_a^b(5x^2 + 3x) = \int_a^b (5x^2 + 3x)$. Answer:

$$S_a^b(5x^2 + 3x) = S_a^b(5x^2) + S_a^b(3x) \quad [\text{by 6.4(b)}]$$
$$= 5 \cdot S_a^b(x^2) + 3 \cdot S_a^b(x) \quad [\text{by 6.4(a)}]$$
$$= 5\int_a^b (x^2) + 3\int_a^b (x) \quad [\text{by parts (b) and (c) of the proposition}]$$
$$= \int_a^b (5x^2 + 3x).$$

6.6. Using the proposition of this section, together with the properties (a) and (b) of exercise 6.4, and *without using* **F3**, prove that

$$S_a^b(7x^3 + 5x^2 + 3x + 1) = \int_a^b (7x^3 + 5x^2 + 3x + 1).$$

(See the solution of the preceding problem.)

6.7. Prove the obvious generalization of the result of exercise 6.6:

   Theorem. *Let f be any cubic function. Then,*

$$S_a^b(f) = \int_a^b f.$$

6.8. Evaluate the integral $\int_{-1}^{3}(7x^3 + 5x^2 + 3x + 1)dx$
   (a) by the fundamental evaluation theorem **F3**.
   (b) by evaluating $S_{-4}^3(7x^3 + 5x^2 + 3x + 1)$ instead and applying the preceding theorem.
   Which method is shorter?

6.9. The integral has the property that if f is a continuous function, then $\int_a^c f = \int_a^b f + \int_b^c f$. Does Simpson's rule have this property? That is, is it true that $S_a^c(f) = S_a^b(f) + S_b^c(f)$? *Hint.* Test the case when $f(x) = x^4$, $a = 0$, $b = 2$, and $c = 4$. Much of your work has already been done in equation (23) and in parts (c) and (d) of exercise 6.1.

6.10. Complete the calculation left unfinished in (25).

6.11. By carrying out the following steps, find an approximation to π. (Carry all your numbers out to at least four decimal places. Use a hand calculator, if available.)
(a) Use **F3** to find $\int_{\pi/6}^{\pi/3} \cos x\, dx$ to several decimal places, using 1.732051 as an approximation to $\sqrt{3}$.
(b) Find $\operatorname{smp}_{\pi/6, \pi/3}(\cos)$ to several decimal places, utilizing decimal approximations to $\sqrt{2}$ and $\sqrt{3}$. ($\sqrt{2} = 1.414214\ldots$.)
(c) Because the interval from $\pi/6$ to $\pi/3$ is rather short, being of length $\pi/6$, and because the cosine function is so nice, we can expect a very close approximation between the integral and its estimation by Simpson's rule:
$$\int_{\pi/6}^{\pi/3} \cos x\, dx \approx S_{\pi/6}^{\pi/3}(\cos) = (\pi/6) \operatorname*{smp}_{\pi/6, \pi/3}(\cos).$$
For the left-hand side of this approximation, substitute your answer to part (a), and substitute your answer to part (b) appropriately on the right. Then deduce an approximation to π.

6.12. Suppose f is continuous at the point a.
(a) What is the limit, as b tends to a, of $f(b)$?
(b) What is the limit, as b tends to a, of $f((a+b)/2)$?
(c) Use your answers to parts (a) and (b) to prove that
$$\operatorname*{Limit}_{b \to a} \left(\operatorname*{smp}_{a,b} f \right) = f(a).$$
(*Begin with the definition given in equation (20).*)
(d) What would you guess is
$$\operatorname*{Limit}_{b \to a} \left(\operatorname*{avg}_{a,b} f \right) ?$$

§7. Quadratures, Mean-Value Theorems, and L'Hôpital's Rule

Mean is another word for *average* and, being shorter, is more often used in mathematics books. Here we shall see some more reasons why the mean (or average) value of a function is of importance. The first reason is that it helps to solve, with ridiculous ease, the problem of *quadratures*.

The problem of quadratures is an ancient problem, dating back to the Greeks. What is this problem? The root *quadra-* means "four", of course, but when the Pythagoreans thought of "four", they thought of

• •

• •

which is a *square*. (That is why a quadratic equation is not an equation of fourth degree, as one might sensibly guess, but rather an equation involving a square.) The problem of quadratures is the problem of exhibiting a square of the same area as a given figure. To construct such a square is to *effect* (*or perform*) *a quadrature* of the figure.

It is trivial to perform a quadrature of a rectangle or a triangle. A quadrature of the rectangle

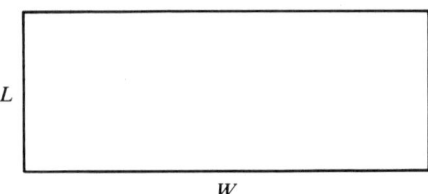

is given by a square whose sides have length $\sqrt{LW}$. (In an exercise to follow, the reader is asked to perform a quadrature of a triangle.) To solve the problem of quadratures, therefore, one need only exhibit a rectangle (or triangle) of the same size as the given figure, since quadratures can readily be performed upon rectangles (or triangles). Thus, in the fifth century B.C., when Hippocrates showed that the indicated "lune" was of the same area as the triangle, he had effectively performed a quadrature of the lune.

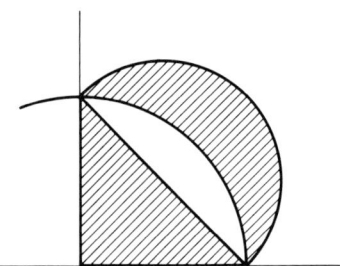

(In a problem at the end of Chapter 2 is sketched a way to see that the lune and the triangle have equal area.)

An outstanding problem was the quadrature of the circle. Archimedes showed (see the problem set at the end of Chapter 2) that a circle of radius r and circumference C is equal in area to a triangle of height r and base C. Archimedes thus was able to perform a quadrature of the circle, assuming that a straight line equal in length to the circumference of the circle could be constructed. This point bears further discussion, but here is not the place for it. The interested reader may consult the appendix on Archimedes.

The problem of quadratures is today handily solved by integrals, together with the notion of the mean, or average, value of a function. Since

$$\int_a^b f(x)\,dx = (b-a) \operatorname*{avg}_{a,b} f,$$

it follows that the area beneath the curve f from a to b is equal to the area of a rectangle of base $b - a$ and height $\operatorname{avg}_{a,b} f$.

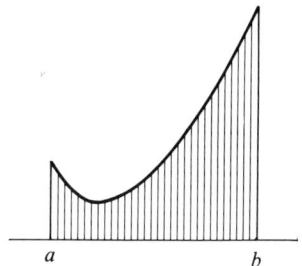
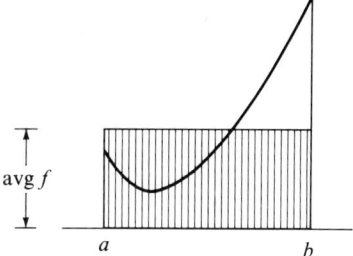

This solves the problem of quadratures for any figure that is the area beneath the graph of any continuous function. A rectangle with the same base but with height equal to the mean value of the function does the trick. Since it is easy to do, we could convert the rectangle into a square of equal size, but for the purposes of the rest of this section, it is better not to do it.

It is better to ponder the figure below, instead, and to see something in it that will prove invaluable. The question to think about is, *what happens as b is taken closer to a?*

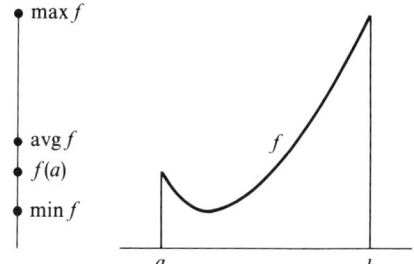
The average, minimum, and maximum of f

To see what happens, note first that avg f must be between the maximum and minimum values of f. This is obvious from the definition of *average*. Thus we have a very useful inequality:

$$\min_{a,b} f \leq \operatorname*{avg}_{a,b} f \leq \max_{a,b} f. \tag{27}$$

What happens to the terms in this inequality as b tends to a? Assuming that f is continuous at a, we can answer this question easily. The maximum and the minimum of f must get closer to each other, and both must get closer to $f(a)$. Thus, as b tends to a, $\operatorname{avg}_{a,b} f$ must tend to $f(a)$, since $\operatorname{avg}_{a,b} f$ is squeezed between two quantities, each of which tends to $f(a)$:

$$\operatorname*{Limit}_{b \to a} \operatorname*{avg}_{a,b} f = f(a) \quad \text{if } f \text{ is continuous at } a. \tag{28}$$

The statement (28) is of great value, as we shall see in the next section. The reader who understands the notion of *continuity* should see that statement (28) is almost obvious. It simply says, in rough terms, that when b is close to a, then the average of f from a to b is expected to be close to $f(a)$. For example, since the cosine function is continuous,

$$\lim_{x \to 0} \operatorname*{avg}_{0,x} (\cos) = \cos 0 = 1. \tag{29}$$

The full importance of statement (28) will be seen in the next section, but the value of it can be indicated right here. We can also illustrate once again the value of doing things backwards.

EXAMPLE 14. Find the limit, as x tends to 0, of $(\sin 2x)/(\sin 3x)$.

The trouble is that both numerator and denominator tend to 0. (This is an example of the so-called indeterminate form $0/0$, which frequently arises.) The fundamental theorem **F3**, used backwards (!), says

$$f(x) - f(0) = \int_0^x f'. \tag{30}$$

When this is applied to $f(x) = \sin 2x$, we have $f(0) = 0$ and $f'(x) = 2\cos 2x$, so that (30) becomes

$$\sin 2x = \int_0^x 2\cos 2t\, dt$$

$$= 2(x - 0) \operatorname*{avg}_{0,x} (\cos 2t).$$

When **F3** is applied backwards to $f(x) = \sin 3x$, then (30) becomes

$$\sin 3x = 3(x - 0) \operatorname*{avg}_{0,x} (\cos 3t).$$

and we can express the quotient we are investigating in terms of mean values:

$$\frac{\sin 2x}{\sin 3x} = \frac{2x \operatorname{avg}_{0,x}(\cos 2t)}{3x \operatorname{avg}_{0,x}(\cos 3t)} = \frac{2}{3} \cdot \frac{\operatorname{avg}_{0,x}(\cos 2t)}{\operatorname{avg}_{0,x}(\cos 3t)} \quad \text{if } x \neq 0;$$

and it is now obvious, with the aid of (28), what the limit is, as x tends to 0. It is

$$\frac{2 \cos 0}{3 \cos 0} = \frac{2}{3},$$

since $\cos 2t$ and $\cos 3t$ are continuous at 0. Hence, $\frac{2}{3}$ is the limit, as x tend to 0, of the quotient $(\sin 2x)/(\sin 3x)$. □

The key to investigating the indeterminate form $0/0$ was to rewrite the quotient in terms of average values. This approach, with the aid of (28), often works, though not always. A shortcut to this approach is to remember this equality: if $f(a) = 0$ and $g(a) = 0$ (so that f/g becomes indeterminate at the point a), then

$$\frac{f(x)}{g(x)} = \frac{\operatorname{avg}_{a,x} f'}{\operatorname{avg}_{a,x} g'} \quad \text{if } x \neq a. \tag{31}$$

7. Quadratures, Mean-Value Theorems, and L'Hôpital's Rule   277

Thus, by (28) we know that the numerator tends to $f'(a)$ and the denominator tends to $g'(a)$, assuming that f' and g' are each continuous at the point a, and we have proved the following rule.

L'Hôpital's Rule. *If the quotient f/g becomes the indeterminate form $\frac{0}{0}$ at the point a, then*

$$\underset{x \to a}{\text{Limit}} \frac{f(x)}{g(x)} = \frac{f'(a)}{g'(a)},$$

provided f' and g' are continuous and $g'(a) \neq 0$.

[The rule should really be named for its discoverer, the Swiss mathematician J. Bernoulli (1667–1748), but it very early became associated with the name of his student L'Hôpital, who wrote one of the first texts on the calculus.] The rule, as stated above, does not work if $g'(a) = 0$. (*Why not?*) A more general rule, to handle this case, is derived in the exercises to follow.

EXAMPLE 15. Find the limit, as x tends to 1, of $(x^4 - x)/(x^3 - x)$.

Solution. When $x = 1$, the quotient becomes the indeterminate form $0/0$. By L'Hôpital's rule, the limit at 1 is equal to

$$\frac{f'(1)}{g'(1)},$$

provided the denominator is not zero. Here we have $f(x) = x^4 - x$ and $g(x) = x^3 - x$, so that

$$\frac{f'(x)}{g'(x)} = \frac{4x^3 - 1}{3x^2 - 1}, \qquad (32)$$

which is equal to $\frac{3}{2}$ when x is 1. The limit is then $\frac{3}{2}$. □

EXAMPLE 16. Find the limit, as x tends to 0, of $(x^4 - x)/(x^3 - x)$.

Solution. By (32), $f'(0)/g'(0) = 1$. The limit is 1, by L'Hôpital's rule. □

"NONEXAMPLE." Find the limit, as x tends to 2, of $(x^4 - x)/(x^3 - x)$.

Solution. When $x = 2$, the quotient does *not* become an indeterminate form. It becomes $(2^4 - 2)/(2^3 - 2) = \frac{7}{3}$ as x tends to 2. L'Hôpital's rule is not applicable, for the quotient is continuous at the point 2. By continuity, its limiting value is equal to the value it takes, namely $\frac{7}{3}$. □

EXAMPLE 17. Find the limit, as x tends to 0, of the quotient $f(x)/g(x) = (1 - \cos x)/x^2$.

Solution. At the point 0, the quotient $f(x)/g(x)$ becomes the indeterminate form $0/0$, so we look at the quotient $f'(x)/g'(x)$, in hopes of applying L'

Hôpital's rule. But,

$$\frac{f'(x)}{g'(x)} = \frac{\sin x}{2x}, \qquad (33)$$

which also becomes an indeterminate form at the point 0! Thus L'Hôpital's rule, as stated above, does not apply. A more general rule is needed to handle this situation, and it is derived in the exercises to follow. It says that if f'/g' is again indeterminate, then the limit of f/g may be obtained by applying L'Hôpital's rule to the indeterminate form f'/g'. Let us do this. Applying L'Hôpital's rule to the quotient (33), we are led to investigate

$$\frac{\cos x}{2},$$

which has a limit of $\frac{1}{2}$ at the point $x = 0$. The limit at 0 of $(1 - \cos x)/x^2$ is $\frac{1}{2}$. □

A result closely related to L'Hôpital's rule, and which will be used to obtain a strengthened form of that rule, is a *mean-value theorem*. It asserts that the mean value of a continuous function on an interval must actually be a value attained by the function. Intuitively, this is obvious, because inequality (27) says that the mean value of f lies between the largest and smallest values of f. Since $f(x)$ varies *continuously* between its largest and smallest values, it must at some point c be equal to its mean value:

$$\operatorname*{avg}_{a,b} f = f(c) \quad \text{for some point } c \text{ between } a \text{ and } b. \qquad (34)$$

A rigorous proof of this is best deferred to a course in analysis. Of course, if f "wiggles" a lot, there may be more than one point where f attains its mean value:

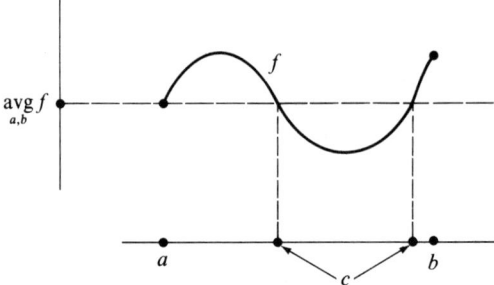

Mean-Value Theorem. *If f is continuous on the interval from a to b, there must be at least one point c, where $a < c < b$, at which $f(c)$ is equal to the mean (or average) value of f.*

The mean-value theorem is of great importance, partly because its content can be rephrased in several (apparently) different ways. There are thus several

7. Quadratures, Mean-Value Theorems, and L'Hôpital's Rule

(apparently) different mean-value theorems, not just one, as we shall note in the exercises to follow.

Let us now see how this enters into a proof of a strengthened form of L'Hôpital's rule. The key is to go back to equation (31) and analyze it a bit further, to show that if f/g is indeterminate at the point a, then

$$\frac{f(b)}{g(b)} = \frac{f'(c)}{g'(c)} \quad \text{for some } c \text{ between } a \text{ and } b. \tag{35}$$

The reader is asked to do this, under suitable hypotheses, in an exercise to follow. Once this is done, a strong form of L'Hôpital's rule is easily forthcoming:

L'Hôpital's Rule (Strong Form). *If the quotient f/g becomes indeterminate at the point a, then, at the point a,*

$$\text{Lim}\frac{f}{g} = \text{Lim}\frac{f'}{g'},$$

provided f' and g' are continuous and $\text{Lim}(f'/g')$ exists at a.

PROOF. [Note that this form of the rule is indeed "stronger" in that it applies even when, as in Example 17, $g'(a) = 0$.] By (35), we have

$$\frac{f(x)}{g(x)} = \frac{f'(c)}{g'(c)} \quad \text{for some } c \text{ between } a \text{ and } x.$$

The choice of c depends, of course, upon the choice of x, but c must lie in between a and x. This means that, as $x \to a$, we automatically have $c \to a$, so

$$\text{Limit}_{x \to a} \frac{f(x)}{g(x)} = \text{Limit}_{c \to a} \frac{f'(c)}{g'(c)} = \text{Lim}\frac{f'}{g'} \quad \text{at } a. \quad \square$$

EXAMPLE 18. Find the limit, as x tends to 0, of $(x - \sin x)/x^3$.

Solution. Since this quotient is indeterminate at 0, the rule just proved says its limit at 0 is the same as the limit of

$$\frac{1 - \cos x}{3x^2}.$$

However, this is again indeterminate, of the form 0/0, at the point 0; so, applying the rule to this indeterminate form, we investigate

$$\frac{\sin x}{6x},$$

for its limit at 0. This is again indeterminate, but one more application of the rule shows that its limit is the limit at 0 of

$$\frac{\cos x}{6},$$

which obviously has a limit of $\frac{1}{6}$ at 0. What we have done is to use repeated applications of the strong form of L'Hôpital's rule to show that, at the point 0,

$$\text{Lim}\frac{x-\sin x}{x^2} = \text{Lim}\frac{1-\cos x}{3x^2} = \text{Lim}\frac{\sin x}{6x} = \text{Lim}\frac{\cos x}{6} = \frac{1}{6},$$

where the final equality comes from the continuity of the cosine function at 0. Therefore,

$$\underset{x\to 0}{\text{Limit}}\frac{x-\sin x}{x^3} = \frac{1}{6}. \qquad \square$$

Exercises

7.1. What size square will form a quadrature of a triangle of base b and height h?

7.2. Show that a rectangle of base π and height $2/\pi$ effects a quadrature of the area beneath the sine function, between 0 and π.

7.3. Using the result of exercise 7.2, using no computation, and using at most two sentences of writing, prove that the area crosshatched vertically is equal in size to the area crosshatched horizontally.

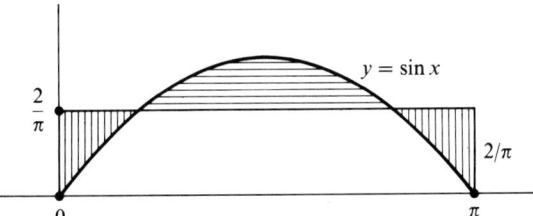

7.4. Perform a quadrature of the indicated area, by finding a rectangle of the same size.

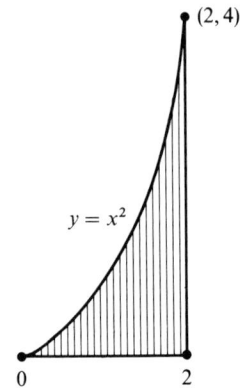

7.5. A student has calculated the average value of the sine function on a certain interval and has found that it is equal to 2. By appealing to an appropriate inequality in this section, show that the student's calculation must be wrong.

7. Quadratures, Mean-Value Theorems, and L'Hôpital's Rule

7.6. Note that if x is not equal to 0, 1, or -1, then

$$\frac{x^4 - x}{x^3 - x} = \frac{x(x^3 - 1)}{x(x^2 - 1)} = \frac{x(x-1)(x^2 + x + 1)}{x(x-1)(x+1)} = \frac{x^2 + x + 1}{x + 1}.$$

(a) Check the results of Examples 15 and 16 by using this equality to find $\text{Lim}(x^4 - x)/(x^3 - x)$ at 1 and at 0.
(b) What is the limit at -1?
(c) What is the limit at -2?

7.7. Find the limit of each of the following, at the point a.
(a) $(x^5 + 1)/(x + 1)$, $a = -1$.
(b) $(\sin x)/x$, $a = 0$.
(c) $(\sin x)/(x - \pi)$, $a = \pi$.
(d) $(x^3 - x)/(x^4 - x)$, $a = -1$.
(e) $(\arctan x)/3x$, $a = 0$.
(f) $(x^2 - 1)/(2x^2 - x - 1)$, $a = 1$.
(g) $(\arcsin x)/(\sin x)$, $a = 0$.
(h) $(\cos 3x)/(x - (\pi/2))$, $a = \pi/2$.

7.8. The mean-value theorem asserts that a *continuous* function must at some point be equal to its mean (or average) value. The "continuous" is vastly different from the "discrete" in this respect. Give an example to show this. (Almost any example you choose will suffice. Think of three test grades, for example, whose average is not itself a grade that was made on a test.)

7.9. Use the mean-value theorem to prove these variant forms of the theorem:
(a) If f is continuous, then there is a point c between a and b such that

$$\int_a^b f(x)\,dx = f(c)(b - a).$$

Hint. In (34), write avg f in terms of an integral.
(b) If F has a continuous derivative F', then there is a point c between a and b such that

$$\frac{F(b) - F(a)}{b - a} = F'(c).$$

Hint. In the equality of exercise 7.9(a), let $f = F'$, and use **F3** to evaluate the integral on the left.
(c) If g has a continuous derivative, then there is a point c between a and x such that

$$g(x) = g(a) + g'(c)(x - a).$$

Hint. In the equality of exercise 7.9(b), let $F = g$ and $b = x$.

7.10. The term *mean-value theorem* is most often applied to the statement in exercise 7.9(b). Use this statement to show that if $P = (a, F(a))$ and $Q = (b, F(b))$ are two points on the continuously differentiable curve F, then there is another point $R = (c, F(c))$ on the curve F, located between P and Q, such that Slope(PQ) is equal to the slope of the tangent line to F at R.

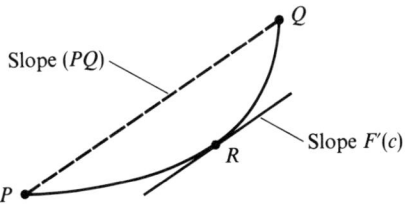

7.11. Assume that f' and g' are continuous and that g' is never equal to zero, except possibly at the point a. Assume that $f(a)$ and $g(a)$ are equal to zero, so that the quotient f/g becomes indeterminate at the point a. By carrying out the following steps, prove statement (35).
   (a) Let $h(x) = f(b)g'(x) - g(b)f'(x)$. Show that $\text{avg}_{a,b}\, h = 0$. *Hint.* Since b is a constant, $\int_a^b h(x)\,dx = f(b) \int_a^b g' - g(b) \int_a^b f'$. Use **F3** to evaluate these integrals and to see that the integral of h, from a to b, is equal to zero.
   (b) Apply the mean-value theorem, together with the result of part (a), to show that $h(c) = 0$ for some c between a and b. Then derive (35).

7.12. Find the limit of each of the following at the indicated point a.
   (a) $(\sin(3x^2))/x^2$, $a = 0$.
   (b) $(x - \tan x)/x$, $a = 0$.
   (c) $(x - \tan x)/x^2$, $a = 0$.
   (d) $(1 - \cos x)/(x \sin x)$, $a = 0$.
   Answers: (a) 3. (b) 0. (c) 0. (d) $\frac{1}{2}$.

7.13. Consider again the three versions of the mean-value theorem established in exercise 7.9.
   (a) Show that the statement in part (a) of exercise 7.9 is true *even if* $b < a$.
   (b) Show that the statement in part (b) of exercise 7.9 is true *even if* $b < a$.
   (c) Show that the statement in part (c) of exercise 7.9 is true *even if* $x < a$.

7.14. Show that statement (35) of this section is true *even if* $b < a$.

§8. Integrals as Antiderivatives

A continuous function does not necessarily have a derivative. Regardless of how complicated the continuous function is, however, it has an antiderivative, expressible in terms of an integral.

Theorem on Integral Representations of Antiderivatives. *Let f be a continuous function defined on a connected domain and let c be any fixed point lying in the domain. Then an antiderivative of f is given by F, where F is defined by the integral expression*

$$F(x) = \int_c^x f. \tag{36}$$

PROOF. Note that the integral expression (36) exists for any point x in the connected domain. The existence is guaranteed by **F2** since f is assumed continuous. To show that $F'(x) = f(x)$, we use Fermat's method.

$$F(x+h) - F(x) = \int_c^{x+h} f - \int_c^x f = \int_x^{x+h} f.$$

Dividing by nonzero h, we get

$$\frac{F(x+h) - F(x)}{h} = \frac{1}{h}\int_x^{x+h} f = \underset{x,x+h}{\text{avg}}\, f. \tag{37}$$

8. Integrals as Antiderivatives

Hence,
$$F'(x) = \underset{h \to 0}{\text{Limit}} \frac{F(x+h) - F(x)}{h}$$
$$= \underset{h \to 0}{\text{Limit}} \underset{x, x+h}{\text{avg}} f \quad [\text{by (37)}]$$
$$= f(x) \quad [\text{by (28)}],$$

since f is continuous at x. Therefore, F is an antiderivative of f. □

The careful reader will want to examine equation (37) to see that it holds for h negative. It does. (See exercise 5.14.)

The expression (36) gives *an* antiderivative of f. It gives *the* antiderivative that takes the value 0 at the point $x = c$, for
$$F(c) = \int_c^c f = 0.$$

By the uniqueness theorem **F1**, any other antiderivative of f differs from (36) by a constant. For example, by the theorem just proved, we know that
$$G(x) = \int_d^x f \tag{38}$$

is also an antidervative of f. (G is the antiderivative that takes the value 0 at the point $x = d$.) The antiderivative G differs from the antiderivative F by a constant. The reader is asked to find this constant in an exercise to follow.

The reader should learn to feel at home with integral representations of functions. Using them gives one the power to write down an antiderivative of any continuous function whatever.

EXAMPLE 19. Find the antiderivative of the function arctan that takes the value

(a) 0 at the point $x = 0$.
(b) 0 at the point $x = \frac{1}{2}$.
(c) 5 at the point $x = 0$.
(d) π at the point $x = \frac{1}{2}$.
(e) $\sqrt{2}$ at the point $x = -\pi$.

This is now child's play, although it would have been formidable indeed prior to the discussion in this section. The answers are, obviously,

(a) $F(x) = \int_0^x \arctan$.
(b) $F(x) = \int_{1/2}^x \arctan$.
(c) $F(x) = 5 + \int_0^x \arctan$.
(d) $F(x) = \pi + \int_{1/2}^x \arctan$.
(e) $F(x) = \sqrt{2} + \int_{-\pi}^x \arctan$.

The answers are unique by the uniqueness theorem **F1**, and they are defined for all x by the existence theorem **F2**, since arctan is everywhere continuous. □

In Example 19 the function in question happened to have a name (arctan). If the function is specified not by name but by an algebraic rule, a dummy variable must be introduced.

EXAMPLE 20. Find the antiderivative of the function $1/x$ that takes the value

(a) 0 at the point $x = 1$.
(b) π at the point $x = 6$.
(c) 17 at the point $x = -4$.

What could be simpler? The answers are

(a) $F(x) = \int_1^x (1/t)\, dt$, defined for $0 < x$.
(b) $F(x) = \pi + \int_6^x (1/t)\, dt$, defined for $0 < x$.
(c) $F(x) = 17 + \int_{-4}^x (1/t)\, dt$, defined for $x < 0$.

By the uniqueness theorem the answers are unique on the specified connected domains. □

A **differential equation** is an equation involving the derivative(s) of a function. To solve a differential equation is to find a function that satisfies the equation. We are in a position to write down the solution to any elementary differential equation of the first order (involving only the first derivative).

Theorem on Elementary First-Order Differential Equations. *Given $f(x)$, the elementary differential equation*

$$\frac{dy}{dx} = f(x), \tag{39}$$

with the boundary condition

$$y(x_0) = y_0, \tag{40}$$

has the unique solution

$$y = y_0 + \int_{x_0}^x f \tag{41}$$

defined on a connected component of the domain of continuity of f.

PROOF. The differential equation (39) simply specifies that y is equal to an antiderivative of f. By the theorem on integral representations of antiderivatives, then,

$$y = \int_{x_0}^x f + \text{some constant,}$$

if x_0 is a point in the domain of the continuous function f. It is clear that condition (40) forces the constant to be y_0, showing that (41) is indeed the unique solution. □

EXAMPLE 21. Find the solution to each of the following differential equations with the given boundary condition.

(a) $dy/dx = \cos x$, $y(0) = 3$.
(b) $dy/dx = x^2$, $y(1) = 2$.
(c) $dy/dx = 1/x$, $y(3) = 7$.

This is a straightforward application of the theorem on differential equations. However, it may take a moment to get used to the way in which the dummy variable enters. The answers are, respectively,

(a) $y = 3 + \int_0^x \cos = 3 + [\sin]_0^x = 3 + \sin x$.
(b) $y = 2 + \int_1^x t^2 \, dt = 2 + [t^3/3]_1^x = 2 + (x^3/3) - \frac{1}{3} = \frac{1}{3}(x^3 + 5)$.
(c) $y = 7 + \int_3^x (1/t) \, dt$.

The answers to (a) and (b) are defined for all x, but the answer to (c) is defined only for $0 < x$. □

The theorem proved in this section is certainly as "fundamental" as **F1**, **F2**, or **F3**. Let us call it **F4** and elevate it to the same exalted status. We may restate it in a striking way:

F4. *If f is continuous, then the derivative of the integral of f is f. That is,*

$$\frac{d}{dx} \int_c^x f = f(x). \tag{42}$$

A plea. Gentle reader, if you remember nothing else about calculus, remember equation (42). There is no possible way to express the connection between differential and integral calculus in a more compact fashion.

EXERCISES

8.1. What is all this fuss about antiderivatives? Doesn't **F3** say that every continuous function has an antiderivative? *Answer*: No. Go back to Chapter 6 and read the proof of **F3**. It says that *if* there is an antiderivative F, then F can be used to evaluate the integral of f.

8.2. Find the antiderivative of the function $\sin^2$ that
(a) takes the value 0 at the point $x = \pi$.
(b) takes the value 4 at the point $x = \pi$.
(c) takes the value $\sqrt{3}$ at the point $x = 0$.
Answer: (c) $F(x) = \sqrt{3} + \int_0^x \sin^2 t \, dt$.

8.3. Fill in properly the indicated places (a) and (b) in each of the tables below.

(i)

x	y	dy/dx
0	5	
x	(a)	$1/(x+3)$
2	(b)	

Answers: (a) $5 + \int_0^x (1/(t+3)) \, dt$. (b) $5 + \int_0^2 (1/(t+3)) \, dt$.

(ii)

x	y	dy/dx
3	$\frac{1}{2}$	
x	(a)	$3x^2$
7	(b)	

Partial answer: (b) 316.5.

(iii)

x	y	dy/dx
1	0	
x	(a)	$1/x$
3	(b)	

Partial answer: (b) $\int_1^3 (1/t) \, dt = 1.0986 \ldots$.

8.4. By the uniqueness theorem expressions (38) and (36) must differ by a constant. Find the constant. Answer: $\int_c^x f - \int_a^x f = \int_c^a f$.

8.5. Each of the following is a continuous function, everywhere defined. Find its antiderivative that takes the value π at the point -5.
  (a) $|x|$.
  (b) $|\sin x|$.
  (c) $\sqrt{3 + |x - 4|}$.
Answer: (b) $F(x) = \pi + \int_{-5}^x |\sin t| \, dt$.

8.6. Use (42) to find the derivative of each of the following:
  (a) $y = 3 + \int_2^x (t + 4t^3) \, dt$.
  (b) $y = \cos x + \int_\pi^x (\arctan t) \, dt$.
  (c) $L = t^2 + \int_0^t (\sin^2 x + 10) \, dx$.
Answers: (a) $dy/dx = x + 4x^3$. (b) $dy/dx = -\sin x + \arctan x$. (c) $dL/dt = 2t + \sin^2 t + 10$.

8.7. Use (42) together with the chain rule to find the derivative of each of the following:
  (a) $y = 3 + \int_2^{r^2} (t + 4t^3) \, dt$.
  (b) $y = \int_\pi^{4r+7} (\arctan t + \cos^3 t) \, dt$.
  (c) $C = \int_0^{\sin x} (\cos^2 s - \tan s) \, ds$.
Answer: (a) Here we have the chain given by the expression in exercise 8.6(a) and the equation $x = r^2$. By the chain rule,
$$\frac{dy}{dr} = \frac{dy}{dx}\frac{dx}{dr} = (x + 4x^3)(2r) = (r^2 + 4r^6)(2r).$$

8.8. Find the derivative of each of the following:
  (a) $y = \int_x^5 \cos^3 t\, dt$.   Hint. $\int_x^5 f = -\int_5^x f$.
  (b) $y = \int_{-x}^x \cos^3 t\, dt$.   Hint. $\int_{-x}^x f = \int_{-x}^0 f + \int_0^x f$.

§9. Integrals as Work and as Distance

The term *work* has a technical meaning in physics and is defined in a very simple way, *provided we have motion against an opposing force that does not vary*. **Work** *is the product of the opposing force and the distance traveled*:

$$W = f \cdot d.$$

For instance, if a person weighs 150 pounds (lb) and he climbs up a rope from a height of 6 feet to a height of 10 feet, then the amount of work done is

$$W = (150 \text{ lb})(4 \text{ ft}) = 600 \text{ ft-lb}. \tag{43}$$

The calculation of 600 foot-pounds (ft-lb) of work is based on the assumption, not entirely valid, that the force of gravity opposing the motion was constant at 150 pounds throughout the motion. Actually, the law of gravity says that the higher one goes, the less one weighs, but surely the weight loss over so short a trip may be ignored here.

Suppose the problem had been to calculate the work done in moving this same person from the earth's surface to a point 500 miles above the earth's surface. Then the weight loss would be substantial and could not be ignored in our calculations. Let us discuss the law of gravity briefly.

According to Newton's law, the weight of an object, which is the force $f(x)$ opposing upward motion, is inversely proportional to the square of the distance x from the earth's center. That is,

$f(x) = $ gravitational force at distance x from earth's center

$$= \frac{k}{x^2},$$

where k is the constant of proportionality. When $x = 4000$ miles (the radius of the earth, according to Eratosthenes' measurement), we then have

$$\text{weight on earth's surface} = f(4000) = \frac{k}{(4000)^2}.$$

Solving this equation for k yields the constant of proportionality:

$$k = (4000)^2 \text{ (weight on earth's surface)},$$

and the force of gravity therefore varies in accordance with the rule

$$f(x) = \frac{(4000)^2}{x^2} \text{(weight on earth's surface), where } x \text{ is in miles.} \tag{44}$$

EXAMPLE 22. How much work is done in taking an object weighing 150 pounds on the earth's surface to a point 500 miles above the earth?

This problem requires a little speculation, for the weight loss is considerable. According to (44), the force of gravity at the end of the trip, when $x = 4500$ miles, is given by

$$f(4500) = \frac{(4000)^2}{(4500)^2}(150) = 118.5 \text{ lb}.$$

How should work be measured, then, when the opposing force varies along the path of travel? The answer, as we shall see, is to measure the work by a certain integral.

Consider the work ΔW done in moving from x to $x + \Delta x$, where Δx is a very small positive number. Since the opposing force at any stage of this short move is between $f(x)$ and $f(x + \Delta x)$, it is clear that the work ΔW lies between the quantities

$$f(x)\,\Delta x \quad \text{and} \quad f(x + \Delta x)\,\Delta x.$$

Dividing by Δx, we see that the difference quotient $\Delta W/\Delta x$ lies between the quantities $f(x)$ and $f(x + \Delta x)$:

$$f(x) \leq \frac{\Delta W}{\Delta x} \leq f(x + \Delta x).$$

What happens here as Δx tends to 0? Since f is continuous, $f(x + \Delta x) \to f(x)$, and the difference quotient $\Delta W/\Delta x$, being squeezed in between, must also tend to $f(x)$. That is,

$$\frac{dW}{dx} = f(x).$$

We then have the information in the following table regarding the relation between the work done and the distance from the earth's center. (*Why is $W = 0$ when $x = 4000$?*)

x	W	dW/dx
4000	0	
x		$f(x)$
4500	?	

To answer the question raised in Example 22 is to fill in the question mark in the table given above. But this is just the sort of thing we have learned to do (see exercise 8.3) by integrals. The work done during the trip from $x = 4000$ to $x = 4500$ is given by

$$\int_{4000}^{4500} f. \tag{45}$$

9. Integrals as Work and as Distance

Since an antiderivative of $1/x^2$ is $-1/x$, we can evaluate this integral by F3. The work done is then, by (44),

$$\int_{4000}^{4500} \frac{(4000)^2(150)}{x^2}\,dx = (4000)^2(150)\left[\frac{-1}{x}\right]_{4000}^{4500}$$
$$\approx 66{,}667 \text{ mi-lb.} \qquad \square$$

The notion of *work* is of great importance in physics because it is closely connected to the notions of *energy* and of *power*. Considerations such as those just made have led physicists to the conclusion that the natural definition of work is in terms of an *integral*, just like (45).

Definition. The **work** done against an opposing force $f(x)$ in moving from position $x = a$ to position $x = b$ is defined as the integral

$$\int_a^b f.$$

It is important to note that this definition of work reduces to the simple definition given at the beginning of this section in case the force $f(x)$ happens to be constant.

EXAMPLE 23. Suppose the force $f(x)$ is constant at 150 pounds. What is the work done in moving from the position $x = 6$ feet to $x = 10$ feet?

According to the definition above, the work done is

$$\int_6^{10} f = \int_6^{10} 150\,dx = 150x\big|_6^{10} = 600 \text{ ft-lb.}$$

This agrees, as it should, with the calculation in (43). $\qquad \square$

The preceding discussion has to do with moving a single object through a fixed distance. If different "parts" of the object go different distances, then the work is again expressed by an integral, but in a different way. This is illustrated below.

EXAMPLE 24. Water weighs 62.4 pounds per cubic foot. Suppose the water contained in an inverted conical container is at a depth of 6 feet, while the container itself is 7 feet high with a circular radius of 5 feet on top. How much work is required to pump all the water out?

The problem here is that all the water doesn't move the same distance. It takes very little work to pump out a cubic foot of water near the top, but more to move the same volume at the bottom. Let W be the work required to pump out all the water if the initial depth of the water is x feet. Our plan is to find dW/dx first, then express W itself as an integral. Fixing x, let Δx be a small positive change. What is the corresponding change ΔW in work?

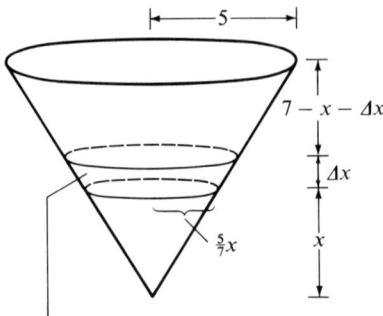

What is the work ΔW required to pump this water to the top?

The weight of the water moved is clearly (see the picture) between

$$62.4\pi\left(\frac{5}{7}x\right)^2 \Delta x \quad \text{and} \quad 62.4\pi\left(\frac{5}{7}(x+\Delta x)\right)^2 \Delta x$$

pounds.

To pump this water to the top of the tank, it must move between

$$7 - x - \Delta x \quad \text{and} \quad 7 - x$$

feet. These facts give us clear bounds on the amount of work ΔW it takes to pump this volume of water to the top:

$$62.4\pi\left(\frac{5}{7}x\right)^2 (7 - x - \Delta x)\Delta x \le \Delta W \le 62.4\pi\left(\frac{5}{7}(x+\Delta x)\right)^2 (7 - x)\Delta x$$

Dividing by Δx we get

$$62.4\pi\left(\frac{5}{7}x\right)^2 (7 - x - \Delta x) \le \frac{\Delta W}{\Delta x} \le 62.4\pi\left(\frac{5}{7}(x+\Delta x)\right)^2 (7 - x).$$

As Δx tends to zero, both sides of this inequality tend to the same limit. Since the difference quotient $\Delta W/\Delta x$ is squeezed in between, it must tend to that limit also. That is,

$$\frac{dW}{dx} = 62.4\pi\left(\frac{5}{7}x\right)^2 (7 - x) = 62.4\pi\left(\frac{25}{49}\right)(7x^2 - x^3). \tag{46}$$

Since $W = 0$ when $x = 0$ (why?), the solution of the differential equation (46) is given by

$$W = \int_0^x 62.4\pi\left(\frac{25}{49}\right)(7t^2 - t^3)\,dt. \tag{47}$$

Equation (47) gives the work W as a function of the initial depth x. If, as in Example 24, the initial depth is 6 feet, then the work required to pump all the water out of the tank is

$$\int_0^6 62.4\pi\left(\frac{25}{49}\right)(7t^2 - t^3)\,dt \text{ ft-lb}. \tag{48}$$

The reader is asked to evaluate this integral in an exercise to follow. □

9. Integrals as Work and as Distance

A less rigorous way of arriving at equation (46) has the advantage of being much quicker. Stare at the figure above and try to make each of the following assertions seem plausible. If Δx is near zero, then all the water of the little slice in question goes approximately $7 - x$ feet. Also, if Δx is near zero the volume of water in the little slice is approximately that of a cylinder of height Δx and radius $\tfrac{5}{7}x$, having a volume of $\pi(\tfrac{5}{7}x)^2 \Delta x$ cubic feet, and a consequent weight of $62.4\pi(\tfrac{5}{7}x)^2 \Delta x$ pounds. The work ΔW is then very close to the product of this approximate weight and this approximate distance:

$$\Delta W \approx 62.4\pi \left(\frac{5}{7}x\right)^2 (7 - x)\, \Delta x.$$

Dividing by Δx and taking the limit as Δx tends to zero leads to equation (46).

Another use of integrals in the physics of motion is in the measurement of distance traveled. This should come as no surprise, as we have seen already the relation between distance traveled and antiderivatives. (See Chapter 5, Section 9.)

Theorem on Integrals and Distance. *Let the speed of a moving particle be given at time t by $v(t)$, where v is a positive continuous function. Then the distance traveled between time $t = a$ and time $t = b$ is given by the integral*

$$\int_a^b v.$$

PROOF. Let s denote the distance traveled by the particle, beginning at time $t = a$. This means that $s = 0$ when $t = a$. Since the speed v is equal to the rate of change of distance, we have $v = ds/dt$. We then have the following information, and we want to fill in the question mark.

t	s	ds/dt
a	0	
t		$v(t)$
b	?	

This is by now a routine problem. The question mark must be filled in by the integral $\int_a^b v$, which therefore gives the distance traveled between times $t = a$ and $t = b$. □

EXAMPLE 25. Suppose the speed at time t of a certain particle is given by $v(t) = 1/t$. How far does the particle go between time $t = 1$ and time $t = 3$? between $t = 4$ and $t = 12$?

Since the speed is positive, the distance traveled is the integral of the speed over the time interval in question. The distance traveled between $t = 1$ and $t = 3$ is

$$\int_1^3 v = \int_1^3 \frac{1}{t}\, dt = 1.0986\ldots \text{ units.}$$

The distance traveled between $t = 4$ and $t = 12$ is then

$$\int_4^{12} v = \int_4^{12} \frac{1}{t} dt.$$

With the aid of equations (7) and (10) at the beginning of this chapter, we see that this integral is equal to

$$\int_4^{12} \frac{1}{t} dt = \int_1^{12} \frac{1}{t} dt - \int_1^4 \frac{1}{t} dt$$

$$= 2.4849\ldots - 1.3863\ldots$$

$$= 1.0986\ldots \text{ units.} \qquad \square$$

EXERCISES

9.1. Find the work done in moving an object from the earth's surface to a point 1000 miles above if the object weighs 100 pounds on the earth's surface. *Answer*: 80,000 mi-lb.

9.2. A satellite weighing 1000 pounds when it is at a point 1000 miles above the surface of the earth is to be pushed out to a point 2000 miles from the earth's surface.
   (a) How much work must be done to accomplish this?
   (b) How much work was required to put the satellite in its original orbit at a distance of 1000 miles from the earth?
   (c) How much did the satellite weigh when it was on the earth's surface?
   Answer: (c) 1562.5 lb.

9.3. Evaluate the integral (48) by
   (a) using **F3**.
   (b) using Simpson's rule. (*Why will Simpson's rule give you the correct answer for this integral?*)

9.4. Find the answer to the question raised in Example 24 if the container is completely full of water. *Answer*: 6370π ft-lb.

9.5. The water contained in an inverted conical container is at a depth of 10 feet. If the container itself is 15 feet high with a circular radius of 9 feet on top, how much work is required to pump all the water out? *Answer*: $62.4\pi(9/25)\int_0^{10}(15t^2 - t^3)dt$ ft-lb. (Evaluate by **F3**.)

9.6. The water contained in a *cylindrical* container is at a depth of 10 feet. If the container itself is 15 feet high with a circular radius of 9 feet on top, how much work is required to pump all the water out? (You can do this by integrals, or by common sense.) *Answer*: $505,440\pi$ ft-lb. (This is the weight of the water times the average distance it has to move.)

9.7. The water contained in a *hemispherical* container is at a depth of 6 feet. The container itself is 7 feet high, which means that it has a circular radius of 7 feet on top. How much work is required to pump all the water out? *Answer*: $62.4\pi \int_1^7 (49t - t^3) dt$ ft-lb.

9.8. Suppose the speed at time t of a certain particle is given by $v(t) = 10/(1 + t^2)$. How far does the particle go between
(a) $t = 0$ and $t = 1$?
(b) $t = 1$ and $t = \sqrt{3}$?
Answer: (b) $5\pi/6$ units.

9.9. Suppose the speed of a certain particle is given at time t by $v(t) = 10/t$. How far does the particle go between
(a) $t = 1$ and $t = 2$?
(b) $t = 2$ and $t = 4$?
(c) $t = 4$ and $t = 8$?

§10. Summary

An integral is *definite* in this sense. If an integral exists, then it can be specified to any degree of accuracy by an approximating sum. This characteristic—of admitting an approximation to an arbitrary degree of closeness—we take to be the principal defining quality of a "real" number. The *number* $\int_a^b f$ may be an average, may effect a quadrature, may measure a quantity of work, or may represent a length of distance traveled. The *function* $\int_c^x f$ represents an antiderivative of f if f is continuous.

Fundamentals are important. It is useful to know theorems guaranteeing the existence of integrals or the uniqueness of solutions to a given equation.

Integrals have other uses that we do not yet know. There are still more mean-value theorems and other versions of L'Hôpital's rule. There are many parts of the house of integrals we have yet to see. But we can begin to feel at home.

Problem Set for Chapter 8

1. (*A discussion question*) Suppose two students are working on the same mathematics problem, in which it is required to find a certain "x", and they give the following answers:

Student A. $x = \arctan 2$.
Student B. $x = \int_0^2 (1/(1 + t^2)) \, dt$.

Both answers are declared to be correct. Assuming that no calculator and no table of tangents and arctangents is available, explain which answer more directly leads to the desired result if
(a) it is desired that the *angle* x be constructed very carefully.
(b) it is desired that the *number* x be calculated very carefully (to many decimal places).

2. Some of the following "calculations" are ridiculous. Find them, and explain why they are in error.
   (a) $\int_0^2 (-2x/(x^2-1)^2)\,dx = (1/(x^2-1))|_0^2 = \frac{4}{3}$.
   (b) $\int_0^{1/2}(-2x/(x^2-1)^2)\,dx = -\frac{1}{3}$.
   (c) $\int_0^{1/2}(1/\sqrt{1-t^2})\,dt = \arcsin\frac{1}{2} - \arcsin 0 = 30 - 0 = 30$.

3. Consider the simple step function given by $f(x) = 4$ if $0 \le x \le 1$ and $f(x) = 6$ if $1 < x \le 2$.
   (a) Sketch the graph of f on the domain $0 \le x \le 2$.
   (b) Apply Eudoxus' method to f on the domain $0 \le x \le 2$, and show that $S_2 = 10$, $S_4 = 10$, $S_6 = 10$.
   (c) Show that $S_n = 10$ whenever n is even.
   (d) Calculate S_3, S_5, and S_7 in Eudoxus' method.
   (e) Does Limit S_n exist?
   (f) Is f continuous throughout the domain $0 \le x \le 2$?
   (g) Does $\int_0^2 f(x)\,dx$ exist?
   (h) Do your answers to parts (f) and (g) contradict **F2**? Explain why not.

4. Consider each of the following integrals. Which of them can be seen to exist by applying **F2**? Which of them are obviously equal to zero?
   (a) $\int_{-1}^{1}(1/t)\,dt$.
   (b) $\int_{-\pi}^{\pi} t^{37}\,dt$.
   (c) $\int_{-2}^{2} \sec^2 t\,dt$.
   (d) $\int_{-a}^{a} \sin 4t \cos 3t\,dt$.
   (e) $\int_{-1/2}^{1/2} \arctan(\arcsin t)\,dt$.
   (f) $\int_{-2}^{2} \arctan(\arcsin t)\,dt$.
   (g) $\int_{-3}^{3} \sqrt{50 - t^3}\,dt$.
   (h) $\int_{-4}^{4} \sqrt{50 - t^3}\,dt$.

5. Approximate the number $\int_1^5 (1/t)\,dt$ by the method illustrated in Section 3, calculating S_4, S_8, and S_{16}. Then make your best guess at it by taking the average of your bounds, as illustrated at the end of Section 3. Would you expect this guess to be slightly high or slightly low?

6. Approximate the number $L(5) = \int_1^5 (1/t)\,dt$ by making the guess (see exercise 2.7) that $L(5) + L(2) = L(10)$, and then solving this equation for $L(5)$ using the information in formulas (5) and (9) of Section 2.

7. Approximate the number $\int_1^5 (1/t)\,dt$ by using Simpson's rule. That is, calculate $S_1^5(1/t)$.

8. (a) Use Simpson's rule to calculate $S_0^{1/2}(1/\sqrt{1-t^2})$. Simplify your answer to $(15 + 16\sqrt{15} + 10\sqrt{3})/180$.
   (b) Explain why $(15 + 16\sqrt{15} + 10\sqrt{3})/30$ ought to be a close approximation to the number π.
   (c) (*Optional*) Use a table of square roots to help evaluate the quotient in part (b) to five decimal places. Compare your answer with 3.14159, the correct expansion of π to five decimal places.

9. A metal bar placed along the x-axis stretches from $x = 2$ to $x = 6$. If its temperature at the point x is given by $2x^3 - x^2 + 3x + 7$, find its *average temperature* by
   (a) using the definition of the average of a function.
   (b) calculating $\mathrm{smp}_{2,6}(2x^3 - x^2 + 3x + 7)$.

10. A rock thrown from the ground has height given by $h = -16t^2 + 64t$ feet at time t seconds after release. Find the rock's *average height*
    (a) during the first 3 seconds of flight.
    (b) during the entire flight. (The rock hits the ground when $t = 4$.)

11. A rock thrown from ground level hits the ground again after T seconds have elapsed. Calculate the *average height* of the rock, treating it as a freely falling body from time $t = 0$ to $t = T$.

12. The weight of an object depends upon its distance x in miles from the center of the earth. The weight of an object is given by $(4000)^2(150)/x^2$ pounds if its weight is 150 pounds on the earth's surface. Find the *average weight* of this object if it is taken at a uniform rate from the earth's surface (radius 4000 miles) to a point
    (a) 500 miles above the earth.
    (b) 1000 miles above the earth.

13. Use L'Hôpital's rule to find the limit, at the point 0, of each of the following indeterminate forms.
    (a) $(\arcsin x)/x$.
    (b) $(\sin \pi x)/3x$.
    (c) $(1 - \cos 2x)/x^2$.
    (d) $(\tan x \sin x)/(\arctan x)$.
    (e) $x^5/(6 \sin x + x^3 - 6x)$.
    (f) $(1 - \cos x)/x$.
    (g) $(1 - \cos 3x)/(x - 5x^2)$.
    (h) $x/\cos(2x - (\pi/2))$.

14. In the problem set at the end of Chapter 7, do problem 3 by applying L'Hôpital's rule.

15. Find each of the following limits.
    (a) $\text{Limit}_{x \to 1}((x^4 - 1)/(x - 1))$.
    (b) $\text{Limit}_{x \to 2}((1/(x - 2))\int_2^x \cos^3 t \, dt)$.
    (c) $\text{Limit}_{x \to \pi}((1/(x - \pi)^2)\int_\pi^x \sin t \, dt)$.
    (d) $\text{Limit}_{x \to 1}\int_1^x (1/t) \, dt$.

16. For each of the following, first specify the domain of all permissible values of x for which the expression makes sense, then find dy/dx.
    (a) $y = 6 + \int_0^x |t| \, dt$.
    (b) $y = \pi + \int_1^x (1/t^2) \, dt$.
    (c) $y = 5x - \int_2^x (1/(t - 4)) \, dt$.
    (d) $y = \int_0^x (1/(t^2 + t + 2)) \, dt$.
    (e) $y = \int_0^x \tan t \, dt$.
    (f) $y = \int_0^{x^2} (1/(t + 2)) \, dt$.
    (g) $y = \int_x^3 \sin^3 t \, dt$.
    (h) $y = \int_0^{x^2} (1/(t^2 + t - 2)) \, dt$.

17. Write down an integral expression for the antiderivative of the absolute value function that takes the value
    (a) 0 at the point 0.
    (b) π at the point 4.

18. Write down an integral expression for the solution of each of the following differential equations, with given boundary condition.
    (a) $dy/dx = x^2$, $y(2) = 4$.
    (b) $dy/dx = 1/x$, $y(3) = 1$.
    (c) $dy/dx = \cos x$, $y(\pi) = 2$.

19. For each of the differential equations in problem 18
    (a) find $y(1)$.
    (b) find $y(5)$.

20. (a) Use the uniqueness theorem **F1** to show that if $F(0) = 0$ and if $F'(x) = 0$ for all x, then $F(x) = 0$ for all x.
    (b) Find $F'(x)$ if $F(x) = \int_{-x}^{x} f(t)\,dt$, assuming f is continuous.
    (c) Combine parts (a) and (b) to give a new proof that the integral $\int_{-a}^{a} f(t)\,dt$ must be zero if f is an odd function, continuous on $-a \leq t \leq a$.

21. (a) Find the derivative of $\int_{1}^{x} (1/t)\,dt$, defined for $0 < x$.
    (b) Find the derivative of $\int_{1}^{\pi x} (1/t)\,dt$, defined for $0 < x$.
    (c) Combine parts (a) and (b) with the uniqueness theorem **F1** to conclude that $\int_{1}^{\pi x} (1/t)\,dt - \int_{1}^{x} (1/t)\,dt$ must be constant. Find that constant. *Hint.* Let $x = 1$.
    (d) Conclude that $L(\pi x) = L(x) + L(\pi)$ if $0 < x$, where $L(x)$ is defined by expression (11), Section 2.

22. In exercise 2.7 we speculated that $L(6) = L(2) + L(3)$, but we were unable to prove equality. Prove that this equation is true. *Hint.* It is easier to prove the more general assertion $L(ax) = L(a) + L(x)$, if x and a are positive. Prove this more general assertion by employing the means of problem 21, but with a replacing π.

23. A snail embarks on a trip at time $t = 0$. His goal is to travel 2 feet. But he gets ever more tired and walks ever more slowly, so that his speed v is given in ft/min by

$$v(t) = 1/(1 + t^2), \qquad 0 \leq t.$$

   (a) What distance does the snail travel between $t = 0$ and $t = 1$?
   (b) How long does it take the snail to travel a distance of $\pi/3$ feet?
   (c) Write down the integral that gives the distance traveled between time $t = 0$ and $t = 100$.
   (d) Explain why the integral in part (c), when evaluated by Eudoxus' method, will not exceed $\pi/2$.
   (e) Does the snail ever come to a complete stop on the (infinite) domain $0 \leq t$?
   (f) Will the snail ever reach his goal?
   (g) "If you set any goal, and move toward it long enough, you will eventually reach it." This is an old saying. Explain why old snails never say it much.

24. Find the amount of work done against the force of earth's gravity if an object weighing 500 pounds on the earth's surface is raised vertically to a point
    (a) 1000 miles above the earth's surface.
    (b) 96,000 miles above the earth's surface.
    (c) L miles above the earth's surface.

25. (a) Explain why the answer to part (c) of problem 24 cannot exceed 2 million mi-lb, regardless of how large L might be.
    (b) "If your capacity for work is limited, there is a bound on how high you can go." This is an old saying. Explain how young astronauts may disprove this saying.

26. (a) Prove that the work done in moving an object is equal to the product of the distance traveled and the average force along the path of travel.
    (b) Prove that the distance traveled is equal to the product of the time traveled and the average speed during this time.

27. A car travels 120 miles in 2 hours. Prove that at some instant during this trip the car's speed is 60 mi/hr. Give a careful argument, illustrating clearly how the mean-value theorem comes into play.

28. Look again at problem 23 at the end of Chapter 6. Show that it is solved by applying the mean-value theorem as stated in exercise 7.10 to the function $f(x) = 4 - x^2$, $-2 \leq x \leq 1$.

29. A container is 5 feet deep and is filled with water (weighing 62.4 pounds per cubic foot) to a depth of 3 feet. Calculate the work required to pump all the water out if the top of the container is a circle of radius 5 feet and the container is shaped like
    (a) a cylinder.
    (b) a cone.
    (c) a hemisphere.

30. A chain 50 feet long weighs 150 pounds, the weight distributed uniformly along its length. If the chain is suspended vertically from the top of a cliff, how much work is required to pull it to the top?

31. Find dy/dx if
$$y = \int_{\arctan x}^{3x^2+4} \cos(xt)\, dt.$$

32. The curve $y = 1/t$, $0 < t$, is of extraordinary interest, as we shall see in the next chapter.
    (a) Show that $S_{1/r}^1(1/t) = S_1^r(1/t)$ if $r > 1$.
    (b) Can one infer, from part (a), that $\int_{1/r}^1 (1/t)\, dt = \int_1^r (1/t)\, dt$?
    (c) Prove that the area indicated by vertical cross-hatching is equal to the area with horizontal cross-hatching.

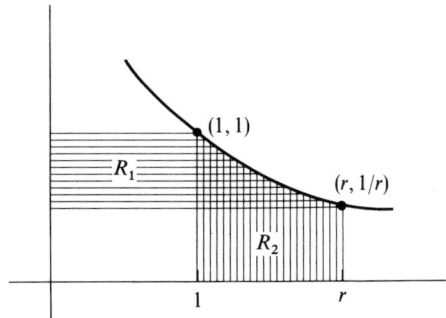

Hint. Begin by showing that the two rectangles R_1 and R_2 have the same area.
    (d) Deduce from part (c) that $\int_{1/r}^1 (1/t)\, dt = \int_1^r (1/t)\, dt$ if $r > 0$.

33. (*A proof of the chain rule*) Suppose we have the chain $y = f(u)$ and $u = g(x)$, where f and g are differentiable functions. By this chain we have
$$y = f(g(x)).$$
A proof of the chain rule is outlined below, under the additional assumption that f' is continuous. (The chain rule can be proved without this assumption, but the proof is more tedious.)

(a) Justify each of these equalities if $g(x+h) \neq g(x)$.

$$f(g(x+h)) - f(g(x)) = \int_{g(x)}^{g(x+h)} f'$$

$$= \left[g(x+h) - g(x) \right] \left[\operatorname*{avg}_{g(x),\, g(x+h)} f' \right].$$

(b) Show that the equalities of part (a) hold *even if* $g(x+h) = g(x)$.
(c) Justify each of the following equalities, thereby proving the chain rule.

$$\frac{dy}{dx} = \operatorname*{Limit}_{h \to 0} \frac{f(g(x+h)) - f(g(x))}{h}$$

$$= \operatorname*{Limit}_{h \to 0} \left[\frac{g(x+h) - g(x)}{h} \right] \left[\operatorname*{avg}_{g(x),\, g(x+h)} f' \right]$$

$$= g'(x) f'(g(x))$$

$$= f'(u) g'(x)$$

$$= \frac{dy}{du} \frac{du}{dx}.$$

34. The second derivative f'' can be derived directly from the function f without taking the first derivative, as follows:

$$f''(x) = \operatorname*{Limit}_{h \to 0} \frac{f(x+h) - 2f(x) + f(x-h)}{h^2},$$

assuming f'' continuous. Prove that this is so, by evaluating this limit with the help of L'Hôpital's rule. *Hint.* It will require two applications of L'Hôpital's rule, and, in each one, you will need to use the chain rule to get derivatives. Remember that x is fixed, and the limit is taken with respect to h.

35. Suppose a cylindrical container of height h and radius r is completely filled with a liquid weighing D pounds per cubic unit.
    (a) Calculate the work required to empty the container by pumping all the liquid to the top.
    (b) Calculate the weight of the liquid in the full container.
    (c) Using your answers to parts (a) and (b), show that this problem could have been solved by pretending that the entire weight of the full container was concentrated at a depth of $h/2$. (See exercise 9.6.)

36. In the situation described in problem 35, suppose the container is an inverted cone instead.
    (a) Calculate the work required to empty the container.
    (b) Calculate the weight of the liquid in the full container.
    (c) Show that this problem could have been solved by pretending that the entire weight of the full container was concentrated at a depth of $3h/4$ (and therefore moves a distance of $h/4$ units).

37. Suppose a hemispherical container of radius r is completely filled with a liquid weighing D pounds per cubic unit. As in problems 35 and 36, find the *centroid* of the liquid. (The **centroid**, or **center of mass**, is an important physical notion. For the purpose of calculating work, the liquid behaves as if all its weight is concentrated at its centroid.)

The Central Height 9

When we began studying trigonometry in Chapter 7, we found the calculus of trigonometric functions to be made easier by the introduction of radian measure. The desire for simplicity brought about the adoption of a "natural" unit of measure—the radian.

We begin this chapter with the study of logarithms. What is the natural way of measuring logarithms in order to simplify the calculus of the logarithmic function? We shall see, and soon.

No longer need we be bound to the lowlands of algebraic and trigonometric functions. We are prepared to conquer the height above. If calculus has a center we may find it here, where logarithms multiply and exponentials grow.

No prior acquaintance with logarithms is necessary in order to read this chapter.

§1. Logarithms

Addition is easier than multiplication. This is one reason why logarithms are valuable. Logarithms can be used to convert problems in multiplication (or division) into problems of addition (or subtraction). How can such a conversion be possible? The principle is really quite simple.

Everybody knows, for example, that

$$2^3 2^7 = 2^{10}.$$

That is, the problem of *multiplying* 2^3 and 2^7 is solved by *adding* 3 and 7 to get 10, and then raising the *base* 2 to the *exponent* 10. The problem of multiplying 8 by 128 can be handled with no multiplication at all:

$$8 \cdot 128 = 2^3 2^7 = 2^{10}.$$

The numbers 3 and 7 here are logarithms, *to the base* 2, of the numbers 8 and 128, respectively. Multiplying the numbers 8 and 128 involves adding their logarithms.

We can speak of logarithms to any base b, provided b is positive and not equal to 1.

Definition. Let b be a positive number, with $b \neq 1$. If a is also positive, then the **logarithm of a to the base b** is denoted

$$\log_b a$$

and is defined as the exponent to which the base b must be raised in order to get a. That is,

$$\log_b a = t \quad \text{means} \quad b^t = a. \tag{1}$$

For example,

$$\log_2 8 = 3 \quad \text{(because } 2^3 = 8\text{),}$$
$$\log_2 128 = 7 \quad \text{(because } 2^7 = 128\text{),}$$
$$\log_2 2 = 1 \quad \text{(because } 2^1 = 2\text{),}$$
$$\log_2 1 = 0 \quad \text{(because } 2^0 = 1\text{),}$$
$$\log_2 \frac{1}{2} = -1 \quad \left(\text{because } 2^{-1} = \frac{1}{2}\right).$$

In the decimal system, the number 10 is the most convenient choice of base. For example,

$$\log_{10} 100 = 2 \quad \text{(because } 10^2 = 100\text{),}$$
$$\log_{10} 10 = 1 \quad \text{(because } 10^1 = 10\text{),}$$
$$\log_{10} 1 = 0 \quad \text{(because } 10^0 = 1\text{),}$$
$$\log_{10} 0.10 = -1 \quad \text{(because } 10^{-1} = 0.10\text{),}$$
$$\log_{10} 0.01 = -2 \quad \text{(because } 10^{-2} = 0.01\text{).}$$

In principle then, it is easy to see how a theory might be constructed to convert multiplication into addition. Just as

$$\log_2 (8 \cdot 128) = \log_2 8 + \log_2 128,$$

so we should expect that the logarithm (to any base) of a product will be equal to the sum of the logarithms of the factors:

$$\text{LOG}(uv) = \text{LOG}(u) + \text{LOG}(v). \tag{2}$$

1. Logarithms

While it is clear in principle that there should be many different logarithmic functions [*that is, functions satisfying condition* (2)], it is not yet clear how any such function might be defined in practice. What, for example, is

$$\log_{10} \pi? \qquad (3)$$

To say $\log_{10} \pi = t$ is to say $10^t = \pi$ [by (1)]. To give a numerical value to expression (3) is to solve the equation

$$10^t = \pi \qquad (4)$$

for t. How can such an equation as this be solved? We shall see. About all we can do now is make guesses. Since

$$10^{1/2} = \sqrt{10} = 3.162 \ldots \approx \pi,$$

we can guess that the solution to (4) is just a little less than $\frac{1}{2}$. Therefore,

$$\log_{10} \pi \approx 0.49.$$

But if we needed $\log_{10} \pi$ to many decimal places, guessing would be a very inefficient procedure. Actually, we have in our hands already the means to overcome all difficulties connected with a practical way of defining a logarithmic function. In the next section we shall begin to move in the proper direction.

EXERCISES

1.1. Evaluate each of the following.
 (a) $\log_2 1024$.
 (b) $\log_2 \frac{1}{4}$.
 (c) $\log_3 27$.
 (d) $\log_{10} 1{,}000{,}000$.
 Answers: (a) 10. (b) -2. (c) 3. (d) 6.

1.2. Evaluate each of the following.
 (a) $\log_{1024} 2$.
 (b) $\log_{1/4} 2$.
 (c) $\log_{27} 3$.
 (d) $\log_{1{,}000{,}000} 10$.
 Partial answers: (a) $\frac{1}{10}$. (b) $-\frac{1}{2}$.

1.3. On the basis of the answers to exercises 1.1 and 1.2 formulate a conjecture about the relation of $\log_a b$ to $\log_b a$.

1.4. For each of the numbers on the left below, make a guess as to the number on the right that most closely approximates it.
 (i) $\log_2 \pi$.   (a) -0.5.
 (ii) $\log_\pi 2$.   (b) -1.5.
 (iii) $\log_{10} 316$.   (c) 1.7.
 (iv) $\log_{10} 3162$.   (d) 0.6.
 (v) $\log_{10} 0.32$.   (e) 3.5.
 (vi) $\log_{10} 0.032$.   (f) 2.5.

1.5. A **logarithmic function** is a function defined on the positive real numbers that satisfies the identity (2). Show that any logarithmic function has the following properties.
   (a) $\text{LOG}(1) = 0$.
   (b) $\text{LOG}(u^2) = 2\,\text{LOG}(u)$.
   (c) $\text{LOG}(u^3) = 3\,\text{LOG}(u)$.
   (d) $\text{LOG}(\sqrt{x}) = \frac{1}{2}\,\text{LOG}(x)$.

   Hint. Let $u = v = 1$ in identity (2) to prove (a); to prove (b), let $v = u$. To prove (d), let $u = \sqrt{x}$ in the equation established in part (b).

1.6. Show that any logarithmic function satisfies the following.
   (a) $\text{LOG}(abc) = \text{LOG}(a) + \text{LOG}(b) + \text{LOG}(c)$ if a, b, and c are positive numbers.
   (b) $\text{LOG}(a_1 a_2 \cdots a_n) = \text{LOG}(a_1) + \text{LOG}(a_2) + \cdots + \text{LOG}(a_n)$ if $a_1, a_2, \ldots, a_n$ are positive numbers.
   (c) $\text{LOG}(c^n) = n\,\text{LOG}(c)$ if c is a positive number and n is a positive integer.

   Hint. Let $u = ab$ and let $v = c$ in identity (2) to prove (a). To prove (c) let $a_1 = a_2 = \cdots = a_n = c$ in the equation established in part (b).

1.7. Given a logarithmic function, suppose that a number e is found satisfying $\text{LOG}(e) = 1$.
   (a) Show, using the result of part (c) of exercise 1.6, that $\text{LOG}(e^n) = n$ for any positive integer n.
   (b) What would you guess that $\text{LOG}(e^x)$ is equal to?

§2. The Natural Log Function

A *logarithmic* (or *log*) *function* is any function that converts multiplication into addition, as in equation (2). We already have such a function in our hands.

Definition. The **natural logarithmic function** ln is defined on the domain $0 < x$ by the equation

$$\ln x = \int_1^x \frac{1}{t}\,dt. \tag{5}$$

Often one sees the expressions

$$\log_e x \quad \text{and} \quad L(x)$$

used in place of $\ln x$ to denote the natural log function. It is easy to prove that $\ln x$ (read "natural log of x" or simply "el-en of x") converts multiplication into addition. By means of this conversion we see that multiplication on the positive real numbers has the same structure as addition defined on all real numbers. A Greek word, *homomorphism* (literally, "same form") is used in modern mathematics to describe this remarkable similarity between the (apparently) unlike operations of multiplication and addition.

The Homomorphism Theorem. *If a and b are positive numbers, then*

$$\ln(ab) = \ln a + \ln b.$$

2. The Natural Log Function

PROOF. By **F4**,

$$\frac{d}{dx}(\ln x) = \frac{d}{dx}\int_1^x \frac{1}{t}\,dt = \frac{1}{x}, \qquad 0 < x. \tag{6}$$

By the chain rule, using the fact that $d(\ln x)/dx = 1/x$, we have

$$\frac{d}{dx}(\ln ax) = \frac{1}{ax}a = \frac{1}{x}, \qquad 0 < x.$$

Therefore, $\ln ax$ is an antiderivative of $1/x$. By **F3**, then,

$$\ln b = \int_1^b \frac{1}{t}\,dt = \ln at\bigg|_1^b = \ln ab - \ln a.$$

Solving this equation for the expression $\ln ab$ yields the desired result. □

The homomorphism theorem was, in essence, proved in problem 22 at the end of Chapter 8, but by a slightly different means. As our familiarity with calculus grows, we shall often find that there is more than one way to see that a given statement is true. For example, from problem 32 at the end of Chapter 8 we know that

$$\int_{1/r}^1 \frac{1}{t}\,dt = \int_1^r \frac{1}{t}\,dt \quad \text{if } 0 < r.$$

We should recognize the integral on the right-hand side here to be $\ln r$, by definition of the natural log function. Thus, if $0 < r$,

$$\ln r = \int_{1/r}^1 \frac{1}{t}\,dt = -\int_1^{1/r} \frac{1}{t}\,dt = -\ln\frac{1}{r}. \tag{7}$$

Theorem on Logarithms of Reciprocals. *If $0 < r$, then*

$$\ln\frac{1}{r} = -\ln r. \tag{8}$$

PROOF. Equation (8) is an immediate consequence of (7), but let us give a different proof, just for fun. A powerful tool for verifying identities, such as (8), is the uniqueness theorem **F1**. By **F1**, two differentiable functions must be identical on a connected domain if they have the same derivative and agree at just one point. (See exercise 1.5 of Chapter 8.) Let us show that $-\ln r$ and $\ln(1/r)$ have the same derivative:

$$\frac{d}{dr}(-\ln r) = -\frac{dr}{dr}(\ln r) = -\frac{1}{r} \quad [\text{by (6)}].$$

By the chain rule,

$$\frac{d}{dr}\left(\ln\frac{1}{r}\right) = \frac{1}{1/r}\cdot\frac{-1}{r^2} = -\frac{1}{r}.$$

We have shown that the two functions on either side of (8) have the same derivative. Since it is obvious that equation (8) holds when $r = 1$, the uniqueness theorem guarantees that equality holds throughout the connected domain $0 < r$. □

Theorem on Logarithms of Numbers Raised to Rational Powers. *Let m and n be integers, with $n \neq 0$. Then for any positive number x,*

$$\ln(x^{m/n}) = \frac{m}{n} \ln x. \tag{9}$$

PROOF. We shall prove the identity (9) by the uniqueness theorem. Obviously equation (9) is true at the point $x = 1$, when both sides of equation (9) are equal to 0.

Taking derivatives, we find that

$$\frac{d}{dx}\left(\frac{m}{n} \ln x\right) = \frac{m}{n} \frac{d}{dx}(\ln x) = \frac{m}{n} x^{-1}.$$

By the chain rule, using the rule* for differentiating rational powers of x,

$$\frac{d}{dx}(\ln x^{m/n}) = \frac{1}{x^{m/n}}\left(\frac{m}{n}\right) x^{m/n - 1} = \frac{m}{n} x^{-1}.$$

By **F1**, equation (9) holds throughout the connected domain $0 < x$. □

To become more familiar with the natural log function, let us set about plotting the graph of

$$L = \ln x, \qquad 0 < x.$$

Taking derivatives, we have

$$\frac{dL}{dx} = \frac{1}{x}, \qquad \frac{d^2 L}{dx^2} = \frac{-1}{x^2}, \qquad 0 < x.$$

Obviously the first derivative is always positive on the domain $0 < x$, and the second derivative is always negative. The curve $L = \ln x$ is therefore continually *rising*, but is *concave downwards*.

We are already familiar with some points on the graph. From equations (5) to (10) of Chapter 8 we have the information in the following table, where numbers are rounded off to three places.

x	$\ln L$
2	0.693
3	1.099
4	1.386
6	1.792
10	2.303
12	2.485

---

* Derived in exercise 8.10 of Chapter 7.

2. The Natural Log Function

From the theorem on logarithms of reciprocals we immediately know twice as much as before:

x	$\ln L$
0.500	−0.693
0.333	−1.099
0.250	−1.386
0.167	−1.792
0.100	−2.303
0.085	−2.485

The graph of the natural log function looks like this:

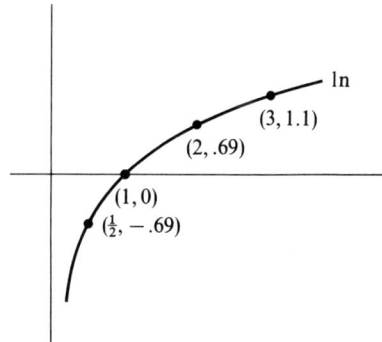

Using the homomorphism theorem and the other results of this section, we can calculate many new values of the natural log function.

EXAMPLE 1. Calculate the natural logarithm of (a) 100, (b) 50, (c) 0.01, (d) $\sqrt{6}$, (e) $\sqrt{6}/4$, and (f) $(100)^{2/3}$.

The following calculations make use of values tabulated above.
(a) $\ln 100 = \ln(10^2) = 2(\ln 10) = 2(2.303) = 4.606$.
(b) $\ln 50 = \ln(\frac{1}{2} \cdot 100) = \ln \frac{1}{2} + \ln 100 = -0.693 + 4.606 = 3.913$.
(c) $\ln 0.01 = \ln \frac{1}{100} = -\ln 100 = -4.606$.
(d) $\ln \sqrt{6} = \ln(6^{1/2}) = \frac{1}{2} \ln 6 = \frac{1}{2}(1.792) = 0.896$.
(e) $\ln(\sqrt{6}/4) = \ln \frac{1}{4}\sqrt{6} = \ln \frac{1}{4} + \ln \sqrt{6} = -1.386 + 0.896 = -0.490$.
(f) $\ln(100^{2/3}) = \frac{2}{3} \ln 100 = \frac{2}{3}(4.606) = 3.071$.

(In each of the calculations above there may be slight error in the last decimal place because the table used is accurate to only three places.)

EXERCISES

2.1. In this section we have two proofs of the fact that the natural log of $1/r$ is the negative of the natural log of r. Give a third proof of this fact, by showing that for any logarithmic function whatever, it is true that $LOG(1/r) = -LOG(r)$. *Hint.* In equation (2), let $u = 1/r$ and let $v = r$.

2.2. Having done exercise 2.1, you now know three different proofs that equation (8) is an identity.
   (a) Which of these proofs might be described as geometric? as algebraic? as analytic?
   (b) (*For students of aesthetics*) Which of these three proofs is the prettiest? Why?
   (c) Which of these proofs illustrates the most powerful general method of proving identities?

2.3. Use the chain rule to find the derivative of each of the following.
   (a) $y = \ln(1 + x^2)$.
   (b) $y = \ln(3 + \cos x)$.
   (c) $y = \ln \pi x$.
   (d) $y = \ln(\arccos x)$.
   (e) $y = \sin(\ln x)$.
   (f) $y = \arctan(\ln x)$.
   (g) $y = \sqrt{3 + (\ln x)^2}$.
   (h) $y = \cos^2(\ln \pi x)$.
   Answers: (a) $2x/(1 + x^2)$. (b) $(-\sin x)/(3 + \cos x)$. (d) $-1/\sqrt{1 - x^2} \arccos x$.
   (e) $(1/x)\cos(\ln x)$. (f) $1/(x + x \ln^2 x)$. (g) $(\ln x)/x\sqrt{3 + \ln^2 x}$.

2.4. Some of the functions defined by the rules in exercise 2.3 do not have an unrestricted domain. Find them and specify the largest domain on which they are defined. (Remember that the natural log function is defined only on the domain of positive real numbers.)
   Answers: (c) $0 < x$. (d) $-1 \leq x < 1$. (e)–(h) $0 < x$.

2.5. Write an equation of the tangent line to the curve $y = \ln x$ at the point
   (a) $(1, 0)$.
   (b) $(\frac{1}{2}, -0.693)$.
   (c) $(4, 1.386)$.
   Answers: (a) $y = x - 1$. (c) $y - 1.386 = 0.250(x - 4)$.

2.6. Prove that the natural log function converts division into subtraction, i.e., $\ln(b/a) = \ln b - \ln a$ if a and b are positive numbers. (Can you give more than one proof?)

2.7. Use only the short table given in this section together with the properties of the natural log function to find the natural logarithm of each of the following.
   (a) 3000.
   (b) $\sqrt{3000}$.
   (c) $\sqrt{3000/100^{2/3}}$.
   (d) 9.
   (e) 0.009.
   (f) 5.
   (g) $2(3/2)^{1/10}$.
   (h) $3\sqrt{2/4}\sqrt{12}$.
   (i) $10(16)^{3/8}$.
   (j) 3^{1000}.
   Answers: (b) 4.003. (c) 0.932. (e) -4.711. (f) 1.609. (g) 0.734. (h) -1.183. (j) 1099.

2.8. The domain of $\ln$ is given by $0 < x$. By carrying out the following steps, establish the range of $\ln$.
   (a) Show that $\ln(3^n) > n$, and $\ln(3^{-n}) < -n$, for any positive integer n. *Hint*. $\ln(3^n) = n \ln 3$.
   (b) Use part (a) to conclude that the range of $\ln$ includes $-n \leq y \leq n$, for any integer n.
   (c) Use part (b) to conclude that the range of $\ln$ is unrestricted.

2.9. Does $\ln x$, $0 < x$, have an inverse $\ln^{-1}$? If so, what is the domain and what is the range of $\ln^{-1}$?

2.10. Justify each of the following equalities, where a, b, and k are positive.

$$\ln b - \ln a = \int_a^b \frac{1}{t} dt = \ln\frac{b}{a} = \int_{ka}^{kb} \frac{1}{t} dt.$$

§3. The Exponential Function

The natural log function has an inverse $\ln^{-1}$ since the curve $y = \ln x$ is always increasing. What is $\ln^{-1}$? It is called the *exponential function* exp, and its place is at the very center of the whole discipline of calculus. *Why?* Because the exponential function, as we shall see, is *its own derivative*!

Definition. The **exponential function** exp is defined as the inverse of the natural log function. That is, exp is the function having the following properties:

$$\ln(\exp x) = x \quad \text{for all } x; \tag{10}$$

$$\exp(\ln x) = x, \quad 0 < x. \tag{11}$$

Equations (10) and (11) should not seem mysterious. If they do, the reader should review problem 26 at the end of Chapter 7. The equations simply say that the function exp has the effect of "undoing" whatever the function ln does. The range of the exponential function is the set of positive real numbers, since this is the domain of the natural log. The domain of exp is unrestricted since (see exercise 2.8) the range of ln is unrestricted.

What does the graph of the exponential function look like? Being the inverse of the natural log, the exponential function has a graph that is the reflection through the line $y = x$ of the graph of ln.

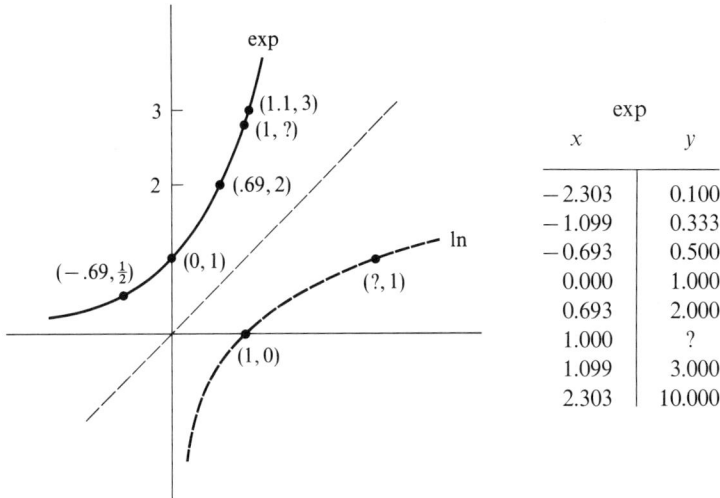

exp	
x	y
−2.303	0.100
−1.099	0.333
−0.693	0.500
0.000	1.000
0.693	2.000
1.000	?
1.099	3.000
2.303	10.000

Since the natural log converts multiplication into addition, it should come as no surprise that the inverse function has the opposite effect:

Homomorphism Theorem. *For all real numbers a and b,*
$$\exp(a + b) = (\exp a)(\exp b). \qquad (12)$$

The proof of this is easy and is left to the reader as an exercise. The most striking property of the exponential function, which gives it a unique status among all functions studied in calculus, is the following:

Theorem on Differentiating Exponentials. *The exponential function is its own derivative:*
$$(\exp)' = \exp; \qquad (13)$$
moreover, if f is any differentiable function then
$$\frac{d}{dx}\exp(f(x)) = f'(x)\exp(f(x)). \qquad (14)$$

PROOF. Equation (10) says
$$\ln(\exp x) = x \quad \text{for all } x.$$
Differentiating both sides yields
$$\frac{1}{\exp x}(\exp)'(x) = 1.$$
Multiplying this equation through by $\exp x$, we get
$$(\exp)'(x) = \exp x,$$
which proves (13).

Equation (14) is an easy consequence of (13) and the chain rule. If $y = \exp(f(x))$, then we have the chain $y = \exp u$ and $u = f(x)$. By the chain rule,
$$\frac{dy}{dx} = \frac{dy}{du}\frac{du}{dx} = (\exp u)f'(x) = f'(x)\exp(f(x)). \qquad \square$$

The application of the rule for differentiating exponentials is straightforward:
$$\frac{d}{dx}\exp x = \exp x,$$
$$\frac{d}{dx}\exp 2x = 2\exp 2x,$$
$$\frac{d}{dx}\exp(x^2) = 2x\exp(x^2),$$
$$\frac{d}{dx}\exp(\sin x) = (\cos x)\exp(\sin x).$$
(15)

3. The Exponential Function

Here are some more examples.

$$\frac{d}{dx}\sqrt{1 + \exp x} = \frac{\exp x}{2\sqrt{1 + \exp x}}, \quad \text{by the square root rule;}$$

$$\frac{d}{dx}(\exp x)^2 = 2(\exp x)(\exp x) = 2(\exp x)^2, \quad \text{by the rule for squares;} \quad (16)$$

$$\frac{d}{dx}\left(\frac{1}{\exp x}\right) = \frac{-1}{(\exp x)^2}(\exp x) = \frac{-1}{\exp x}, \quad \text{by the reciprocal rule.} \quad \square$$

Look again at the graph of the exponential function. *What is* exp 1? From the graph it appears to be somewhere between 2.5 and 3. Later on we shall see how this number can be calculated to any desired degree of accuracy. As we shall see in Section 9,

$$\exp 1 = 2.71828\ldots. \quad (17)$$

It is convenient to introduce a special symbol to stand for this number, which plays a central role in the theory of exponentials. It is called e.

Definition. The number e is defined by the equation

$$e = \exp 1. \quad (18)$$

Equivalently, e is that number satisfying the equation

$$\ln e = 1. \quad (19)$$

By (17) and (18), $e = 2.71828\ldots$. The number e is very special. Like π, it pops up in the most unexpected places.

Why is the number e so important to the exponential function? Consider the following. For any rational number m/n we have

$$\ln(e^{m/n}) = \frac{m}{n}\ln e \quad \text{[by (9)]}$$

$$= \frac{m}{n} \quad \text{[by (19)]}.$$

Therefore, $m/n = \ln(e^{m/n})$, so

$$\exp\frac{m}{n} = \exp(\ln(e^{m/n}))$$

$$= e^{m/n} \quad \text{[by (11)]}.$$

This shows that

$$\exp x = e^x, \quad (20)$$

whenever $x = m/n = $ a rational number. The exponential function exp is now seen to be aptly named. By (20) it is indeed an exponential function, and its base is e.

EXERCISES

3.1. Suppose that $\ln s = \ln t$, where s is a positive number and t is a positive number. Prove that $s = t$ either
   (a) by an informal argument referring to the rising graph of the natural log function; or
   (b) by writing
   $$0 = \ln t - \ln s = \int_s^t \frac{1}{x} dx$$
   and arguing that the only way the integral can be zero is for s and t to be equal.

3.2. (a) Show that $\ln(\exp(a+b)) = a + b$.
   (b) Show that $\ln[(\exp a)(\exp b)] = a + b$ also, by using the homomorphism property of the natural log, together with property (10).
   (c) Use the results of parts (a) and (b), together with exercise 3.1, to prove the identity (12).

3.3. Check to see which of the following are identities. (*Take the natural log of both sides in each equation. By exercise 3.1, positive numbers are equal if and only if they have the same natural logarithm.*)
   (a) $\exp(x^2) = 2 \exp x$.
   (b) $(\exp x)^2 = \exp 2x$.
   (c) $(\exp x)^n = \exp nx$.
   (d) $(\exp a)/(\exp b) = \exp(a/b)$.
   (e) $(\exp a)/(\exp b) = \exp(a - b)$.
   Partial answer: Equation (a) is not an identity. The natural log of the left-hand side is x^2, while the natural log of the right-hand side is $\ln 2 + x$.

3.4. Write an equation of the tangent line to the graph of $y = \exp x$ at the point
   (a) $(0, 1)$.
   (b) $(1, e)$.
   Answers: (a) $y = x + 1$. (b) $y - e = e(x - 1)$.

3.5. Find the second derivative of the exponential function. Then find the 37-th derivative of it.

3.6. Find the derivative of each of the following.
   (a) $x \exp x$.
   (b) $\exp(\tan x)$.
   (c) $\exp \pi x$.
   (d) $(\ln x)(\exp x)$.
   (e) $(\sin x)(\exp 3x)$.
   (f) $\sin(\exp 3x)$.
   Answers: (a) $\exp x + x \exp x$. (b) $(\sec^2 x) \exp(\tan x)$. (c) $\pi \exp \pi x$. (d) $(\ln x)(\exp x) + (1/x)(\exp x)$.

3.7. The exponential function is continuous everywhere (because it is differentiable everywhere). Use the continuity of exp to answer the following.
   (a) What is $\text{Limit}_{x \to \pi} \exp x$?
   (b) Does the integral $\int_0^2 \exp t \, dt$ exist?
   (c) What is $d(\int_\pi^x \exp t \, dt)/dx$?
   (d) What is $\text{Limit}(\exp(1/n))$ as n increases without bound?
   Answers: (a) $\exp \pi$. (d) 1.

3. The Exponential Function

3.8. Since the exponential function is its own derivative, the exponential function is also its own antiderivative. Use this to evaluate the following integrals by **F3**.
   (a) $\int_0^2 \exp t \, dt$.
   (b) $\int_0^1 \exp t \, dt$.
   (c) $\int_1^\pi \exp t \, dt$.
   Answers: (a) $e^2 - 1$. (b) $e - 1$. (c) $\exp \pi - e$.

3.9. Is exp an even function? an odd function? (Look at its graph to tell.) Write the exponential function as the sum of an even function and an odd function. [See exercise 5.9, parts (i)–(k), in Chapter 8.]
   Answer: See problem 5 at the end of this chapter.

3.10. (*Only for those with a Pythagorean love of numbers*) The following is pure play with numbers, for which only a little justification is offered. By serendipity we are able to approximate the number *e* very nicely.
   (a) Simpson's rule is expected to work better and better over smaller and smaller intervals. Convince yourself that the following approximation ought to be quite good if *n* is a large integer.

$$e^{1/n} - e^{-1/n} = \int_{-1/n}^{1/n} \exp$$

$$\approx S_{-1/n}^{1/n} \exp = \frac{1}{3n}(e^{-1/n} + 4 + e^{1/n}).$$

   (b) Multiplying both sides of an approximation by a large positive integer *n* offers no guarantee that the approximation will still be close. Nevertheless, multiply the approximation of part (a) so as to obtain

$$(3n - 1)e^{2/n} - 4e^{1/n} - (3n + 1) \approx 0.$$

   (c) The expression on the left in part (b) may be regarded as a quadratic in $e^{1/n}$. Apply the quadratic formula, throwing away (*why?*) the negative root, and obtain

$$e^{1/n} \approx \frac{2 + \sqrt{3 + 9n^2}}{3n - 1}.$$

   (d) Even if the approximation in part (c) is very close there is no guarantee it will still be close if both sides are raised to the *n*-th power. Nevertheless, do this hopefully, writing

$$e \approx \left[\frac{2 + \sqrt{3 + 9n^2}}{3n - 1} \right]^n.$$

   (e) The approximation in part (d) is expected to get better as *n* is taken larger and larger. Evaluate the expression on the right, using a table of square roots, for
   (i) $n = 1$.
   (ii) $n = 2$.
   (iii) $n = 4$.
   (iv) $n = 8$.
   Answers: (i) 2.732. (ii) 2.719. (iii) 2.7183. (iv) 2.718286.

§4. Exotic Numbers

Everyone today has heard the phrase "exponential growth", especially in connection with the alarming rate at which the world's population is rising. The phrase refers, of course, to the rapid rise of the exponential function's graph. The exponential function also pops up in studying the decay of radioactive material and in the theory of compound interest. We shall study these "external" applications of the exponential function shortly. Right now, let us study how the exponential function is applied "internally" to mathematics itself, rather than to the external world.

Consider such an exotic expression as

$$e^{\pi}. \tag{21}$$

What could this mean? What should it mean? Is it a real number? Can we make sense out of such expressions as (21), or, for example,

$$2^{\pi}, \quad e^{\sqrt{3}}, \quad 3^{\sqrt{2}}, \quad \text{and} \quad 6^{e}? \tag{22}$$

How can a number be raised to an *irrational* power?

As we have done so often in the past, we must call on our old friend Lim. Consider the expression (21). Let us first say what it should *not* mean. It should not mean any of the following numbers:

$$e^{3}, \quad e^{3.1}, \quad e^{3.14}, \quad e^{3.141}, \quad e^{3.1415}, \ldots$$

The expression e^{π} should not denote any of these numbers, but should denote their *limit*. That is, e^{π} should be defined as

$$e^{\pi} = \lim_{x \to \pi} e^{x},$$

where x tends to π through rational values, like 3, 3.1, 3.14, etc., in the sequence given above. But $e^{x} = \exp x$ if x is rational, by (20). Therefore the definition just given implies that

$$e^{\pi} = \lim_{x \to \pi} \exp x = \exp \pi,$$

by the continuity of the exponential function. This answers our question about raising a number to an irrational power, if that number happens to be e. By the same token as above, it is natural to define e^{t} as follows:

$$e^{t} = \exp t$$

for any real number t, rational or irrational. For example,

$$e^{\sqrt{3}} = \exp \sqrt{3},$$
$$e^{0} = \exp 0 = 1,$$
$$e^{\ln 2} = \exp(\ln 2) = 2,$$
$$e^{\ln a} = \exp(\ln a) = a,$$
$$e^{-1} = \exp(-1) = \frac{1}{e}.$$

4. Exotic Numbers

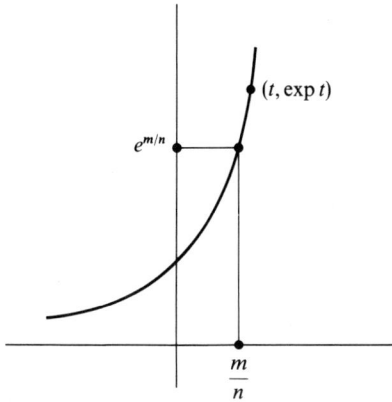

Since e^t and $\exp t$ agree when t is rational, it is natural to define e^t as $\exp t$ for all t

Having seen how to define e^t for any real number t, we can apply similar considerations to define a^t, where a is any positive number. From equation (9) we already know that the following identity holds for every rational number:

$$\ln(a^{m/n}) = \frac{m}{n} \ln a.$$

It is natural to define the number a^t in such a way that the equation

$$\ln a^t = t \ln a \tag{23}$$

becomes an identity, holding for all real numbers t. By exponentiating both sides, we see that (23) is equivalent to

$$a^t = \exp(t \ln a) = e^{t \ln a}. \tag{24}$$

We are thus led to this definition.

Definition. Let a be a positive number and let t be any real number. Then a^t is defined by equation (24).

The exotic numbers listed in (22) are then rendered as follows:

$$2^\pi = \exp(\pi \ln 2) \approx \exp(3.14(0.69)) = \exp 2.17 = e^{2.17},$$
$$e^{\sqrt{3}} = \exp(\sqrt{3} \ln e) = \exp(\sqrt{3}) \approx \exp(1.73) = e^{1.73},$$
$$3^{\sqrt{2}} = \exp(\sqrt{2} \ln 3) \approx \exp(1.41(1.10)) = \exp 1.55 = e^{1.55},$$
$$6^e = \exp(e \ln 6) \approx \exp(2.72(1.80)) = \exp 4.90 = e^{4.90}.$$

Any positive real number can be expressed as a power of e. The problem of calculating such powers as $e^{2.17}$, $e^{1.73}$, etc., to several decimal places is discussed in a later section.

We should check that this definition of raising numbers to powers agrees with our familiar notion in cases where the situation is familiar. For example,

$$2^{-1} = \exp((-1)\ln 2) = \exp(-0.69\ldots) = 0.500.$$

$$\left(\frac{1}{2}\right)^2 = \exp\left(2\ln\frac{1}{2}\right) = \exp(2(-0.69\ldots)) = \exp(-1.38\ldots) = 0.250.$$

$$\left(\frac{1}{4}\right)^n = \exp\left(n\ln\frac{1}{4}\right) = \exp(n(-\ln 4)) = \exp(-n\ln 4) = \exp(\ln 4^{-n}) = 4^{-n}.$$

What about the laws of exponents that we are used to? Do they still hold? Is $(a^s)^t$ equal to a^{st}, for example? By definition we have

$$(a^s)^t = \exp(t \ln a^s), \tag{25}$$
$$a^{st} = \exp(st \ln a). \tag{26}$$

The following theorem answers some questions.

Theorem on Exponentials. *Let a be a positive number, and let s and t be any real numbers. Then*

$$(a^s)^t = a^{st}; \tag{27}$$
$$a^s a^t = a^{s+t}. \tag{28}$$

PROOF. To prove (27) first note that

$$\ln a^s = s \ln a \tag{29}$$

by (23). Therefore,

$$(a^s)^t = \exp(t \ln a^s) \qquad [\text{by (25)}]$$
$$= \exp(st \ln a) \qquad [\text{by (29)}]$$
$$= a^{st}. \qquad [\text{by (26)}]$$

A proof of (28) is given by the following chain of equalities. The first and last of these equalities are by definition. The crucial step involves the homomorphism theorem [identity (12) of Section 3].

$$a^s a^t = [\exp(s \ln a)][\exp(t \ln a)] = \exp((s+t)\ln a) = a^{s+t}. \qquad \square$$

At last we can prove a rule of differentiation that the reader probably was able to guess in Chapter 4, problem 32.

General Rule for Powers. *If $f(x) = x^t$, with domain $0 < x$, then f has a derivative given by*

$$f'(x) = tx^{t-1}.$$

PROOF. By definition of x^t we have

$$f(x) = \exp(t \ln x).$$

4. Exotic Numbers

By (14),
$$f'(x) = \frac{t}{x}\exp(t \ln x) = \frac{t}{x}x^t = tx^{t-1}.$$ □

For example,

if $f(x) = x^\pi$, then $f'(x) = \pi x^{\pi-1}$;
if $f(x) = x^e$, then $f'(x) = ex^{e-1}$;
if $f(x) = x^{-1}$, then $f'(x) = -x^{-2}$.

The reader should note the sharp contrast between powers and exponentials.

General Rule for Exponentials. *Let a be a positive number and let $f(x) = a^x$. Then f has a derivative given by*
$$f'(x) = (\ln a)a^x.$$

PROOF. By definition of a^x, we have
$$f(x) = \exp(x \ln a).$$
By (14),
$$f'(x) = (\ln a)\exp(x \ln a) = (\ln a)a^x.$$ □

For example,

if $f(x) = \pi^x$, then $f'(x) = (\ln \pi)\pi^x$;
if $f(x) = 2^x$, then $f'(x) = (\ln 2)2^x$;
if $f(x) = e^x$, then $f'(x) = (\ln e)e^x = e^x$.

As seen in these examples, *the derivative of any exponential function is proportional to the function itself*. If $f(x) = a^x$, then $f'(x) = (\ln a)f(x)$, by the rule given above. The constant of proportionality is $\ln a$. The natural base for an exponential function is e. Only in this case is the constant of proportionality equal to one.

EXERCISES

4.1. Formulas (27) and (28) of this section were proved by the "direct" method. Give indirect proofs of these formulas by showing that both sides of each formula have the same natural log. Then appeal to the result of exercise 3.1.

4.2. Find the derivative of each of the following.
   (a) $x^\pi e^x$.
   (b) $x^{\sqrt{2}} \ln x$.
   (c) 10^x.
   (d) x^{10}.
   (e) π^e.
   Answers: (a) $x^\pi e^x + \pi x^{\pi-1}e^x$. (b) $x^{\sqrt{2}-1}(1 + \sqrt{2}\ln x)$. (c) $(\ln 10)10^x$. (d) $10x^9$. (e) 0.

4.3. Find the derivative of each of the following.
  (a) e^x.
  (b) e^{2x}.
  (c) e^{x^2}.
  (d) $e^{\sin x}$.
  (e) $\sqrt{1 + e^x}$.
  (f) $(e^x)^2$.
  (g) $1/e^x$.
  Answer: (The answers are given in formulas (15) and (16), where the notation $\exp x$ is used in place of e^x.)

4.4. Find the derivative of each of the following.
  (a) 3^x.
  (b) 3^{2x}.
  (c) 3^{x^2}.
  (d) $3^{\sin x}$.
  (e) $\sqrt{1 + 3^x}$.
  (f) $(3^x)^2$.
  (g) $1/3^x$.
  Answers: (b) $2(\ln 3)3^{2x}$. (c) $2x(\ln 3)3^{x^2}$. (d) $(\cos x)(\ln 3)3^{\sin x}$.

4.5. Which of the following are identities?
  (a) $3^{xy} = 3^{y^x}$.
  (b) $(3^x)^y = (3^y)^x$.
  (c) $5^x 5^y = 5^{xy}$.
  (d) $10^t = e^{10 \ln t}$.
  (e) $10^t = e^{t \ln 10}$.
  (f) $\sqrt{e^x} = e^{x/2}$.
  (g) $e^{-2x} = (1/e^x)^2$.
  (h) $e^x e^y = e^{x+y}$.
  (i) $e^{xy} = e^x e^y$.
  (j) $a^x/b^x = (a/b)^x$.
  Hint. One way to test for equality of numbers is to see if they have the same natural log.

4.6. Find an antiderivative of the function f given by $f(x) = x^t$, $0 < x$. Be careful. Answer: If $t \neq -1$, an antiderivative is given by $F(x) = x^{t+1}/(t + 1)$; if $t = -1$, an antiderivative is $F(x) = \ln x$.

4.7. (a) Use **F3** together with the result of exercise 4.6 to evaluate the integral $\int_1^a x^t \, dx$, where $t \neq -1$.
  (b) Evaluate $\int_1^a x^t \, dx$, where $t = -1$.
  (c) On the basis of your answers to parts (a) and (b) explain why it is plausible to conjecture that
  $$\underset{t \to -1}{\text{Limit}} \frac{a^{t+1} - 1}{t + 1} = \ln a \quad \text{if } 0 < a.$$
  (d) Use L'Hôpital's rule to evaluate the limit in part (c). Is the conjecture true?

4.8. Consider the problem of finding $\text{Limit}_{h \to 0}((a^h - 1)/h)$, where a is some positive number. Do this in three ways:
  (a) By letting $h = t + 1$, justify each of the following steps:
  $$\underset{h \to 0}{\text{Limit}} \frac{a^h - 1}{h} = \underset{t \to -1}{\text{Limit}} \frac{a^{t+1} - 1}{t + 1} = \ln a.$$
  (b) Use the derivative to find the slope of the tangent line to the curve $f(x) = a^x$ at the point $(0, 1)$. Equate this number with the limit that results from applying Fermat's method to find the slope of this tangent line.
  (c) Evaluate the limit directly by an application of L'Hôpital's rule.

4.9. Consider the problem of finding $\text{Limit}_{h \to 0}((\ln(1 + hk))/h)$, where k is some real number. Do this in three ways.
(a) First justify the following equalities:

$$\frac{1}{h}\ln(1 + hk) = \frac{1}{h}\int_1^{1+hk} \frac{1}{t} dt$$

$$= k\frac{1}{hk}\int_1^{1+hk} \frac{1}{t} dt = k\left[\operatorname*{avg}_{1,\,1+hk}\left(\frac{1}{t}\right)\right].$$

Then find the limit as h tends to zero by using the continuity of the function $1/t$ at the point $t = 1$.
(b) Use the derivative to find the slope of the tangent line to the curve $f(x) = \ln(1 + kx)$ at the point $(0, 0)$. Equate this number with the limit that arises by applying Fermat's method to find the slope of this tangent line.
(c) Evaluate the limit by applying L'Hôpital's rule.

4.10. (a) Give an example of a function whose rate of growth is proportional to itself. That is, find an example of a function f such that if $y = f(t)$, then $dy/dt = ky$, where k is some constant of proportionality.
(b) Can you think of any quantity from the "real world" that might be expected to grow always at a rate proportional to its size?
Hints. (a) Read the last paragraph in this section.
(b) Read Sections 6, 7, and 8.

§5. Uses of Logarithms; Logarithmic Differentiation

A table of logarithms may be found in an appendix to this book. Both natural logs and logarithms to the base 10 are given. While it is true that the natural log function ln makes calculus simpler, nevertheless our decimal representation of real numbers gives some advantage to the function $\log_{10}$. How is ln related to $\log_{10}$? Recall from Section 1 that

$$\log_{10} a = t \quad \text{means} \quad 10^t = a.$$

It is now easy to solve the equation $10^t = a$ for t. Taking the natural log of both sides gives

$$t \ln 10 = \ln a,$$

$$t = \frac{\ln a}{\ln 10}.$$

Therefore,

$$\log_{10} a = \frac{\ln a}{\ln 10} \approx \frac{\ln a}{2.302585} = 0.43429 \ln a. \tag{30}$$

This shows that $\log_{10}$ is simply a multiple of the natural log function. The same is true of logarithms to any base b whatever.

Theorem on Logarithms to Arbitrary Bases. *Let b be a positive number, with* $b \neq 1$. *Then*

$$\log_b x = \frac{\ln x}{\ln b}, \quad 0 < x. \tag{31}$$

In particular,
$$\log_e x = \ln x. \tag{32}$$

PROOF. To prove (31), let
$$t = \log_b x. \tag{33}$$

By (1), then,
$$b^t = x.$$

Therefore,
$$t \ln b = \ln x,$$

$$t = \frac{\ln x}{\ln b}. \tag{34}$$

Putting (33) and (34) together proves (31). Letting $b = e$ in equation (31) proves (32), which shows that *the natural log function is the logarithmic function to the base e.* □

A logarithmic function to any base is then just a multiple of the natural log function. By virtue of this, a logarithmic function to any base inherits the homomorphism property already proved for the natural log, plus all the properties that follow from the homomorphism property. Just as e^x gives the inverse function for the natural log, so does b^x give the inverse function for $\log_b$, provided $b \neq 1$. Whether one wishes to work with logarithms to base e, base 10, or to some other base is largely a matter of convenience.

EXAMPLE 2. Use a table of logarithms to calculate each of the following numbers to several decimal places:

(a) $\sqrt{12}$.
(b) $10e^{0.12}$.
(c) $100(1.06)^{20}$.

The method is first to find the logarithm of the number in question, then to use the inverse log function to find the number itself. We shall do each of these in two ways, first using natural logs, then logs to base 10. Tables are in an appendix.

(a) Let $x = \sqrt{12}$. Then, using natural logs, we have $\ln x = \frac{1}{2} \ln 12 \approx 1.242$.
By the inverse log function,
$$x \approx e^{1.242} \approx 3.46.$$

Using logs to base 10 instead, we have $\log_{10} x = \frac{1}{2} \log_{10} 12 \approx 0.5396$.
By the inverse log function,
$$x \approx 10^{0.5396} \approx 3.46.$$

5. Uses of Logarithms; Logarithmic Differentiation

(b) Let $x = 10e^{0.12}$. Using natural logs we have
$$\ln x = \ln 10 + 0.12 \ln e = \ln 10 + 0.12 \approx 2.423.$$
By the inverse log function,
$$x \approx e^{2.423} \approx 11.28.$$
Using logs to base 10 instead, we have
$$\log_{10} x = \log_{10} 10 + 0.12 \log_{10} e \approx 1 + 0.0521.$$
By the inverse log function,
$$x \approx 10^{1.0521} \approx 11.28.$$

(c) Let $x = 100(1.06)^{20}$. Using natural logs we have
$$\ln x = \ln 100 + 20 \ln 1.06 \approx 4.605 + 1.165 = 5.770.$$
By the inverse log function,
$$x \approx e^{5.770} \approx 320.7.$$
Using logs to base 10, we have
$$\log_{10} x = \log_{10} 100 + 20 \log_{10} 1.06 \approx 2 + 0.5061.$$
By the inverse log function,
$$x \approx 10^{2.5061} \approx 320.7. \qquad \square$$

Whenever an expression involves products, quotients, or powers, it is often easier to deal with the logarithm of the expression than with the expression itself.

EXAMPLE 3. Find the limit, as h tends to 0, of each of the following:

(a) $\exp((\sin h)/h)$.
(b) $(1 + h)^{1/h}$.
(c) $(1 + hk)^{1/h}$.

Each of these expressions is rather complicated, but its log is not so complicated. Let us first find the limits of the logs of these expressions. Once this is done, it will be easy to find the limits of the expressions themselves.

(a) Let $y = \exp((\sin h)/h)$. Then $\ln y = (\sin h)/h$. By L'Hôpital's rule,
$$\underset{h \to 0}{\text{Limit}} \frac{\sin h}{h} = \underset{h \to 0}{\text{Limit}} \frac{\cos h}{1} = \cos 0 = 1.$$
This shows that $\ln y$ tends to 1. This immediately implies that y tends to e. *Reason*: By continuity of the exponential function,
$$y = \exp(\ln y) \to \exp 1 = e.$$

(b) Let $y = (1 + h)^{1/h}$. Then $\ln y = (1/h)\ln(1 + h)$. By L'Hôpital's rule,

$$\underset{h \to 0}{\text{Limit}} \frac{\ln(1 + h)}{h} = \underset{h \to 0}{\text{Limit}} \frac{1/(1 + h)}{1} = 1.$$

This shows that $\ln y$ tends to 1, which immediately implies that y tends to e.

(c) Let $y = (1 + hk)^{1/h}$. Then $\ln y = (1/h)\ln(1 + hk)$. By L'Hôpital's rule,

$$\underset{h \to 0}{\text{Limit}} \frac{\ln(1 + hk)}{h} = \underset{h \to 0}{\text{Limit}} \frac{k/(1 + hk)}{1} = k.$$

This shows that $\ln y$ tends to k, which immediately implies (*why?*) that y tends to e^k. □

EXAMPLE 4. Find dy/dx, given each of the following:

(a) $y = 2^x$.
(b) $y = x^x$.
(c) $y = x^2/(1 + x^2)^{3/2}$.

Each of these expressions is fairly complicated, but its log is somewhat simpler to deal with. Let us first find $\ln y$, then use implicit differentiation. (The technique of using implicit differentiation after applying the logarithmic function is known as **logarithmic differentiation**.)

(a) If $y = 2^x$, then $\ln y = x \ln 2$. Differentiation of both sides with respect to x yields

$$\frac{1}{y}\frac{dy}{dx} = \ln 2.$$

Therefore,

$$\frac{dy}{dx} = y \ln 2 = 2^x \ln 2.$$

(b) If $y = x^x$, then $\ln y = x \ln x$. Differentiation of both sides with respect to x yields

$$\frac{1}{y}\frac{dy}{dx} = x\left(\frac{1}{x}\right) + \ln x = 1 + \ln x.$$

Therefore,

$$\frac{dy}{dx} = y(1 + \ln x) = x^x(1 + \ln x).$$

(c) If $y = x^2/(1 + x^2)^{3/2}$, then $\ln y = 2 \ln x - (3/2)\ln(1 + x^2)$. Differentiation of both sides with respect to x yields

$$\frac{1}{y}\frac{dy}{dx} = 2\left(\frac{1}{x}\right) - \left(\frac{3}{2}\right)\frac{2x}{1 + x^2} = \frac{2}{x} - \frac{3x}{1 + x^2}.$$

5. Uses of Logarithms; Logarithmic Differentiation

Therefore,
$$\frac{dy}{dx} = y\left(\frac{2}{x} - \frac{3x}{1+x^2}\right) = \frac{x^2}{(1+x^2)^{3/2}}\left(\frac{2}{x} - \frac{3x}{1+x^2}\right). \qquad \square$$

The simplicity achieved by the use of logarithmic differentiation is obvious. A seeming drawback to this technique is that we can take the logarithm only of *positive* quantities. Can the technique be made applicable to differentiate negative quantities? The answer is *yes, by taking absolute values first*. The following theorem explains why.

Theorem on Logarithmic Differentiation. *If f is a differentiable function (whether positive or negative), then*

$$\frac{d}{dx}\ln|f(x)| = \frac{f'(x)}{f(x)} \quad \text{if } f(x) \neq 0. \qquad (35)$$

PROOF. We must consider separately the cases when $f(x)$ is positive and when $f(x)$ is negative, and verify that the formula (35) holds in each case. This is easy to do.

If $f(x)$ is positive, then formula (35) asserts that

$$\frac{d}{dx}\ln f(x) = \frac{f'(x)}{f(x)},$$

which is easily seen to be true by a straightforward application of the chain rule.

If $f(x)$ is negative, then $|f(x)| = -f(x)$ by definition of the absolute value of a negative quantity. In this case the assertion of formula (35) is that

$$\frac{d}{dx}\ln(-f(x)) = \frac{f'(x)}{f(x)}. \qquad (36)$$

This again is easily seen to be true by the chain rule. The reader is asked to verify (36) as an exercise. $\square$

EXAMPLE 5. Use logarithmic differentiation to find the derivative of each of the following:

(a) $(x+1)/(x-1)$.
(b) $x(x^2+3)^{1/4}/(x-6)$.
(c) $f(x)g(x)$.

In each case we take absolute values first, since we cannot take the logarithm of a negative quantity. The formulas we obtain will be valid everywhere except for a few points that we shall note at the end.

(a) Let $y = (x+1)/(x-1)$. Then $|y| = |x+1|/|x-1|$, so

$$\ln|y| = \ln|x+1| - \ln|x-1|. \qquad (37)$$

By the theorem on logarithmic differentiation,

$$\frac{y'}{y} = \frac{1}{x+1} - \frac{1}{x-1}$$

$$y' = y\left(\frac{1}{x+1} - \frac{1}{x-1}\right) = \left(\frac{x+1}{x-1}\right)\left(\frac{1}{x+1} - \frac{1}{x-1}\right).$$

(b) Let $y = x(x^2+3)^{1/4}/(x-6)$. Then $|y| = |x||x^2+3|^{1/4}/|x-6|$, so

$$\ln|y| = \ln|x| + \frac{1}{4}\ln|x^2+3| - \ln|x-6|. \tag{38}$$

Logarithmic differentiation therefore shows that

$$\frac{y'}{y} = \frac{1}{x} + \left(\frac{1}{4}\right)\frac{2x}{x^2+3} - \frac{1}{x-6}.$$

Multiplying through by y, we obtain

$$y' = \frac{x(x^2+3)^{1/4}}{x-6}\left(\frac{1}{x} + \frac{\frac{1}{2}x}{x^2+3} - \frac{1}{x-6}\right).$$

(c) Let $y = f(x)g(x)$. Then $|y| = |f(x)||g(x)|$, so

$$\ln|y| = \ln|f(x)| + \ln|g(x)|. \tag{39}$$

Assuming f and g to be differentiable, we obtain

$$\frac{y'}{y} = \frac{f'(x)}{f(x)} + \frac{g'(x)}{g(x)}.$$

Multiplying through by $y = f(x)g(x)$, we derive the product rule

$$y' = f(x)g(x)\left(\frac{f'(x)}{f(x)} + \frac{g'(x)}{g(x)}\right) = g(x)f'(x) + f(x)g'(x). \quad \square$$

There is one slight drawback to the method of logarithmic differentiation. The number 0 is not in the domain of the natural log function. Therefore equation (37) does not hold when $x = 1$ or when $x = -1$. Equation (38) fails when $x = 0$ or when $x = 6$. And equation (39) is not true when either f or g takes the value zero. The advantages of logarithmic differentiation far outweigh this slight drawback.

EXERCISES

5.1. We may speak of logarithms to any base b, provided b is positive and *not equal to one*. Why do we not speak of "logarithms to base 1"?

5.2. Use (31) to prove each of the following.
   (a) $\log_b a = 1/\log_a b$ if a and b are both positive and neither is equal to 1.
   (b) $(\log_b a)(\log_c b) = \log_c a$ if a, b, and c are positive and neither is equal to 1.
   (c) $\log_b xy = \log_b x + \log_b y$ if b, x, and y are positive and $b \neq 1$.

5. Uses of Logarithms; Logarithmic Differentiation

5.3. We have seen in this section that a logarithmic function to any base is simply a certain multiple of the natural log function. In a similar vein, show that an exponential function to any base is simply a certain *power* of the exponential function e^x. Answer: $b^x = (e^x)^{\ln b}$.

5.4. In exercise 1.4 we made guesses about approximations to logarithms of certain numbers to certain bases. Use tables, together with formula (31) if needed, to find these logarithms correct to several decimal places.

5.5. Find each of the following logarithms to several decimal places, by any means.
  (a) $\log_{10} \pi$.
  (b) $\log_\pi 10$.
  (c) $\log_e 3$.
  (d) $\log_3 e$.
  (e) $\log_2 3$.
  (f) $\log_{1/2} 3$.
  (g) $\log_{1/4} 10$.
  (h) $\log_4 10$.
  Answers: (a) 0.497. (e) 1.586. (f) -1.586. (h) 1.661.

5.6. Find the limit as h tends to zero:
  (a) $\exp((\cos h - 1)/h)$.
  (b) $(1 + h)^h$.
  (c) $(1 + h)^{1/h}$.
  (d) $(1 + 0.06h)^{1/h}$.
  (e) $\exp(\ln h)$.
  (f) $(1 - h)^{1/h}$.
  (g) $(1 + h)^{-1/h}$.
  (h) $(1 + h^2)^{1/h}$.
  Answers: (a) 1. (b) 1. (c) e. (d) $e^{0.06}$. Hint. (e) Use identity (11).

5.7. Prove equation (36) by the chain rule.

5.8. Use logarithmic differentiation to find the derivative of each of the following.
  (a) 10^x.
  (b) $(1 + x)^x$.
  (c) $|x|$.
  (d) $(x|x + 3|^{1/4})/(8 + x^3)$.
  (e) $1/g(x)$.
  (f) $\sqrt{g(x)}$.
  (g) $(f(x))^2$.
  (h) $f(x)/g(x)$.
  Answers: (b) $(1 + x)^x((x/(1 + x)) + \ln(1 + x))$, $-1 < x$. (c) $|x|/x, x \neq 0$.
  (d) $((x|x + 3|^{1/4})/(8 + x^3))((1/x) + (1/(4x + 12)) - (3x^2/(8 + x^3)))$, $x \neq 0, -2, -3$.

5.9. Use formula (35) to find antiderivatives of each of the following.
  (a) $1/(1 + x)$.
  (b) $-3/(5 - 3x)$.
  (c) $1/(5 - 3x)$.
  (d) $2x/(1 + x^2)$.
  (e) $x/(1 + 3x^2)$.
  (f) $(-\sin x)/(\cos x)$.
  Answers: (c) $(-1/3) \ln|5 - 3x|$. (f) $\ln|\cos x|$.

5.10. Use **F3** to evaluate each of the following integrals, after first checking to see if the integral exists.
  (a) $\int_1^3 (1/(1 + x))\,dx$.
  (b) $\int_{-1}^0 (-3/(5 - 3x))\,dx$.
  (c) $\int_{-1}^1 (1/x)\,dx$.
  (d) $\int_0^4 (2x/(1 + x^2))\,dx$.
  (e) $\int_0^3 \tan x\,dx$.
  (f) $\int_0^{\pi/4} \tan x\,dx$.
  Answers: (a) $\ln|1 + x|\big|_1^3 = \ln 4 - \ln 2 = \ln(4/2) = \ln 2$.
  (b) $\ln|5 - 3x|\big|_{-1}^0 = \ln 5 - \ln 8$. (c) Does not exist. (d) $\ln|1 + x^2|\big|_0^4 = \ln 17$.
  (e) Does not exist. (f) $-\ln|\cos x|\big|_0^{\pi/4} = -\ln(\sqrt{2}/2) = \frac{1}{2}\ln 2$.

5.11. Do exercise 4.4 by the method of logarithmic differentiation. Is it easier by this method?

§6. Compound Interest; Continuous Interest

The reader is probably familiar with the notion of compound interest. If not, then the following example should make clear the notion of interest *compounded annually, compounded semiannually, compounded quarterly*, etc.

EXAMPLE 6. A principal of $1,000 is deposited in a savings account to grow at an interest rate of 12% per year. Find the value (principal plus interest) of the savings account at the end of one year if interest is compounded

(a) annually.
(b) semiannually (that is, at 6-month intervals).
(c) quarterly (that is, at 3-month intervals).
(d) monthly.
(e) daily.
(f) n times (that is, at intervals of length $1/n$ years).

All of these are really instances of the last. In (a) the interest is added to the principal once, in (b) twice, in (c) four times, in (d) 12 times, and in (e) 365 times. Thus, (a) through (e) are really instances of (f), where n is equal to 1, 2, 4, 12, and 365, respectively. Nevertheless, let us consider (a) through (f) in order.

(a) At the end of 1 year an interest of 12% per year on a principal of $1,000 will produce an income of $1000(0.12) = 120$ dollars. Adding this to the original principal gives

$$1000 + 1000(0.12) = 1000(1 + 0.12) = 1000(1.12),$$

or $1,120.

(b) Here the interest of 12% per year is compounded twice, that is, by adding 6% interest at the end of each half-year. At the end of the first half-year the balance is

$$1000 + 1000(0.06) = 1000(1 + 0.06) = 1000(1.06) \qquad (40)$$

dollars. Equation (40) says that compounding interest semiannually has the effect of increasing the account's balance by a factor of 1.06 at the end of each half-year. The balance (40), after another half-year, will therefore be increased to

$$1000(1.06)(1.06) = 1000(1.06)^2,$$

or $1,123.60.

(c) Here the interest of 12% per year is compounded four times, that is, by adding 3% interest to the existing balance at the end of each quarter. This means that the account's value is multiplied by a factor of 1.03 each quarter, so that after 1 year the account is worth

$$1000(1.03)^4$$

dollars. This is equal to $1,125.51.

6. Compound Interest; Continuous Interest

(d) Here the interest of 12% per year is compounded 12 times, that is, by adding 1% interest to the existing balance at the end of each month. This means that the principal is multiplied by a factor of 1.01 twelve times in the course of a year, making the account worth

$$1000(1.01)^{12} \qquad (41)$$

dollars at the end of the year. The reader is asked to use logarithms to evaluate (41).

(e) Here the interest of 12% per year is compounded 365 times, that is, by adding 12/365% interest each day. At the end of the year the account is worth

$$1000\left(1 + \frac{1}{365}(0.12)\right)^{365} \qquad (42)$$

dollars. The reader is asked to use logarithms to evaluate (42), and to show that it is equal to $1,127.44.

(f) If the interest of 12% per year is compounded n times, that is, by adding $(12/n)$% interest to the existing balance each time, then the value of the account at the end of one year is

$$1000\left(1 + \left(\frac{1}{n}\right)(0.12)\right)^n \qquad (43)$$

dollars. □

The values calculated in Example 6 are tabulated here.

n	The year-end value of $1000 investment at 12% per year interest compounded n times
1	$1120.00
2	$1123.60
4	$1125.51
12	$1126.82
365	$1127.44
n	$1000(1 + (0.12/n))^n$

As n is taken larger and larger, that is, as the interest is compounded more and more often, we approach what might be called *continuous interest*, or *interest compounded continuously*. In Example 6 the value of the account at the end of 1 year with interest compounded continuously is the limit of (43), as n increases without bound. What is the limit of (43)? Letting $h = 1/n$, so that $h \to 0$ as n increases, we see that it is

$$\operatorname*{Limit}_{h \to 0} 1000(1 + h(0.12))^{1/h} = 1000e^{0.12},$$

by part (c) of Example 3 in the preceding section. Tables [or see equation (67) of Section 9] give $e^{0.12} \approx 1.12749$, so that continuous interest in Example 6 will produce a year-end value of

$$\$1{,}127.49.$$

As might be expected, this is only slightly larger than the value tabulated for interest compounded daily.

The example just considered was for a principal $A_0 = 1000$ with a rate of growth $k = 0.12$ over a period of time $t = 1$. Continuous interest at a rate k on an initial investment A_0 over a period of time t units is defined by applying exactly the same considerations. The result is described as follows.

Theorem on Continuous Interest. *Let k denote a rate of growth of value of an investment, and let A_0 denote the investment's initial value. Then its value A at time t is given by*

$$A = A_0 e^{kt}$$

if the growth is continuous.

PROOF. Because of the multiplicity of symbols in the calculation that follows, things may seem much more complicated than they are. Remember that A_0, k, and t are simply constants in the calculation, representing the principal, the rate of interest, and the period of time over which the investment is carried.

By compounding n times over the period t the value of the initial investment becomes

$$A_0\left(1 + \left(\frac{t}{n}\right)k\right)^n. \tag{44}$$

We are interested in the limit of (44) as n is taken larger and larger, that is, as interest is compounded more and more often. How can we calculate the limit of (44) in order to get the value of the investment under continuous interest? If we let $h = t/n$ (so that $n = t/h$), we see that the limit of (44) is given by

$$\underset{h \to 0}{\text{Limit}}\ A_0(1 + hk)^{t/h} = \underset{h \to 0}{\text{Limit}}\ A_0[(1 + hk)^{1/h}]^t$$
$$= A_0[e^k]^t$$
$$= A_0 e^{kt},$$

where we have used once again the result of part (c) of Example 3. □

EXAMPLE 7. At an interest rate of 6% per year, how much will an initial investment of $1,000 be worth after 5 years under

(a) annual interest?
(b) continuous interest?

(a) For annual interest the investment's value is multiplied each year by a a factor of 1.06. After 5 years the investment will be worth

$$1000(1.06)^5 = 1338.23$$

dollars.

(b) Applying the theorem on continuous interest with $A_0 = 1000$, $k = 0.06$, and $t = 5$, we see that the investment will be worth

$$1000e^{0.06(5)} = 1000e^{0.3} = 1349.86$$

dollars after 5 years. □

EXAMPLE 8. A bank advertises an interest rate on savings at 7.5% per year, to be compounded continuously. What amount of interest compounded annually is equivalent to this?

After 1 year at the advertised rate, an investment A_0 would be worth $A_0 e^{0.075}$ by the theorem on continuous interest. After 1 year at an annual rate k, the same investment would be worth $A_0(1 + k)$. For these to be equivalent we must have

$$A_0(1 + k) = A_0 e^{0.075},$$
$$1 + k = e^{0.075},$$
$$k = e^{0.075} - 1. \quad (45)$$

From a table, $e^{0.075} \approx 1.0779$, so $k \approx 0.0779$ by (45). An annual rate of approximately 7.79% is then equivalent to interest compounded continuously at 7.5%. □

At interest compounded annually, quarterly, or daily, the value of the initial investment grows by discrete steps. With interest compounded continuously the value A grows *continuously* at a rate proportional to A. The following theorem should not be surprising.

Theorem on Exponential Growth. *The function given by*

$$A(t) = A_0 e^{kt} \quad (46)$$

describes a quantity A whose initial value is A_0 and whose rate of growth is always proportional to itself, that is,

$$\frac{dA}{dt} = kA. \quad (47)$$

PROOF. Obviously from (46) we have $A(0) = A_0$. All that we need to do is verify that the function defined in (46) satisfies the "growth law" given by the differential equation (47). This is straightforward. If (46) holds, then

$$\frac{dA}{dt} = \frac{d}{dt} A_0 e^{kt} = A_0 k e^{kt} \quad [\text{by (14)}]$$
$$= kA \quad [\text{by (46)}]. \quad □$$

The theorem just proved may seem insignificant. We shall see, however, that the growth law expressed by the differential equation (47) is of particular interest. Equation (47) is one of the most important differential equations ever written. And the theorem just proved tells us exactly how to solve it. A solution to (47) is given by the function expressed in equation (46).

EXERCISES

6.1. Use logarithms to evaluate expressions (41) and (42).

6.2. A principal of $100 is deposited for 20 years at an interest rate of 6% per year. How much will this account be worth at the end of 20 years if the interest is compounded
(a) annually?
(b) continuously?
Answers: (a) About $321, by Example 2, part (c). (b) About $332, by the theorem on continuous interest.

6.3. At an interest rate of 12% per year, how much will an initial investment of $1,000 be worth after 5 years under
(a) annual interest?
(b) continuous interest?

6.4. A bank advertises an interest rate on savings at 8% per year, to be compounded continuously. What amount of interest compounded annually is equivalent? *Answer*: 8.33%.

6.5. A bank advertises an interest rate on savings at 8% per year, to be compounded annually. What amount of interest compounded continuously is equivalent? *Answer*: 7.70%.

6.6. What amount of interest compounded continuously will double the initial value of an investment after 5 years? *Answer*: We must find the growth rate k that satisfies $2A_0 = A(5) = A_0 e^{5k}$. Dividing by A_0 we obtain $2 = e^{5k}$, implying that $\ln 2 = 5k$, or $k = (\ln 2)/5 = 0.1386$. An interest rate of 13.86% per year will suffice.

6.7. What amount of interest compounded continuously will double the initial value of an investment after 10 years?

6.8. What amount of interest compounded annually will double the value of an investment after 5 years? *Answer*: 14.87% per year.

6.9. Money is deposited at 6% interest per year to be compounded continuously. How long will it take for the account to double in value?

6.10. A quantity A grows in accordance with the law $A = A_0 e^{kt}$, where k is positive. How long does it take for the quantity A to double in size? *Answer*: Let T be the required time so that $A(t + T) = 2A(t)$. That is,

$$A_0 e^{k(t+T)} = 2A_0 e^{kt}.$$

Solving by logarithms gives $T = (\ln 2)/k$.

6.11. Consider the differential equation $dA/dt = 2A$.
(a) Show that $A = e^{2t}$ satisfies this differential equation.
(b) Show that $A = e^{2t} + C$ does not satisfy this differential equation unless $C = 0$.
(c) Show that $A = Ce^{2t}$ does satisfy this differential equation, where C can be any constant.

6.12. Consider the differential equation $dA/dt = 2t$. (Note the difference between this equation and the equation of exercise 6.11.) Solve this equation. *Answer*: $A = t^2 + C$.

6.13. Solve each of the following differential equations with given initial conditions.
(a) $dA/dt = 2A$, $A(0) = 5$.
(b) $dA/dt = \pi A$, $A(0) = 10$.
(c) $dA/dt = -2A$, $A(0) = 3$.
Answer: (c) $A = 3e^{-2t}$, by the theorem on exponential growth.

6.14. The exponential function has a special feature in connection with the product rule which makes it quite useful in solving certain types of differential equations. To see this property, take the derivative of the product of each of the following, as indicated. Note that the exponential function factors out of your answer.
(a) $(t^2 e^t)'$.
(b) $(t^4 e^{3t})'$.
(c) $(e^{kt} \sin t)'$.
(d) $(e^{F(t)} g(t))'$.
Answers: (b) $e^{3t}(4t^3 + 3t^4)$. (c) $e^{kt}(\cos t + k \sin t)$. (d) $e^{F(t)}(g'(t) + f(t)g(t))$, where $F' = f$.

6.15. Try to combine careful observation with a little luck to guess an *antiderivative* of each of the following. Use the property of the exponential function illustrated in the preceding exercise.
(a) $e^t(1 + t)$.
(b) $e^t(\cos t + \sin t)$.
(c) $e^{kt}(\cos t + k \sin t)$.
(d) $e^{-t}((1/t) - \ln t)$.
(e) $e^{-kt}(\cos t - k \sin t)$.
(f) $e^{-kt}((dy/dt) - ky)$.
Answers: (a) te^t. (e) $e^{-kt} \sin t$. (f) $e^{-kt} y$.

§7. Population Growth

Is there a law governing the way the population N of some species varies with time? It is hard to conceive of a law that might be valid if one is prone to worry about violent upheaval, massive drought, stalking pestilence, nuclear holocaust, apocalyptic cataclysm, and the like. In the absence of such catastrophes, however, the rate of growth of population at any time should be proportional to the population at that time. The **growth law** is

$$\frac{dN}{dt} = kN, \qquad (48)$$

where k is some constant of proportionality characteristic of the species in question. The larger k is, the faster the population multiplies.

The growth law (48) should not be accepted as absolutely accurate. It represents a simpleminded way of making a mathematical model to predict population growth. This model has the great virtue of simplicity, but it also has its drawbacks, just as the simple model for studying freely falling bodies has drawbacks. A slight drawback here is that the population N is treated as a *continuously* increasing quantity in (48). In fact, of course, every population changes by *discrete* steps with each birth and death. However, if the population N is large then the discrete steps will occur so close together as to approximate the continuous growth in (48) to great accuracy. (Recall from the preceding section how little difference there is between interest compounded daily and interest compounded continuously.) Thus we may expect the growth law (48) to make fairly accurate predictions when applied to large populations, in the absence of catastrophic events.

Equation (48) is of the same form as the differential equation (47) considered in the theorem on exponential growth. By that theorem we know that

$$N = N_0 e^{kt} \tag{49}$$

is a solution, where N_0 is the initial population, i.e., the population when $t = 0$.

Knowing that (49) is a solution of the growth law (48) is certainly helpful, but we cannot proceed with security until we investigate whether (49) is the *only* solution of (48). It is, as we shall see below. In addition to proving the uniqueness of the solution (49), we shall introduce the idea of an *integrating factor*, which is quite a useful tool in attacking differential equations in general.

The possibility of finding an integrating factor is increased by a curious, but simple, property of the exponential function that has been illustrated in exercise 6.14. The property is this. If an exponential function is multiplied by a second function, and the derivative of the product is taken, then the exponential function can be factored out of the derivative. It is easy to check that this is true. For any differentiable function $N(t)$, the product rule yields

$$(e^t N(t))' = e^t \frac{dN}{dt} + N(t)e^t = e^t \left(\frac{dN}{dt} + N(t) \right),$$

showing how e^t factors out of the derivative of the product. The same game can be played with e^{kt} (or even with $e^{F(t)}$, as investigated in the exercises). For example, the product rule yields

$$(e^{-kt} N(t))' = e^{-kt} \left(\frac{dN}{dt} - kN(t) \right). \tag{50}$$

How could equation (50) be of any use? It becomes useful when you read it backwards (!), as in the proof below.

Existence and Uniqueness Theorem for the Growth Law. *For the differential equation* (48) *there exists a unique solution, given by* (49).

PROOF. We have already seen that there exists a solution, for the function (49) satisfies (48). To prove uniqueness we shall go in the reverse direction. That is, given (48), which can be rewritten

$$\frac{dN}{dt} - kN = 0, \tag{51}$$

7. Population Growth

we shall derive (49). To do this multiply equation (51) through by the integrating factor e^{-kt} to get

$$e^{-kt}\left(\frac{dN}{dt} - kN\right) = 0.$$

By (50) we then have

$$(e^{-kt}N)' = 0,$$

so by the fundamental principle **F1**, $e^{-kt}N$ must be constant:

$$e^{-kt}N = C.$$

Multiplying through by e^{kt} yields

$$N = Ce^{kt}. \tag{52}$$

What is C? Letting $t = 0$ in (52) we obtain $N_0 = Ce^0 = C$. Therefore $C = N_0$ and (49) follows. □

The trick of multiplying through (51) by the integrating factor e^{-kt} doubtless appears at first as if pulled out of a hat by magic. But in fact anybody can do it who works the exercises below. A fair number of differential equations succumb to this trick, when used in conjunction with the fundamental principles of calculus.

The technique of dealing with examples illustrating population growth is much the same as in the examples illustrating interest compounded continuously.

EXAMPLE 9. Assume that the population of a certain planet grows from 2 billion to 3 billion in 10 years.

(a) How long will it take the population to double in size?
(b) When will the population reach 10 billion?

If we assume the growth law valid and begin measuring time when the population N is 2 billion, then by (49) we have

$$N = 2e^{kt}, \tag{53}$$

where N is measured in billions, t in years. We can find k by using the fact that $N = 3$ when $t = 10$, so that (53) implies

$$3 = 2e^{10k},$$
$$\ln 3 = \ln 2 + 10k.$$

Solving for k we get (rounding off to four decimal places)

$$k = \frac{\ln 3 - \ln 2}{10} = 0.0405,$$

and (53) becomes

$$N = 2e^{0.0405t}. \tag{54}$$

Once we have the population expressed as a function of time, as in (54), we can easily answer questions (a) and (b).

(a) The population doubles when $N = 4$, i.e., when
$$4 = 2e^{0.0405t}.$$
Solving for t by logarithms yields $t = 17.1$ years.

(b) The population reaches 10 billion when
$$10 = 2e^{0.0405t}.$$
Solving for t by logarithms yields $t = 39.7$ years. □

In Example 9 there is an alternate way of expressing equation (54) that has the advantage of not requiring the use of log tables. The trick is to solve for e^k instead of k in the equation $3 = 2e^{10k}$. This is easy:

$$e^{10k} = \frac{3}{2},$$

$$e^k = \left(\frac{3}{2}\right)^{1/10}.$$

Therefore $e^{kt} = (3/2)^{t/10}$ by (27), and (53) becomes

$$N = 2\left(\frac{3}{2}\right)^{t/10}. \tag{55}$$

Both (54) and (55) represent the same function giving population in terms of time.

Exercises

7.1. The population of a chicken farm grows from 1000 to 1200 in 1 month. Assuming the growth law (48) valid, find an expression for the population in terms of time. Then find
  (a) how long it takes for the population to double.
  (b) when the population will reach 2500.
  Partial answer: $N = 1000(\frac{6}{5})^t$, where t is in months.

7.2. In Example 9 find the average population over the 10-year period in which the population rises from 2 billion to 3 billion. *Hint*. An antiderivative of e^{kt} is given by e^{kt}/k. *Answer*: From (54) the average value of the population N is
$$\frac{1}{10}\int_0^{10} 2e^{0.0405t}\,dt = \frac{1}{5}\left[\frac{e^{0.0405t}}{0.0405}\right]_0^{10} \approx 2.47 \text{ billion.}$$

7.3. An experiment in biology begins with a colony of 3 million bacteria. After 8 hours the population of the colony is 15 million. Assume the growth law valid.
  (a) Find an expression for the population in terms of time.
  (b) How long does it take the colony to double in size?

7. Population Growth

7.4. The population of a certain country was 151 million in 1950. Twenty years later it was 203 million. Assuming the growth law valid, find
 (a) how long it takes the population to double.
 (b) the population in 1960.
 (c) the population in 1930.
 (d) the population in 1990.
 (e) when the population will reach 400 million.
 Hint. Start measuring time in 1950. Then parts (b), (c), and (d) correspond to $t = 10$, -20, and 40, respectively.

7.5. Consider the general differential equation $dy/dx = f(x)y$, where $f(x)$ is some given function. Let F be an antiderivative of f. Derive the solution $y = Ce^{F(x)}$. *Answer:* Rewrite the differential equation as $(dy/dx) - f(x)y = 0$ and multiply through by the integrating factor $e^{-F(x)}$ to obtain

$$e^{-F(x)}\left(\frac{dy}{dx} - f(x)y\right) = 0$$

$$(e^{-F(x)}y)' = 0,$$

$$e^{-F(x)}y = C,$$

$$y = Ce^{F(x)}.$$

7.6. As in exercise 7.5, find an integrating factor to aid in solving each of the following differential equations. *Do not simply write down the answer. Get used to going through all the steps as in exercise 7.5.*
 (a) $(dy/dx) - 2y = 0$. (b) $(dy/dx) - (\sin x)y = 0$.
 (c) $(dy/dx) + 3x^2 y = 0$. (d) $(dy/dx) - (\sec^2 x)y = 0$.
 (e) $(dy/dx) + (1/x^2)y = 0$. (f) $(dy/dx) - (1/x)y = 0$.
 Answers: (a) $y = Ce^{2x}$. (b) $y = Ce^{-\cos x}$. (d) $y = Ce^{\tan x}$. (f) $y = Cx$.

7.7. (Read again Section 8 of Chapter 8 before doing this problem.) Use an integrating factor in each of the following differential equations to find the solution that takes the value 3 when x is 2.
 (a) $(dy/dx) - 2y = 0$. (b) $(dy/dx) - 2y = x$.
 (c) $(dy/dx) + 3x^2 y = \cos x$. (d) $(dy/dx) + (y/x^2) = \tan x$.
 (e) $(dy/dx) - 2xy = 4$. (f) $(dy/dx) + 2xy = e^x$.
 Answers:
 (b) Multiplying through by the integrating factor e^{-2x}, we obtain

$$e^{-2x}\left(\frac{dy}{dx} - 2y\right) = xe^{-2x}$$

$$(e^{-2x}y)' = xe^{-2x}.$$

It follows (see Chapter 8, Section 8) that for some C,

$$e^{-2x}y = C + \int_2^x te^{-2t}\,dt.$$

When x is 2, y must equal 3, which implies $C = 3e^{-4}$ and therefore

$$y = 3e^{-4+2x} + e^{2x}\int_2^x te^{-2t}\,dt.$$

 (c) $y = 3e^{8-x^3} + e^{-x^3}\int_2^x e^{t^3}\cos t\,dt$.
 (d) $y = (3e^{1/x}/\sqrt{e}) + e^{1/x}\int_2^x e^{-1/t}\tan t\,dt$.

§8. Radioactive Decay

We have so far considered the growth law $dA/dt = kA$ only in case k is *positive*. In this case, when an initial positive value A_0 is prescribed, dA/dt is always positive, which means the quantity A is continually *increasing* as time goes on. We have seen that A must grow in an exponential fashion in accordance with the equation $A = A_0 e^{kt}$.

One of the features of this growth is that no matter what size the quantity is, there is a fixed time T it takes for the size to double. As seen in exercise 6.10, this "doubling time" is given by

$$T = \frac{\ln 2}{k}.$$

It has been observed for some time that radioactive substances behave in rather the opposite way. A radioactive substance *decays* (that is, its rate of growth is *negative*) in such a way that there is a fixed length of time it takes to reduce itself to half its original size. Suppose, for example, that this half-life is 5 years. This means that if we begin, say, with 16 grams of the radioactive substance, its size will be cut in half every 5 years. Thus we shall have 8 grams left after 5 years, 4 grams after 10 years, 2 grams after 15 years. In general we shall have $16(\frac{1}{2})^n$ grams left after n half-lives have elapsed.

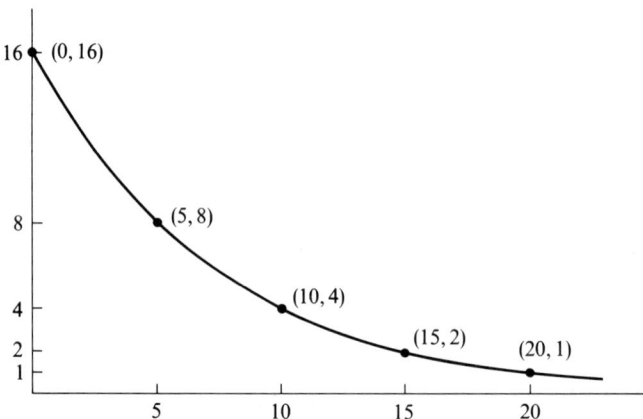

Radioactive decay with 16 grams initially and a half-life of 5 years. The curve has equation (57). It also has equation (58).

How do we fill in the rest of the points here? It seems reasonable to assume that the rate of decay of a radioactive substance should be proportional to its size at any given instant. That is,

$$\frac{dA}{dt} = kA,$$

8. Radioactive Decay

where k is negative. (Remember that dA/dt represents the rate of *increase* of A.) By our theorem of the previous section, which is valid whether k is positive or negative, we conclude that

$$A = A_0 e^{kt}.$$

For the curve pictured above, we have $A_0 = 16$, so that $A = 16e^{kt}$. What is k? Since $A = 8$ when $t = 5$, we have

$$8 = 16e^{5k},$$

$$\frac{1}{2} = e^{5k}, \tag{56}$$

$$-\ln 2 = 5k,$$

$$k = \frac{-\ln 2}{5} = -0.139,$$

where we have rounded off to three places. The curve given above then has the equation

$$A = 16e^{-(0.139)t}. \tag{57}$$

The curve also may be represented by a different equation, derived as follows. Taking the fifth root of both sides of (56) we have

$$e^k = \left(\frac{1}{2}\right)^{1/5}.$$

Therefore $e^{kt} = (\frac{1}{2})^{t/5}$ and it follows that

$$A = 16\left(\frac{1}{2}\right)^{t/5}. \tag{58}$$

Equation (58) is equivalent to (57), but the form of (58) makes it quite obvious that we have what we were seeking. Only a moment's scrutiny of equation (58) will convince the reader that it describes a quantity A which takes the value 16 when $t = 0$, and which is cut in half with every passage of 5 units of time.

The proof of the following theorem is left to the reader. The essential idea of the proof may be found in the concrete example just discussed.

Theorem on Radioactive Decay. *Let A denote the amount of a certain radioactive substance present at time t, and let A_0 denote the amount present when $t = 0$. If the rate of decay is proportional at each instant to the amount present, then*

$$A = A_0 \exp\left(\frac{-t \ln 2}{T}\right) = A_0 \left(\frac{1}{2}\right)^{t/T}, \tag{59}$$

where T is the half-life of the substance.

Here are some examples dealing with radioactive decay.

EXAMPLE 10. Five grams of a certain radioactive substance decays to 3 grams in 10 days. Find the half-life T of the substance.

If we begin measuring time when 5 grams is present, then by (59) we have $A = 5\exp((-t\ln 2)/T)$. Since $A = 3$ when $t = 10$, we get

$$3 = 5\exp\left(\frac{-10\ln 2}{T}\right),$$

$$\ln 3 = \ln 5 - \frac{10\ln 2}{T}.$$

Solving for T we get

$$T = \frac{10\ln 2}{\ln 5 - \ln 3} \approx 13.6 \text{ days}. \qquad \square$$

EXAMPLE 11. Five grams of a certain radioactive substance decays to 3 grams in 10 days. Find the average amount present during these 10 days.

Beginning to measure time when 5 grams is present, we have

$$A = 5\exp\left(\frac{-t\ln 2}{13.6}\right) = 5e^{-(0.051)t}$$

from the work of Example 10. The average of A between $t = 0$ and $t = 10$ is given by

$$\frac{1}{10}\int_0^{10} 5e^{-(0.051)t}\,dt = \frac{1}{2}\int_0^{10} e^{-(0.051)t}\,dt$$

$$= \left[\frac{-1}{0.102}e^{-(0.051)t}\right]_0^{10}$$

$$\approx 3.9 \text{ grams}. \qquad \square$$

EXERCISES

8.1. During a period of 1 year, 15 grams of a certain radioactive substance decays until only 12 grams is left. Find the half-life T of the substance. Answer: $T \approx 3.11$ years.

8.2. In exercise 8.1 find the average amount of the radioactive substance present during the year.

8.3. Write out a careful proof of the theorem on radioactive decay stated in this section. That is, derive equation (59) from the growth law $dA/dt = kA$, where k is negative.

8.4. Suppose that initially there are 15 grams of a radioactive substance X and 24 grams of a radioactive substance Y. After 1 year there are 12 grams of X and 18 grams of Y. When will there be the same amount of X and Y? Answer: When $15(\frac{4}{5})^t = 24(\frac{3}{4})^t$, or when (solving by logs) $t \approx 7.3$ years.

9. Algebraic Approximations to Transcendental Functions   337

8.5. Radium has a half-life of 1620 years. If 10 grams of radium are present in a certain piece of material today, find
(a) a formula for the amount of radium present as a function of time.
(b) the rate of radioactive decay today.
(c) the amount of radium present in this material at the time of Pythagoras (when $t = -2500$ years).
Answers: (a) $A = 10(\tfrac{1}{2})^{t/1620}$, where A is the amount present in grams and t is time in years from today. (b) $dA/dt|_{t=0} \approx -0.0043$. This gives rate of *growth*. The rate of of decay today is 0.0043 grams per year, approximately. (c) $10(\tfrac{1}{2})^{-2500/1620} \approx 29.1$ grams.

8.6. During a certain period of time, 20 grams of a radioactive substance decays to 15 grams. Find the average amount of radioactive substance present during this period if the length of the period is
(a) 1 year.
(b) 100 years.
Answer: Approximately 17.4 grams, in either case.

8.7. During a certain period of time, A grams of a radioactive substance decays to B grams. Find the average amount of radioactive substance present during this period.
Answer: $(A - B)/(\ln A - \ln B)$ grams (regardless of the length of the period).

§9. Algebraic Approximations to Transcendental Functions

A transcendental function (exp, sin, ln, arctan, etc.) has a rule that is not expressible in terms of algebraic operations. This tends to make transcendental functions seem more removed from us who have known algebraic functions so much longer. If a function is not algebraic, then we naturally ask whether it might be approximated closely by an algebraic function.

There is an obvious line of attack which can be illustrated by taking a concrete example. Let us consider the function exp near a particular point on its graph. The most convenient point to choose is $(0, 1)$, although the considerations that follow apply equally well to any point on the graph. Near the point $(0, 1)$ the tangent line, having equation

$$y = x + 1, \tag{60}$$

is the line that most closely approximates the curve exp. We have in equation (60) an algebraic approximation to the transcendental curve exp near the point $(0, 1)$. Obviously the approximation

$$\exp x \approx 1 + x$$

is a good one only if $x \approx 0$. For instance, we expect that

$$\exp 0.12 \approx 1 + 0.12, \tag{61}$$

since 0.12 ≈ 0. However, if x is relatively far from 0, then the approximation is not so good. When $x = 1$ for instance, the algebraic rule $1 + x$ takes the value 2, yet the exponential function takes the value $e \approx 2.72$.

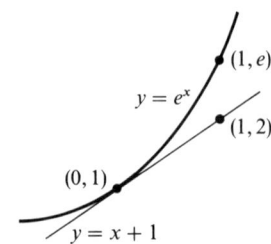

Linear approximation to exp near (0, 1)

It is natural to say that at the point $x = 0$ the transcendental function exp and the algebraic rule given by $1 + x$ agree to **first order**. That is, at $x = 0$ they have the same value and the same first derivative:

x	y	y'
0	1	1

They do not agree to any higher order, however, for all higher derivatives of the linear function $1 + x$ are equal to zero; whereas all higher derivatives of exp are equal to exp, which takes the value 1 when $x = 0$. The function exp has the following "signature" at the point $x = 0$:

x	y	y'	y''	y'''	...
0	1	1	1	1	...

The linear approximation (60) agrees with exp to first order at $x = 0$. Let us seek a *quadratic* approximation that agrees with the signature of exp to **second order**, i.e., that satisfies this table:

x	y	y'	y''
0	1	1	1

This is easy. We can easily guess (and we check this guess below) that the linear part of the quadratic we seek is given in (60). Thus we need only find the right coefficient a of the second degree term in

$$y = ax^2 + x + 1. \qquad (62)$$

From (62) we derive

$$y' = 2ax + 1, \qquad (63)$$

$$y'' = 2a. \qquad (64)$$

9. Algebraic Approximations to Transcendental Functions

From (62) and (63) we see that $y = 1$ and $y' = 1$ when $x = 0$, regardless of the value of a. From (64) we see that in order to have $y'' = 1$ we must have $a = \frac{1}{2}$. Thus the quadratic function given by

$$y = 1 + x + \frac{1}{2}x^2 \qquad (65)$$

agrees with the exponential function to second order at the point $x = 0$. We expect that the approximation

$$\exp x \approx 1 + x + \frac{1}{2}x^2$$

will be better than our former attempt. For instance,

$$\exp 0.12 \approx 1 + 0.12 + \frac{1}{2}(0.12)^2 = 1.1272 \qquad (66)$$

ought to be better than the approximation in (61). And when $x = 1$, the approximation

$$\exp(1) \approx 1 + 1 + \frac{1}{2}(1)^2 = 2.5$$

is respectably close to e.

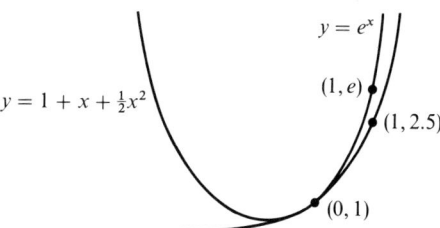

Quadratic approximation to exp near $(0, 1)$

A cubic approximation of exp to **third order** suggests itself. It ought to be of the form

$$y = 1 + x + \frac{1}{2}x^2 + ax^3$$

for some a, in order that its quadratic part agree with (65). Upon taking three derivatives we find

$$y''' = 6a.$$

In order to have $y''' = 1$, we must have $a = \frac{1}{6}$. A third-order approximation to exp at the point $x = 0$ is then given by

$$\exp x \approx 1 + x + \frac{1}{2}x^2 + \frac{1}{6}x^3.$$

Using this third-order approximation we have

$$\exp 0.12 \approx 1 + 0.12 + \frac{1}{2}(0.12)^2 + \frac{1}{6}(0.12)^3 = 1.127488, \qquad (67)$$

$$\exp 1 \approx 1 + 1 + \frac{1}{2}(1)^2 + \frac{1}{6}(1)^3 \approx 2.67. \qquad (68)$$

We shall find that (68) is a pretty good approximation to e, while (67) is of such accuracy as to please all but the most complete perfectionist.

We may continue this process to any order we please. The reader will find that a **fourth-order** approximation at the point $x = 0$ is given by

$$\exp x \approx 1 + x + \frac{1}{2}x^2 + \frac{1}{6}x^3 + \frac{1}{24}x^4,$$

and, in general, an **n-th–order** approximation is

$$\exp x \approx 1 + x + \frac{1}{2}x^2 + \cdots + \frac{1}{n!}x^n, \qquad (69)$$

where $n!$ (read "n factorial") is defined as the product of the integers 1 through n:

$$n! = 1 \cdot 2 \cdot 3 \cdots n.$$

Setting $x = 1$ in (69) leads to the approximation

$$e \approx 1 + 1 + \frac{1}{2} + \frac{1}{6} + \frac{1}{24} + \frac{1}{120} + \cdots + \frac{1}{n!}, \qquad (70)$$

where the approximation, one hopes, becomes more and more accurate with increasing n. In summation notation, (70) is written

$$e \approx 1 + \sum_{k=1}^{n} \frac{1}{k!}. \qquad (71)$$

For $n = 3$ this approximation is evaluated in (68). In the exercises the reader is asked to evaluate it for larger n.

Approximations become more useful if one has a good idea of their accuracy. How close is the approximation in (71)? As we shall see in Section 10, the error in the approximation (71) cannot possibly exceed $3/(n + 1)!$, showing that the approximation does indeed become as accurate as we please by taking n sufficiently large. Recall that to define a real number it is necessary to specify it to an arbitrary degree of accuracy. Since the approximation given in (71) does this, the number e is sometimes *defined* as the limit of the approximating sums in (71).

EXAMPLE 12. Find an algebraic approximation to the curve $y = e^{-x}$ near $(0, 1)$.

9. Algebraic Approximations to Transcendental Functions 341

Let us find the signature of this function at the point $x = 0$. This is made easy by the fact that successive derivatives of e^{-x} are $-e^{-x}$, e^{-x}, $-e^{-x}$, e^{-x}, etc., forming an alternating pattern. At $x = 0$ the successive derivatives are then $-1, 1, -1, 1, -1$, etc.

x	y	y'	y''	y'''	$\cdots$
0	1	-1	1	-1	$\cdots$

The reader can verify that this gives rise to the algebraic approximation

$$e^{-x} \approx 1 - x + \frac{1}{2}x^2 - \frac{1}{6}x^3 + \frac{1}{24}x^4 - \cdots. \tag{72}$$

There is also another way to arrive at (72). Just consider the approximation (69), which holds if x is near zero. It must therefore hold if x is replaced by $-x$, which results in (72). □

EXAMPLE 13. Find an algebraic approximation to the sine function that is valid near the point $x = 0$.

Let us find the signature of this function at the point $x = 0$. This is made easy by the fact that the successive derivatives of sin repeat in patterns of four: cos, $-\sin$, $-\cos$, sin, cos, $-\sin$, $-\cos$, sin, etc. At the point $x = 0$ the successive derivatives are then $1, 0, -1, 0, 1, 0, -1, 0$, etc., repeating in patterns of four.

x	y	y'	y''	y'''	$y^{(4)}$	$y^{(5)}$	$\cdots$
0	0	1	0	-1	0	1	$\cdots$

The reader can verify that this gives rise to the approximation

$$\sin x \approx x - \frac{1}{6}x^3 + \frac{1}{120}x^5 - \cdots. \tag{73}$$

Note that only odd powers of x appear. This is to be expected because sin is an odd function. □

EXAMPLE 14. Find an algebraic approximation to $\ln(1 + x)$ that is valid near $x = 0$.

If $y = \ln(1 + x)$, the successive derivatives are given by

$$y' = (1 + x)^{-1},$$
$$y'' = -(1 + x)^{-2},$$
$$y''' = 2(1 + x)^{-3},$$
$$y^{(4)} = -6(1 + x)^{-4},$$

etc., where the power rule for taking derivatives has been used repeatedly. Plugging in $x = 0$, we see that the successive derivatives take the values 1, $-1, 2, -6, 24, -120$, etc.

x	y	y'	y''	y'''	$y^{(4)}$	...
0	0	1	-1	2	-6	...

This gives rise to the approximation

$$\ln(1 + x) \approx x - \frac{1}{2}x^2 + \frac{1}{3}x^3 - \frac{1}{4}x^4 + \cdots \qquad (74)$$

□

EXERCISES

9.1. Find a fourth-order approximation to the exponential function at the point $x = 0$. Then plug in $x = 1$ to get an approximation for e.

9.2. The approximation (71) becomes more accurate with larger values of n. Evaluate the right-hand side of (71) for $n = 5, 6$, and 7. *Partial answer*: For $n = 7$ the approximation in (71) becomes 2.71826 ....

9.3. It is shown in Section 10 that the approximation (71) is accurate to within $3/(n + 1)!$ at least. Use this "error estimation" to determine what value of n in (71) will ensure an accuracy of at least

(a) $\frac{1}{10}$.

(b) $\frac{1}{100}$.

(c) $\frac{1}{1000}$.

(d) $\frac{1}{10,000}$.

Answer: (b) In order that the error be less than $\frac{1}{100}$, we want to choose n so that $3/(n + 1)!$ is less than $\frac{1}{100}$. We can find n by trial-and-error methods. When $n = 5$, then $3/(n + 1)!$ is equal to $\frac{3}{6!} = \frac{3}{720} = \frac{1}{240}$. Thus the accuracy in the approximation (71) is well within $\frac{1}{100}$, or 0.01, when n is 5.

9.4. Estimate the value of $1/e$ to several decimal places, either by plugging in $x = -1$ in (69) or by plugging in $x = 1$ in (72). (Stop after taking six or seven terms in the series.)

9.5. Estimate the square root of e. [Plug in $x = \frac{1}{2}$ in (69).]

9.6. Estimate the sine of an angle of 1 radian, or 57.3 ... degrees. [Plug in $x = 1$ in the approximation (73).]

9.7. Estimate the natural log of 1.1. [Plug in $x = 0.1$ in the approximation (74).]

9.8. Estimate the natural log of $\frac{1}{2}$. [Plug in $x = -\frac{1}{2}$ in (74).]

9.9. The natural log function is undefined for negative numbers. We might expect, therefore, that the right-hand side of (74) does not tend to a limit if $x \le -1$.
  (a) Plug in $x = -1$ and see what happens.
  (b) Plug in $x = 1$ and see what happens.
  (c) Plug in $x = 2$ and see what happens.

9. Algebraic Approximations to Transcendental Functions 343

Partial answer: The approximation (74) works as we would hope only if $-1 < x \le 1$.

9.10. Consider the cosine function, as its graph passes through the point (0, 1). Find its signature and deduce an algebraic approximation to the cosine function that is valid near the point $x = 0$. *Hint*. This is similar to Example 13 except that only even powers will appear.

9.11. The derivative of the sine is the cosine. Differentiate the right-hand side of (73) term by term. Do you get agreement with your answer to exercise 9.10? You should.

9.12. Consider the function given by $f(x) = \sqrt{100 + x}$. Find an algebraic approximation that agrees with f near the point $x = 0$
   (a) to first order.
   (b) to second order.
   (c) to third order.
   Answer: (c) $\sqrt{100 + x} \approx 10 + \frac{1}{20}x - \frac{1}{8000}x^2 + \frac{1}{1,600,000}x^3$.

9.13. By plugging in appropriate values of x in part (c) of exercise 9.12, estimate
   (a) $\sqrt{101}$.
   (b) $\sqrt{104}$.
   (c) $\sqrt{90}$.
   (d) $\sqrt{110}$.
   (e) $\sqrt{121}$.

9.14. Consider the following signature.

x	y	y'	y''	y'''	...
a	b	c	d	e	...

   (a) Find a linear approximation to first order.
   (b) Find a quadratic approximation to second order.
   (c) Find a cubic approximation to third order.
   Answers: (a) $y = b + c(x - a)$. (b) $y = b + c(x - a) + (d/2)(x - a)^2$. (c) $y = b + c(x - a) + (d/2)(x - a)^2 + (e/6)(x - a)^3$.

9.15. Find a cubic approximation to the exponential function that agrees with it to third order near the point $x = 1$. *Answer*: The exponential function has the following signature at the point $x = 1$:

x	y	y'	y''	y'''	...
1	e	e	e	e	...

$y = e + e(x - 1) + (e/2)(x - 1)^2 + (e/6)(x - 1)^3$ is the required cubic, by part (c) of exercise 9.14.

9.16. Find a cubic approximation to the square root function that agrees with it to third order near the point
(a) $x = 100$.
(b) $x = 25$.
(c) $x = 1$.

Answer: (a) The square root function has the following signature at the point $x = 100$:

x	y	y'	y''	y'''	...
100	10	$\frac{1}{20}$	$\frac{-1}{4000}$	$\frac{3}{800,000}$	...

$y = 10 + \frac{1}{20}(x - 100) - \frac{1}{8000}(x - 100)^2 + \frac{1}{1,600,000}(x - 100)^3$ is the required cubic, by part (c) of exercise 9.14. [The same answer could have been obtained by replacing x with $x - 100$ in the answer to part (c) of exercise 9.12.]

9.17. Verify that the following approximations agree to third order.
(a) $\tan x \approx x + \frac{1}{3}x^3$.
(b) $\ln(1 - x) \approx -x - \frac{1}{2}x^2 - \frac{1}{3}x^3$.
(c) $e^x/(1 - x) \approx 1 + 2x + \frac{5}{2}x^2 + \frac{8}{3}x^3$.
(d) $e^x \tan x \approx x + x^2 + \frac{5}{6}x^3$.

9.18. (*For use in Section* 10) The exponential function e^x is the solution to the differential equation $dy/dx = y$ that takes the value 1 when x is 0. Its algebraic approximations (69) "try" to do this as well as they can.
(a) Show that the first-order approximation $y = 1 + x$ satisfies the differential equation $dy/dx = y - x$.
(b) Show that the second-order approximation $y = 1 + x + \frac{1}{2}x^2$ satisfies $dy/dx = y - \frac{1}{2}x^2$.
(c) Show that the third-order approximation satisfies $dy/dx = y - \frac{1}{6}x^3$.
(d) Show that the n-th–order approximation (69) satisfies $dy/dx = y - (1/n!)x^n$.

§10. Where Do We Go from Here?

In this chapter we have been able to look at some previous work in a new light by introducing the natural log function. We have learned a little more about properties of the real number system, and we have seen the central role played in calculus by the exponential function. We have reached a certain plateau in our study of calculus, and it is time for this chapter to come to an end.

The reader should not think, however, that the plateau we have reached marks the height of modern knowledge about calculus. In fact it marks the beginning, for very little has been described here that was not commonly known by eighteenth-century mathematicians. We have barely touched upon the contributions of the nineteenth and twentieth centuries.

10. Where Do We Go from Here?

What is left to study in calculus? Some indication may be gotten from all the loose strings left hanging in the preceding section. Can we really bring transcendental functions down to earth so easily? Will such algebraic approximations always work? Can we differentiate and integrate transcendental functions by performing these operations upon their algebraic approximations instead? Can we learn to solve complicated differential equations?

More fundamental considerations are these. Can we put our discipline on a firmer foundation? Can we give a precise definition of limit so that we do not have to use such vague terms as *tends to* or *approaches* or *approximates*. Can our arguments about functions and about real numbers be made to rest upon something more solid than the pictures we sketch? If so, can we apply ourselves to the study of situations we cannot picture? Instead of studying functions of a single variable, as we have done so far, can we study functions whose rule of correspondence depends upon two, three, or more variables? How can we make our way into higher dimensions?

These questions and many more have been attacked by modern mathematicians with great success. We can indeed explore spaces of high dimension with the same security we have in the two-dimensional plane. We can indeed be more precise about what we say.

Here is a small sample of ideas typical of nineteenth-century analysis. In the preceding section our hopes were raised that e could be approximated as closely as we please by a number of the form

$$s_n = 1 + \sum_{k=1}^{n} \frac{1}{k!}. \tag{75}$$

That is, we think it is probably true that $e = \text{Limit } s_n$ as n increases without bound. Analysis cannot be content with assertions that are "probably true". We ask about the size of the difference between e and s_n, in order to be *sure* that the difference $e - s_n$ becomes arbitrarily small.

Theorem on the Approximation of e. *If s_n is defined by (75), then for each positive integer n,*

$$0 < e - s_n < \frac{3}{(n+1)!}. \tag{76}$$

PROOF. First note for later use that

$$e < 3. \tag{77}$$

This follows because $e = e^1 < e^{1.09} < e^{\ln 3} = 3$.

Let n be a particular (fixed) positive integer, and define the function y by

$$y(x) = 1 + \sum_{k=1}^{n} \frac{x^k}{k!}. \tag{78}$$

It follows (see exercise 9.18) that the function y satisfies the differential equation

$$\frac{dy}{dx} = y - \frac{x^n}{n!}. \tag{79}$$

If we compare the equation (79) with the differential equation $dy/dx = y$ satisfied by the exponential function, we see that they are identical except for the expression $x^n/n!$. This expression ought therefore to be related somehow to the difference between $y(x)$ and e^x. It is, as we see below in equation (80).

Let us rewrite (79) as

$$\frac{dy}{dx} - y = -\frac{x^n}{n!}$$

and multiply through by the integrating factor e^{-x}:

$$e^{-x}\left(\frac{dy}{dx} - y\right) = -\frac{x^n}{n!}e^{-x}$$

$$(e^{-x}y)' = -\frac{x^n}{n!}e^{-x}.$$

Therefore $-e^{-x}y(x)$ is an antiderivative of $(x^n/n!)e^{-x}$. By **F3** we have, for any t,

$$\int_0^t \frac{x^n}{n!}e^{-x}\,dx = -e^{-x}y(x)\Big|_0^t$$

$$= -e^{-t}y(t) + 1,$$

since $y(0) = 1$. This shows that

$$1 - e^{-t}y(t) = \int_0^t \frac{x^n}{n!}e^{-x}\,dx$$

for any real number t. Multiplying through by e^t yields

$$e^t - y(t) = e^t \int_0^t \frac{x^n}{n!}e^{-x}\,dx. \tag{80}$$

Equation (80) holds for all t. If we let $t = 1$, we have

$$e - s_n = e \int_0^1 \frac{x^n}{n!}e^{-x}\,dx. \tag{81}$$

Since the right-hand side of (81) is positive (*why?*), it is clear that

$$0 < e - s_n,$$

which proves the first half of inequality (76). To prove the other half we must show that the right-hand side of (81) does not exceed $3/(n+1)!$.

Let us first consider the integral on the right-hand side of (81). On the domain $0 \leq x \leq 1$ it is clear that e^{-x} never exceeds 1. Therefore,

$$\int_0^1 \frac{x^n}{n!} e^{-x} dx \leq \int_0^1 \frac{x^n}{n!}(1) dx = \frac{1}{n!} \int_0^1 x^n dx. \tag{82}$$

By the fundamental theorem of calculus we have

$$\frac{1}{n!} \int_0^1 x^n dx = \frac{1}{n!} \left[\frac{x^{n+1}}{n+1} \right]_0^1 = \frac{1}{1 \cdot 2 \cdots n} \left(\frac{1}{n+1} \right) = \frac{1}{(n+1)!}. \tag{83}$$

From (82) and (83) we deduce the inequality

$$\int_0^1 \frac{x^n}{n!} e^{-x} dx \leq \frac{1}{(n+1)!}. \tag{84}$$

Finally,

$$e - s_n \leq e \frac{1}{(n+1)!} \quad \text{[by (81) and (84)]}$$

$$< \frac{3}{(n+1)!} \quad \text{[by (77)]}. \qquad \square$$

Problem Set for Chapter 9

1. Find the indicated area:

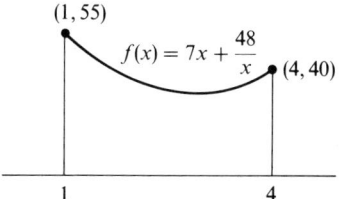

2. Find the derivative of each of the following.
   (a) $e^{2x} \ln x$.
   (b) $\ln(\ln x)$.
   (c) $\exp(e^x)$.
   (d) $x^2 e^x \sin x$.
   (e) $(e^x)^x$.
   (f) $(2 + \sin x)^x$.
   (g) $\log_{10} x$.
   (h) $\log_x 10$.

3. Use L'Hôpital's rule to find the following limits.
   (a) $\text{Limit}_{x \to 1}((\ln x)/(x - 1))$.
   (b) $\text{Limit}_{x \to 0}((e^x - 1)/x)$.
   (c) $\text{Limit}_{x \to 0}((10^x - 1)/x)$.
   (d) $\text{Limit}_{x \to 1}((\log_{10} x)/(x - 1))$.
   (e) $\text{Limit}_{x \to 0}((e^x - 1 - x)/x^2)$.
   (f) $\text{Limit}_{x \to 1} \ln((x^5 - 1)/(x - 1))$.
   (g) $\text{Limit}_{x \to 2}((1 - \log_2 x)/(x - 2))$.
   (h) $\text{Limit}_{x \to 2}((1 - \log_x 2)/(x - 2))$.

4. Evaluate each of the following integrals.
   (a) $\int_0^1 e^{kx} dx$.
   (b) $\int_0^1 2^x dx$.
   (c) $\int_0^1 (3x^2/(1 + x^3)) dx$.
   (d) $\int_0^{\pi/3} \tan x \, dx$.
   (e) $\int_{-1}^1 (2/(4 + 2x)) dx$.
   (f) $\int_0^3 e^{x^2}(2x) dx$.

5. The **hyperbolic sine** and **hyperbolic cosine** are defined as follows:
$$\sinh x = \frac{e^x - e^{-x}}{2}; \quad \cosh x = \frac{e^x + e^{-x}}{2}.$$
They have properties roughly analogous to the sine and cosine functions. Establish the following.
   (a) The hyperbolic sine is an odd function, the hyperbolic cosine is an even function, and their sum is the exponential function.
   (b) $\cosh^2 x - \sinh^2 x = 1$, for all x.
   (c) $d(\sinh x)/dx = \cosh x$; $d(\cosh x)/dx = \sinh x$.
   (d) $\sinh(x + y) = (\sinh x)(\cosh y) + (\cosh x)(\sinh y)$ for all x and y.

6. Show that $(\cosh x + \sinh x)^n = \cosh nx + \sinh nx$.

7. Find the critical points and inflection points, if any, of the function f given by $f(x) = xe^x$, and use this information to help sketch the graph of f.

8. Make a rough sketch of the graphs of $y = x^2$ and $y = 2^x$. They intersect at $(2, 4)$ and $(4, 16)$. Find the area enclosed by the curves between these two points of intersection.

9. Which number is larger: e^π or π^e?

10. (a) Given $y = fgh$, the product of three positive functions, use logarithmic differentiation to find y'.
    (b) Guess a *generalized product rule*. If $y = f_1 f_2 \cdots f_n$ is the product of n functions, what is y'?

11. Use logarithmic differentiation to find the derivatives of the following.
    (a) $100(\frac{1}{2})^{t/3.5}$.
    (b) $\sqrt{2 + x^4}/(x^2 + 6)^{3/2}$.
    (c) $(3 + \cos t)^{5t}$.
    (d) $x^2(5 + \sin x)^3/e^x(2 - \arctan x)^\pi$.

12. Find the limit of each of the following, as h tends to 0.
    (a) $(1 + 3h)^{2/h}$.
    (b) $(1 - \pi h)^{-1/h}$.
    (c) $(1 + h^2)^{1/h^2}$.
    (d) $(1 - 3h^3)^{1/h}$.

13. For each of the following, find the volume of the solid of revolution generated by revolving about the x-axis the area beneath f from $x = 0$ to $x = 1$.
    (a) $f(x) = e^x$.
    (b) $f(x) = e^x - 1$.
    (c) $f(x) = 1/\sqrt{1 + x}$.
    (d) $f(x) = 10^x$.

14. Assume that a function f satisfies the identity
$$f(x + h) = f(x)f(h). \qquad (*)$$
Assume also that f is not always equal to zero.
    (a) Apply Fermat's method to f and deduce that if f has a derivative, then $f'(x) = kf(x)$, where
$$k = \underset{h \to 0}{\text{Limit}} \frac{f(h) - 1}{h}.$$

(b) By setting both x and h equal to zero in (*) deduce that $f(0)$ is either 0 or 1. Deduce further that $f(0)$ cannot be 0, and therefore $f(0) = 1$, and therefore (why?) $k = f'(0)$.

(c) Use parts (a) and (b) to show that any differentiable function f satisfying the homomorphism property (*) is either identically zero or else of the form
$$f(x) = e^{kx},$$
where $k = f'(0)$.

15. A principal of $1,000 is deposited in a savings account and left to grow at 7% interest per year, to be compounded continuously.
    (a) How much is the account worth after 10 years?
    (b) What is the average balance in the account during its first 10 years?
    (c) When will the account be worth $3,000?

16. A snake farm increases its population from 10,000 to 13,000 in 1 month.
    (a) Assuming the growth law (48) valid, find how long it takes for the population to double.
    (c) When will the population reach 30,000?

17. An unknown quantity of a new breed of fish is introduced into a lake. After 3 years the population of this breed is estimated at 2000, and after 5 years at 4000. Estimate how many fish were originally introduced, assuming the growth law is valid.

18. A certain radioactive substance takes a week to decay from 10 grams to 9 grams. What is its half-life?

19. Initially there are 10 grams of a radioactive substance X and 20 grams of a radioactive substance Y. After 3 days there are 9 grams of X and 15 grams of Y. When will there be equal amounts of X and Y?

20. Carbon-14 has a half-life of 5568 years.
    (a) How long does it take for two-thirds of carbon-14 to decay?
    (b) In a sample of material left undisturbed for millennia, 10 grams of carbon-14 are present today. How much carbon-14 was present in this sample at the time of Pythagoras?

21. A fast-decaying radioactive substance goes from 1000 grams to 1 gram in 1 year. Find the average amount present during this year.

22. Solve each of the following differential equations through the use of an appropriate integrating factor.
    (a) $(dy/dx) - 3y = 0$.
    (b) $(dy/dx) - 3y = e^x$.
    (c) $(dy/dx) + 2y = x$.
    (d) $(dy/dx) - y = e^{-x}$.
    (e) $(dy/dx) + 3x^2 y = 0$.
    (f) $(dy/dx) + (y/x) = 3$.

23. Estimate the square root of 28 by carrying out the following steps.
    (a) Find a cubic approximation, near the point $x = 0$, to the function $\sqrt{25 + x}$.
    (b) Let $x = 3$ in your cubic.

24. Consider the function given by $f(x) = 1/(1 - x)$. Find linear, quadratic, cubic, and n-th-order approximations valid near the point $x = 0$. Compare your answer with the approximation to $1/(1 - x)$ obtained in the appendix on sums and limits.

25. In your answer to problem 24 replace x by $-x$ and thus obtain linear, quadratic, cubic, and n-th-order approximations to $1/(1 + x)$. Is your answer the same as that obtained by differentiating both sides of the approximation (74)?

26. (a) In your answer to problem 25 replace x by x^2 to obtain
$$\frac{1}{1+x^2} \approx 1 - x^2 + x^4 - x^6 + \cdots$$

(b) Integrate both sides of the approximation in part (a) from $x = 0$ to $x = 1$ to obtain
$$\frac{\pi}{4} \approx 1 - \frac{1}{3} + \frac{1}{5} - \frac{1}{7} + \frac{1}{9} - \cdots$$

(c) Deduce *Leibniz's approximation for* π:
$$\pi \approx 4 - \frac{4}{3} + \frac{4}{5} - \frac{4}{7} + \frac{4}{9} - \frac{4}{11} + \cdots$$

(d) Add up a few of the numbers on the right-hand side of the approximation in part (c). Does the sum seem to be getting closer and closer to π as more terms are taken? (Actually, while this is a very beautiful result obtained by Leibniz, the series converges much too slowly to be of great practical value in approximating π.)

27. *Newton's law of cooling* deals with the temperature change in a heated body placed in a large medium kept at constant temperature. If y is the temperature of the body, Newton's law states that the rate of change of y is always proportional to the difference between y and the constant temperature of the surrounding medium. (The hotter the body, the faster it will cool.) Suppose a small metal ball is heated to a temperature of 180 degrees Celsius and placed at time $t = 0$ in a large tub of water kept at a constant temperature of 20 degrees Celsius.

(a) Write down Newton's law in the form of a differential equation.
(b) Use an appropriate integrating factor to solve the differential equation of part (a).
(c) Suppose that at time $t = 2$ minutes the temperature y of the metal ball is 90 degrees Celsius. Use this information to adjust the constant appearing in your answer to part (b).
(d) What will be the temperature of the ball at time $t = 4$, assuming Newton's law valid?
(e) When will the ball's temperature be 25 degrees Celsius, assuming all of the above?

28. (*For ambitious students*) Our model for dealing with freely falling bodies began with the assumption that the rate of change of speed is -32 ft/sec per second. This does not take into account the fact that air friction becomes more significant as the speed v gets larger. Accordingly, instead of assuming $dv/dt = -32$, as we did in Chapter 5, assume
$$\frac{dv}{dt} = -32 - kv,$$
where k is some constant of proportionality that depends upon the shape of the falling body. Find the height h as a function of time. Does your answer approach the height formula
$$h = -16t^2 + v_0 t + h_0$$
of Chapter 5 as k tends to zero?

29. Approximations are more valuable if one can estimate the size of error in them. Prove that the error in the approximation (69) can be represented as an integral:

$$\exp(t) - \left(1 + \sum_{k=1}^{n} \frac{t^k}{k!}\right) = \int_0^t \frac{x^n}{n!} \exp(t-x)\, dx.$$

(This is a special case of *Taylor's theorem*, one of the most valuable theorems in analysis.) *Hint*. Begin with equation (80).

10 Romance in Reason

Our story of the development of calculus began with the "Greek miracle", when the spirit of modern mathematics was first felt. That spirit then lay dormant for so long that its rejuvenation in the Renaissance seems almost like a second miracle. Except for the Arabs, it might never have happened.

Arabic culture, as early as A.D. 1000, placed great value upon learning and scholarship. The Arabs even produced, in Omar Khayyám (1043?–1123?), a man who was both poet and mathematician. In studying how to find the roots of a cubic equation, Omar took significant steps in algebra, the chief mathematical discipline cultivated by the Arabs. The word *algebra* is Arabic.

However, our debt to the Arabs is not primarily for their development of algebra, which bears no comparison in depth to the development of geometry by the Greeks. Nor is our principal debt to their rejection of the preposterous Roman numerals and the introduction of the Arabic system we use today. Our real debt is rather to their love of learning, which compelled them to preserve and translate the ancient classical writings, at a time in history when these might have been irretrievably lost.

For several hundred years the Arabs cradled such works as those of Euclid, Apollonius, and Archimedes. The rediscovery by western Europe of these works helped rekindle the flame of mathematics in the Renaissance. Neoclassicism, a movement to revive or to adapt the classical style, arose in mathematics just as it arose in literature, art, and music.

As every student of history knows, all this led in time to the Age of Reason, the Enlightenment, and eventually to the Romantic Movement. And every student of the liberal arts will know something of the way these historical movements are reflected in literature, art, and music.

Let us not leave mathematics out of the liberal arts. How does mathematics enter into this scheme of things? It is obvious that the development of calculus helped bring about the rise of modern science, to which the Enlightenment pointed with such pride. However, let us not be content with such an obvious remark. While we may not learn "the true position of mathematics as an element in the history of thought", we may yet learn something by musing about the nature of mathematics.

The key word of the discussion in this chapter is *tension*. It has been contended that the life in any work of art derives from the creation and resolution of tension, where "tension" is understood in a rather broad sense. Certainly the vitality of mathematics springs from a kind of tension. Mathematics itself, being in residence between the humanities and the sciences, is stretched in many directions: toward beauty, form, and vision on the one hand; and toward utility, function, and rationality on the other. And these are only a few of the struggles taking place within mathematics:

> Mathematics as an expression of the human mind reflects the active will, the contemplative reason, and the desire for aesthetic perfection. Its basic elements are logic and intuition, analysis and construction, generality and individuality. Though different traditions may emphasize different aspects, it is only the interplay of these antithetic forces and the struggle for their synthesis that constitute the life, usefulness, and supreme value of mathematical science.
>
> R. Courant and H. Robbins*

§1. Guessing versus Reasoning

Mathematics has always been associated with "reason", or "rational thought". The word *rational* still carries the connotation of "measurement" or "calculation". A rational man measures, or calculates, the effect of his activity.

The mathematician, in the act of making a discovery, is hardly "rational", however. The view that mathematicians employ only a cold, unexcited, strictly logical approach to their calling is somewhat distorted. Archimedes, struggling with a perplexing problem, once had such an exciting idea that he jumped from his bath to run naked and screaming down the streets of Syracuse. And Newton's greatness as a mathematician seems not to have been due primarily to his ability to reason correctly:

> I fancy his pre-eminence is due to his muscles of intuition being the strongest and most enduring with which a man has ever been gifted. Anyone who has ever attempted pure scientific or philosophical thought knows how one can hold a problem momentarily in one's mind and apply all one's powers of concentration to piercing through it, and how it will dissolve and escape

---

* *What is Mathematics?* Oxford University Press, New York, 1941, p. xv.

and you will find that what you are surveying is a blank. I believe that Newton could hold a problem in his mind for hours and days and weeks until it surrendered to him its secret. Then being a supreme mathematical technician he could dress it up, how you will, for purposes of exposition, but it was his intuition which was pre-eminently extraordinary—'so happy in his conjectures', said de Morgan, 'as to seem to know more than he could possibly have any means of proving' . . . .

John Maynard Keynes[*]

Conjectures, or guesses, play a largely unrecognized role in mathematics. We have seen in this book some instances of how they work. Early in Chapter 3 we guessed that the slope of a certain tangent line was 2, yet it took a while to find a reason why. Early in Chapter 6 we guessed that a certain area was 6 square units, but reasoned justification for that guess could come only much later. "Humble thyself, impotent reason!" exhorted Pascal, who almost discovered the calculus himself, before Newton. While reason may demonstrate the truth of a guess, reason alone rarely discovers anything of significance.

> Mathematics is regarded as a demonstrative science. Yet this is only one of its aspects. Finished mathematics presented in a finished form appears as purely demonstrative, consisting of proofs only. Yet mathematics in the making resembles any other human knowledge in the making. You have to guess a mathematical theorem before you prove it; you have to guess the idea of the proof before you carry through the details. You have to combine observations and follow analogies; you have to try and try again. The result of a mathematician's creative work is demonstrative reasoning, a proof; but the proof is discovered by plausible reasoning, by guessing.

G. Pólya[†]

Let there be no doubt of the existence, in mathematics, of knowledge acquired in nonrational ways:

> I have had my solutions for a long time, but I do not yet know how I am to arrive at them.

Gauss

> It's plain to me by the fountain I draw from, though I will not undertake to prove it to others.

Newton

---

[*] "Newton, the Man" from *Essays in Biography*, Horizon Press Inc., New York, 1951, p. 312. (This essay appears also in *Newton Tercentenary Celebrations*, Cambridge University Press, 1947.)

[†] *Induction and Analogy in Mathematics*, Princeton University Press, 1954, p. vi.

2. Atomism versus Common Sense

> Certain things first became clear to me by a mechanical method, although they had to be demonstrated by geometry afterwards, because investigation by the said method did not furnish an actual demonstration.
>
> Archimedes

Gauss*, Newton, and Archimedes stand in a class above all other mathematicians. We thus have it on the highest authority that imagination plays a role in mathematics at least rivaling, and perhaps surpassing, the role of reason. Great mathematicians have both gifts, in great degree.

EXERCISES

1.1. Make a guess as to the formula for the surface area S of a sphere of radius r. Reason by analogy with a circle, which is a "sphere" in the plane: for a circle, $A = \pi r^2$ and $C = 2\pi r$; for a sphere, $V = \frac{4}{3}\pi r^3$ and $S = ?$

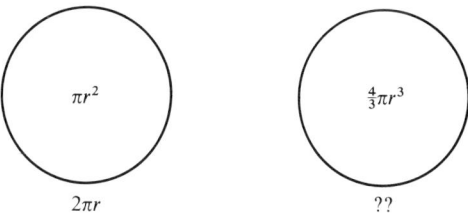

1.2. Give the approximate dates and general characteristics of the Renaissance, the Age of Reason, and the Romantic Movement. For help, consult an encyclopedia or a history of Western civilization.

1.3. Leaf through Volume 1 of *Mathematics and Plausible Reasoning*, G. Pólya, Princeton University Press, Princeton, N.J., 1954, for a fuller understanding of the art of guessing.

§2. Atomism versus Common Sense

How is nonrational, or intuitive, knowledge possible? What sorts of tricks were used by the developers of the calculus to tell them which way to go? Many tricks stem from the "atoms of Democritus". The Greek Democritus (ca. 460–370 B.C.) supported the doctrine of *atomism*, which holds that bodies are made up of *atoms*, or indivisible units. An atomist would raise no objection to thinking of a line as the sum of its points, or of an area as the sum of its

---

* Carl Friedrich Gauss (1777–1855), preeminent German mathematician.

vertical line segments. Atomism regards time as being made up of instants, an instant being a "point" in time.

Like most philosophical doctrines, atomism has its drawbacks. Common sense seems to tell us that time, like a pencil point moving smoothly along a line, is a "flowing", or "continuous", kind of thing. How can a *continuous* entity like time be made up of *discrete* instants? How can an atomist answer the Arrow Paradox of Zeno (ca. 495–435 B.C.)?

> Consider an arrow flying through the air. At each instant the arrow is motionless. How can the arrow move if it is motionless at each instant?

The same general sort of "paradox" is not uncommon in mathematics:

> How can a line segment have nonzero length, if each of its points has length zero?

> How can a planar figure have nonzero area, if each of its vertical line segments has area zero?

The inadequacy of atomism is evident. The atoms of a body apparently need not reflect all the properties of that body: whereas a line has length, its points do not. The whole may be something more than the sum of its atoms.

What good, then, is atomism? In mathematics it is often an aid to the intuition. Democritus used it to make an inspired guess about the proper formula for the volume of a cone or pyramid (one-third the area of the base times the height, in either case). Democritus had the imagination to guess the correct answer, but was never able to offer any rational justification for that answer. It was Eudoxus whose method provided the demonstrative proof. Both deserve credit: Democritus as seer, and Eudoxus as sage.

> ... in the case of the theorems the proof of which Eudoxus was the first to discover, namely that the cone is a third part of the cylinder, and the pyramid of the prism, having the same base and equal height, we should give no small share of the credit to Democritus who was the first to make the assertion ... but did not prove it.
>
> Archimedes

The passage above, as well as the quotation from Archimedes given earlier, is taken from a letter addressed to Eratosthenes. Archimedes goes on in this letter to describe *discovery* and *proof* as complementary aspects of mathematics. He then describes how he used atomism, in a novel way, to conjecture the truth of some of his most celebrated theorems, which he proved later by a masterful use of Eudoxus' method. The means of discovery and the means of proof were completely different.

3. Seer versus Sage

EXERCISES

2.1. Archimedes' letter to Eratosthenes is discussed briefly in an appendix to this book. Read the appendix on Archimedes.

2.2. (*The purpose of this exercise is to give a clue as to how Democritus might have used atomism to guess that the volume of a pyramid is one-third the area of the base times the height.*) Consider a bunch of cannonballs (to be thought of as large atoms) arranged in a pyramid with square base.

(a) Find the number of cannonballs in a pyramid if the base is
   (i) two by two.
   (ii) three by three.
   (iii) four by four. *Answer*: 30.
   (iv) n by n. *Hint*. See appendix on sums.

(b) Find the number of cannonballs in a *cube* if the base is
   (i) two by two.
   (ii) three by three.
   (iii) four by four. *Answer*: 64.
   (iv) n by n.

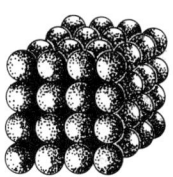

(c) Find the ratio of the number of cannonballs in a pyramid with square base to the number of cannonballs in a cube with the same base, if the base is n by n. *Answer*: $n(n+1)(2n+1)/6n^3$.
(d) The ratio in part (c) has 1/3 as a limit. How might this have helped Democritus?

§3. Seer versus Sage

It has never been a secret that the pursuit of mathematics requires more than the power of deductive reasoning. Even a rationalist allows this possibility.

> There are only two ways open to man for attaining certain knowledge of truth: clear intuition and necessary deduction.
>
> Descartes

The *seer* who discovers is just as much a mathematician as the *sage* who proves. Some mathematicians, like Archimedes, are coequally seer and sage, and no mathematician is wholly one or the other. Nevertheless the distinction is useful. The seer has the gift of vision—intuition, divination, or imaginative insight. Whereas the sage is blessed with wisdom—sound judgment, good taste, and reason. The seer points his hand to the sky while his eye darts around the heavens as if to see everything at a single instant. The sage, however, plants his feet squarely on the ground, his gaze fixed upon his object, and marks his world with a steady eye.

The distinction between seer and sage is of interest when one examines the philosophies of Plato and Aristotle, insofar as they pertain to mathematics. Platonism has often been seen to animate speculation, the searching and re-searching for undiscovered truths lying just beyond our ken. Seers are often disciples of Plato.

If Plato animates speculation, says Whitehead, then Aristotle animates scholarship. Aristotle emphasized the consolidation of knowledge, through reason, into a coherent system. Aristotle's influence may be seen in the form in which Euclid's *Elements* was cast, even though the content of the *Elements* owes its existence to the spirit of speculation. Aristotle's influence was great, and Greek mathematics appears almost always in finished form, cold, unexcited, and with strict logic, as if it might have been written by a sage alone.

The fact that classical Greek texts presented only proofs became a source of some annoyance later. One might think that the Greeks, in a wondrous plot, had all agreed to conspire against the seventeenth century by refusing to divulge their means of discovery.

> ... like some artisans who conceal their secret, they feared, perhaps, that the ease and simplicity of their [hidden] method, if become popular, would diminish its importance, and they preferred to make themselves admired by leaving to us, as the product of their art, certain barren truths deduced with subtlety, rather than to teach us that art itself, the knowledge of which would end our admiration.
>
> Descartes

Descartes could not have known that the "art itself", the seer's vision, had been freely given by Archimedes in his letter* to Eratosthenes mentioned earlier. And, having revealed his secret method (discussed in an appendix to this book), Archimedes wrote,

> I am persuaded that this method will be of no little service to mathematics. For I foresee that this method, once understood, will be used to discover other theorems which have not yet occurred to me, by other mathematicians, now living or yet unborn.

---

* See *The Method of Archimedes*, a Supplement to *The Works of Archimedes*, edited by T. L. Heath, Cambridge University Press, 1912 (also available in paperback by Dover Publications).

3. Seer versus Sage 359

Figure 3.* Plato and Aristotle in Raphael's "School of Athens". *Even now there is a very wavering grasp of the true position of mathematics as an element in the history of thought.* —Whitehead.

* Reproduced, with permission, from *Raphael*, by Oskar Fischel, translated by Bernard Rackham, Routledge & Kegan Paul Ltd., London, 1948.

Unfortunately for Descartes, the contents of Archimedes' letter had been lost for centuries, and were found again only in 1906, in Turkey, by the Danish philologist J. L. Heiberg.

The preference expressed by Descartes for the seer as opposed to the sage was typical of seventeenth-(and eighteenth-) century thought in mathematics. The influence of Aristotle was at an ebb, and Platonism was once again ascendent. It is curious that, in the Age of Reason, the climate was such as to permit a lapse of rigor in the reasoning used by mathematicians. The "idea itself", the means of discovery, became more important than the rigorous logical demonstration. The happy acceptance of vague, but intuitively suggestive remarks as a valid proof was not unusual. A reaction set in against the "over-precise" manner of the Greeks, which could only impede the progress of seventeenth-century mathematics.

An illustration of this is seen in Newton's *Principia*, written in 1687. In composing this greatest of works Newton attempted to emulate the rigorous Archimedean style; but, by doing so, he only made the *Principia* more difficult for modern minds to comprehend:

> The ponderous instrument of synthesis (Archimedism), so effective in his hands, has never since been grasped by one who could use it for such purposes; and we gaze at it with admiring curiosity, as on some gigantic implement of war, which stands idle among the memorials of ancient days, and makes us wonder what manner of man he was who could wield as a weapon what we can hardly lift as a burden.
>
> William Whewell

The passage above is from a nineteenth-century book on the history of science. In the seventeenth century, one might well have heard of Newton's *Principia* what King James had earlier said of Francis Bacon's *Novum Organum*, that "it was like the peace of God, which passeth all understanding."

Exercises

3.1. "If logic is the hygiene of the mathematician, it is not his source of food." The twentieth-century mathematician André Weil said this. What does Weil mean?

3.2. On the whole, has the spirit of Plato or of Aristotle been more conducive to the progress of science?

3.3. Find Newton's *Principia* in a library. The full title in English is *Mathematical Principles of Natural Philosophy*.
   (a) What did Newton mean by "natural philosophy"?
   (b) Why did Newton write in Latin?
   (c) Newton's book is generally acknowledged as the greatest single work ever written on science. Why?
   (d) Why is Newton's book so little read today?

3.4. (*For more ambitious students*) "The world will again sink into the boredom of a drab detail of rational thought, unless we retain in the sky some reflection of light from the sun of Hellenism." Read Chapter 7, "Laws of Nature", from *Adventures of Ideas*, A. N. Whitehead, Macmillan, New York, 1933, then tell what is meant by Whitehead's warning.

§4. The Discrete versus the Continuous

The attraction of atomism is probably the emphasis it places upon the *discrete*, a notion that seems quite transparent to the intuition. Certainly the discrete is easier to comprehend than the continuous. It is easier to think about a stationary pebble than about flowing sand. It is easier to think about a stationary instant than about flowing time.

The temptation is great to attempt to explain the continuous in terms of the discrete. Newton tried to explain light this way. Although common sense tells us that light is a "continuous" phenomenon, Newton spoke of "particles of light", as if a light ray was made up of a huge number of discrete units. Newton's description was not really taken seriously until the development of quantum theory in the twentieth century.

Within mathematics itself, the tension between the discrete and the continuous is profound.

> The whole of mathematical history may be interpreted as a battle for supremacy between these two concepts. This conflict may be but an echo of the older strife so prominent in early Greek philosophy, the struggle of the One to subdue the Many. But the image of a battle is not wholly appropriate, in mathematics at least, as the continuous and the discrete have frequently helped one another to progress.
>
> E. T. Bell[*]

The development of calculus given in this book has been based upon the notion of *limit*, which is intimately related to the idea of *continuity*. Thus we have placed much more emphasis upon the continuous than upon the discrete. It must now be confessed that the seventeenth century attempted a discrete approach to the calculus as well, the description of which makes up a remarkable chapter in the history of ideas.

An attempt will now be made to describe this discrete approach. If the reader finds this approach slightly incomprehensible, there is a good reason for it. The description is kept at a very intuitive level in order to ignore any difficulties that might be seen by close logical scrutiny. This is the way some things were done in the mathematics of the seventeenth century, and much

---

[*] E. T. Bell, *The Development of Mathematics*, McGraw-Hill, 1945, p. 13.

of the great progress made then is undoubtedly due to this approach. Had there not been a lapse in emphasis upon logical rigor, Newton and Leibniz might have feared to put some of their speculations into print. It is helpful to remember that the tone of seventeenth-century mathematics contrasts greatly with the classical Greek. Seventeenth-century mathematics often reads as if written by seer alone.

With our eyes "in a fine frenzy rolling", let us seek the seer's vision. Consider the following question:

> What does the difference Δx become, as Δx tends to zero, but is never allowed to equal zero?

The answer given by Leibniz might run something like the following:

> The difference Δx becomes a quantity of infinitesimal size, to be denoted by "dx" and called the *differential of x*. To say that dx is an infinitesimal is to say that dx is not zero, but is smaller than any positive number.

The differential of y, where y is a function of x, is "defined" in a similar way: the differential dy is what the difference Δy becomes, as Δx tends to zero, but is not allowed to equal zero. Leibniz thought of the derivative as an actual quotient of the differentials dy and dx. Thinking this way leads one to discover the chain rule. Newton thought in an intuitive way along much the same lines, but his terminology was different. He spoke of *fluents*, *fluxions*, and their *moments*, because he thought of a variable as a flowing quantity.

In reply to the question given above, many of us would say that the question is simply ill-posed and has no answer. The difference Δx does not "become" anything, because it is never zero. It just "keeps on changing". This would be the reply of a sage. Calculus was not discovered by a sage.

It is easy to criticize the notion of an infinitesimal, if it is regarded as a fixed quantity somehow squeezed between zero and the positive numbers. Where could it be on the number line? There is no place for it:

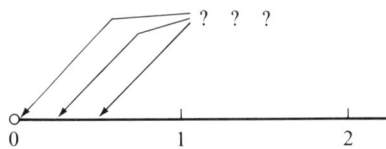

If there is such a thing as an infinitesimal, it must be a new kind of quantity, for it cannot be pictured as a point on the number line. The notion of the *infinitesimal* is one of the most elusive ideas ever conceived. Attempts to describe it, as with the adjectives *nascent* and *evanescent*, bordered upon the the comic. The first adjective means "just born", the second means "just vanishing".

However, we should not laugh at this seventeenth-century version of atomism called *infinitesimal analysis*. Though it all seems so vague, it was really a noble attempt to reconcile, through a rather mystical notion, the two great cooperating opposites of mathematics. An infinitesimal was supposed to be a discrete entity that retained qualities of the continuous.

EXERCISES

4.1. Prove that there is no positive number lying "next" to zero. That is, show that between any positive number and zero lies another number. *Hint.* Halving a non-zero quantity always results in a new quantity that is closer to zero.

4.2. Criticize the following statement. "The tangent line to a curve at a point is the line through the point and the next point on the curve."

4.3. (*For more ambitious students*) Read in a philosophy book about Leibniz's theory of monads. Write a paper explaining this theory, and explaining how it may be related to Leibniz's theory of differentials.

§5. The Infinitesimal Calculus

Let us continue in the spirit of the preceding section, agreeing to pretend that we know what an infinitesimal is. Let us also agree to accept the romantic notion that logic is unimportant, that something is true as soon as it is felt. This is the setting for discussing the remarkable theory of the infinitesimal calculus, in the spirit of seventeenth-century mathematical thought.

To start off with a bang, let us consider the fundamental theorem of calculus:

$$\int_a^b f(x)\,dx = F(b) - F(a) \quad \text{if } F' = f.$$

Infinitesimal calculus is better done in Leibniz's notation, which does not name functions, only variables. To put the fundamental theorem in this notation, let $y = F(x)$, so that $dy/dx = F'(x) = f(x)$, $y(b) = F(b)$, and $y(a) = F(a)$.

The Fundamental Theorem of Caculus. $\int_a^b (dy/dx)\,dx = y(b) - y(a)$.

"PROOF"! Canceling the differential dx we have

$$\int_a^b \frac{dy}{dx}\,dx = \int_a^b dy. \tag{1}$$

Now $\int_a^b dy$ is simply the sum, from a to b, of all the infinitesimal changes in y! This will obviously add up to the *total* change in the function y, as x runs from a to b:

$$\int_a^b dy = y(b) - y(a).$$

The fundamental theorem is simply the result of putting equations (1) and (2) together! (?)

EXAMPLE 1. Use infinitesimal calculus to find the area between the curves $y = 2 + 2x - x^2$ and $y = x - x^2$, between $x = 0$ and $x = 2$.

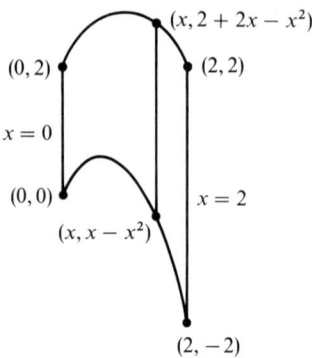

Let x be any number between 0 and 2. Consider the vertical segment through x whose width is infinitesimal! The length of this segment is

$$2 + 2x - x^2 - (x - x^2) = 2 + x.$$

The width of this segment is the infinitesimal dx! Its area is therefore $(2 + x)\,dx$! The entire area between the curves is the sum of these infinitesimal areas $(2 + x)\,dx$, as x runs from 0 to 2! The entire area is then

$$\int_0^2 (2 + x)\,dx = 6 \text{ square units,}$$

by the fundamental theorem of calculus! (?)

EXAMPLE 2. Use infinitesimal calculus to find dA/dt, where A is the area beneath the curve y, between $x = a$ and $x = t$.

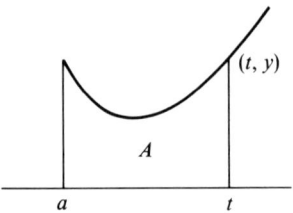

This is easy! As any transcendental eye can see, the infinitesimal change dA in area is clearly given by a rectangle of height y and width dt!

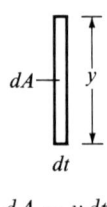

$$dA = y\,dt.$$

Dividing by dt, we see that $dA/dt = y$. Thus the area beneath a curve y yields an antiderivative of y. (?)

EXAMPLE 3 Use infinitesimal calculus to find the volume of a solid generated by revolving the area beneath a curve about the x-axis.

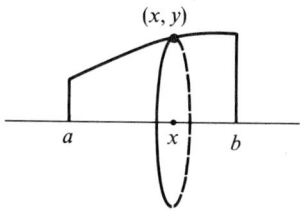

Let us consider the solid of revolution generated by the curve y from $x = a$ to $x = b$. If x is any number between a and b, consider the volume of the indicated slice through x of infinitesimal width!

Its infinitesimal volume dV is given by

$$dV = \pi y^2 \, dx,$$

by the well-known formula for the volume of an infinitesimal cylinder! The total volume V of the solid is the sum of all these infinitesimal dV's, as x runs from a to b!

$$V = \int_{x=a}^{x=b} dV = \int_a^b \pi y^2 \, dx.$$

This is the formula for the volume of a solid of revolution! (?)

EXAMPLE 4. Suppose a solid of revolution is generated by revolving the area beneath a curve about the y-axis. Use infinitesimal calculus to find the formula for the solid.

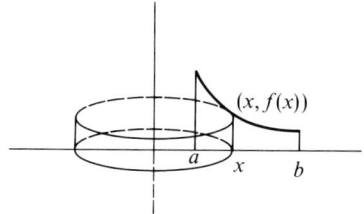

Let x be any number between a and b, and consider what happens to the vertical segment through x, as it is revolved about the y-axis. The surface of a cylinder is obtained, the cylinder being of height $f(x)$ and radius x. Let us find the infinitesimal volume of this surface, whose thickness is infinitesimal! When the surface is flattened out, a rectangular solid is obtained, dimensions $2\pi x$, $f(x)$, and dx.

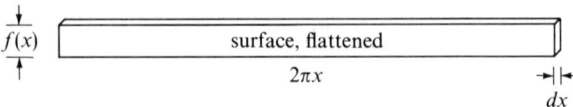

Its infinitesimal volume dV is then given by

$$dV = 2\pi x f(x)\, dx.$$

The total volume V is the sum of these infinitesimals dV, as x runs from a to b!

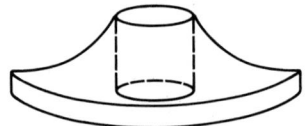

$$V = \int_{x=a}^{x=b} dV = \int_{a}^{b} 2\pi x f(x)\, dx$$

is the required formula! (?)

EXAMPLE 5. Use infinitesimal calculus to find the derivative of the squaring function.

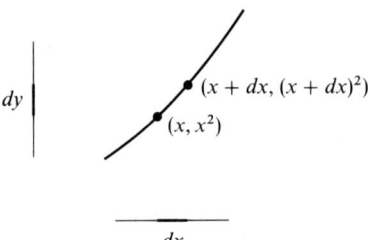

The two points virtually coincide

On the curve $y = x^2$ consider the point (x, x^2). If x is changed by an infinitesimal amount dx, then

$$dy = (x + dx)^2 - x^2 = 2x\, dx + (dx)^2.$$

Dividing by dx, we obtain

$$\frac{dy}{dx} = 2x + dx. \tag{3}$$

5. The Infinitesimal Calculus

Figure 4.* Engraving from Euler's *Introduction in Analysin Infinitorum* (1748), the first great treatise on analysis. Every textbook on calculus today borrows, more or less, from Euler.

* This illustration serves as frontispiece for Abraham Robinson's historic *Non-standard Analysis*, North-Holland, Amsterdam, 1966. Reproduced by kind permission of Elsevier-North Holland.

Since dx is infinitesimal, equation (3) becomes

$$\frac{dy}{dx} = 2x. \quad (?) \tag{4}$$

Example 5 leads to an interesting question for anyone who professes to understand the "reasoning" in it. *How can $2x + dx$ be equal to $2x$ unless dx is zero? And if dx is zero then how do you justify dividing by it to arrive at equation (3)?* One way to avoid embarrassment is to discard the notion that an infinitesimal is fixed and to think of it instead as a variable tending to zero. In that sense equation (3) does "become" equation (4). This leads to the discarding of fixed infinitesimals in favor of the notion of a *limit*.

EXERCISES

5.1. Discuss the following quotations of Bertrand Russell.
  (a) "It is a peculiar fact about the genesis and growth of new disciplines that too much rigour too early imposed stifles the imagination and stultifies invention. A certain freedom from the strictures of sustained formality tends to promote the development of a subject in its early stages, even if this means the risk of a certain amount of error." (*Wisdom of the West*, Rathbone Books Limited, London, 1959, p. 280.)
  (b) "Instinct, intuition, or insight is what first leads to the beliefs which subsequent reason confirms or confutes ... Reason is a harmonising, controlling force rather than a creative one. Even in the most purely logical realms, it is insight that first arrives at what is new." (*Our Knowledge of the External World*, W. W. Norton & Company, Inc., 1929, p. 22.)

5.2. In the spirit of this section write out a "proof" of each of the following, using infinitesimals. If ever you feel yourself getting into logical difficulties, adopt the visionary's style of reasoning by making exclamations instead of statements.
  (a) The distance traveled is the integral of the speed function.
  (b) $dy/dt = (dy/dx)(dx/dt)$.
  (c) $ds/dt = (dt/ds)^{-1}$.
  Answer: (c) *What else could it be!*

5.3. Consider a curve f. What is its *length* from the point $(a, f(a))$ to $(b, f(b))$?

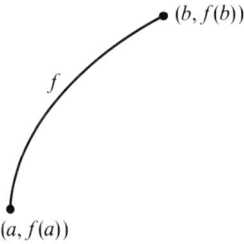

Use infinitesimals to try to guess a formula for the length of a curve. (Try to express it as an integral.) After you have made your guess, check to see if it works for a straight line whose length you are sure of. Then (if you have read Chapter 7) see if it checks out for some arc of the unit circle whose length you are sure of.

5.4. By carrying out the following steps, check to see if the formula of Example 4 works in this simple case considered here.
   (a) Use the known formula (derived in Chapter 6) for the volume of a cone to find the volume of a cone of radius 2, height 4.
   (b) Consider the cone generated by revolving the area beneath the line $y = 4 - 2x$, $0 \le x \le 2$, about the y-axis.

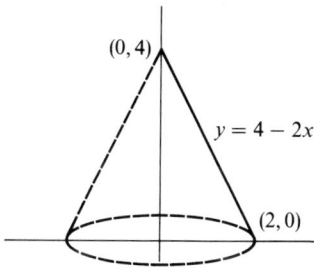

Apply the formula of Example 4 to this situation. Does it give the proper answer as found in part (a)?

5.5. (*Only for those who have read Chapter 9*) Work exercise 7.6 of Chapter 9 by throwing around infinitesimals with reckless abandon. *Answer*: (a) If $(dy/dx) - 2y = 0$, then $dy/dx = 2y$. Therefore (!),

$$\frac{1}{y} dy = 2\, dx,$$

$$\int_{x=0}^{x=t} \frac{1}{y} dy = \int_{x=0}^{x=t} 2\, dx,$$

$$\ln y(x) \Big|_0^t = 2x \Big|_0^t,$$

$$\ln y(t) - \ln y(0) = 2t,$$

$$\ln y(t) = \ln y(0) + 2t,$$

$$y(t) = y(0) e^{2t}.$$

5.6. (*How do you measure surface area?*) In exercise 1.1 you were asked to make a guess as to the formula for the surface area S of a sphere of radius r. Consider the following way one might reason by the use of infinitesimals. The sphere of radius r is obtained by revolving the graph of $y = \sqrt{r^2 - x^2}$, $r \le x \le r$, about the x-axis.

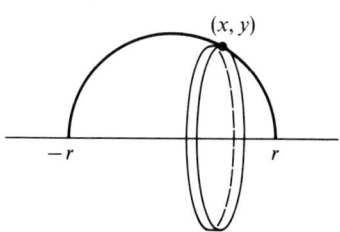

When the point (x, y) is revolved we get the surface of an infinitesimal cylinder of radius y and height dx! This surface, when flattened out into a rectangle, is seen to have an area of $2\pi y\, dx$. Therefore (!), the total surface area is given by

$$S = \int_{-r}^{r} 2\pi y\, dx = 2\pi \int_{-r}^{r} y\, dx = 2\pi \begin{pmatrix} \text{the area beneath the curve } y \\ \text{from } -r \text{ to } r \end{pmatrix}.$$

Since the area beneath the semicircle is $\pi r^2/2$, it follows that $S = 2\pi(\pi r^2/2) = \pi^2 r^2.$ (?)

(a) This probably disagrees with the guess you made in exercise 1.1. Do you still have more confidence in your guess, or do you accept the infinitesimal analysis just given? *Why?*

(b) Are you willing to accept all the infinitesimal analysis given in this section? If not, how do you decide what to accept and what to reject as invalid?

Hint. Archimedes showed that the surface area of a sphere of radius r is given by $S = 4\pi r^2$.

§6. Analysis versus Modern Developments

As Leibniz grew older, he began to move away from infinitesimals and toward the notion of *limit*. The following excerpt from a letter written in 1702 is seen as evidence for this.

> One must remember ... that incomparably small quantities ... are by no means constant and determined. On the contrary, since they may be made as small as we like, they play the same part in geometric reasoning as the infinitely small in the strict sense. For if an antagonist denies the correctness of our theorems, our calculations show that the error is smaller than any given quantity, since it is in our power to decrease the incomparably small ... as much as is necessary for our purpose.
>
> Leibniz*

This passage can be construed as saying, in effect, that the use of infinitesimals can be regarded as a shortcut means of taking a limit. The objectionable reasoning in Example 5, Leibniz seems to say, is merely a quick way of getting the result that was derived with a little more care in Chapter 5, Section 1.

It was doubly difficult for Leibniz to discard the infinitesimal, for this conception had inspired both his mathematics and, in his theory of monads, his philosophy. In deciding whether to disown his brainchild Leibniz must have experienced quite a struggle between mind and heart. His indecision is understandable. It was easier for Newton, who renounced the infinitely

---

* From a letter to Varignon, as given in *Ways of Thought of Great Mathematicians*, by Herbert Meschkowski, Holden-Day, 1964, p. 58.

6. Analysis versus Modern Developments

small in his later work, in favor of an intuitive understanding of the limit notion. Analysis in mathematics became based upon limits.

Atomism continues to survive today, though, and so do infinitesimals. Though sometimes, as in exercise 5.6, infinitesimals lead one astray, generally they point one in the proper direction like magic. Their success led Leo Tolstoy to seek an historical adaptation.

> [Infinitesimal calculus], unknown to the ancients, when dealing with problems of motion admits the conception of the infinitely small, and so conforms to the chief condition of motion (absolute continuity) and thereby corrects the inevitable error which the human mind cannot avoid when it deals with separate elements of motion instead of examining continuous motion.
>
> In seeking the laws of historical movement just the same thing happens. The movement of humanity, arising as it does from innumerable arbitrary human wills, is continuous....
>
> Only by taking infinitesimally small units for observation (the differential of history, that is, the individual tendencies of men) and attaining to the art of integrating them (that is, finding the sum of these infinitesimals) can we hope to arrive at the laws of history.
>
> Tolstoy*

These words were written in the middle of the nineteenth century, showing that infinitesimals were alive and kicking then. Even in the mid-twentieth century infinitesimals were used in calculus, though only, it was supposed to be emphasized, as a shortcut means of deriving what was done more rigorously by means of limits.

Analysis, a branch of mathematics growing out of calculus to develop a precise notion of limit, had in the late nineteenth century given calculus a firm foundation. Analysis tried to do for the calculus what Euclid had attempted to do for geometry: base the entire structure upon a few simple general principles. The "ε-δ" definition of a limit (the discussion of which we defer) has given a precise meaning to that notion. (The reader will have observed that the discussion of limits so far offered in this book has been completely intuitive in character.)

Quite recently mathematics has seen the exciting development of *nonstandard analysis*, which makes real sense out of infinitesimals. This work was pioneered in the 1960s by Abraham Robinson (1918–1974), who used sophisticated modern mathematical ideas to capture the intuitive notion of an infinitesimal. The discrete approach to the calculus, thought for so long to have been a heroic failure, may yet be a success after all.

At the same time, a movement in quite the opposite direction has been born. Errett Bishop (1928–   ) and his followers have developed an approach to the calculus that is more down-to-earth and constructive in nature than traditional analysis. It will be interesting to see how calculus looks in the year 2015, on its 350-th birthday.

---

* *War and Peace*, translated by Louise and Aylmer Maude, Simon and Schuster, New York, 1942, p. 918.

EXERCISES

6.1. Read Chapter 7, "The Beginning of Modern Mathematics, 1637–1687", in *The Development of Mathematics*, E. T. Bell, McGraw-Hill, New York, 1945.

6.2. Read the first section of Book Eleven of *War and Peace*, from which Tolstoy's quotation above is taken.

6.3. Read pp. 1–2 of *Non-standard Analysis*, Abraham Robinson, North-Holland, Amsterdam, 1966. Compare it with "A Constructivist Manifesto", pp. 1–10 of *Foundations of Constructive Analysis*, Errett Bishop, McGraw-Hill, New York, 1967.

§7. Faith versus Reason

Having had a very brief view of what has happened to calculus recently, let us get back to the early eighteenth century, where the old conflict between faith and reason still raged. A minor, but revealing incident in this conflict concerns infinitesimals, which became the ammunition for a skirmish between Edmund Halley, the astronomer, and George Berkeley (pronounced BARK-ly), the philosopher.

Halley, so the story goes, had persuaded a friend of Berkeley's to become skeptical about his religious beliefs, whereupon they were rejected on the grounds that theologian's claims could not be justified so soundly as the claims of mathematicians. This infuriated the Irishman Berkeley, who had just been made a bishop in the Church of England. His outrage was so great that he sought not to shore up the foundations of theology, but to undermine those of mathematics. The result was an extraordinary essay, *The Analyst*, "a discourse addressed to an infidel mathematician".

> Whereas then it is supposed that you apprehend more distinctly, consider more closely, infer more justly, and conclude more accurately than other men, and that you are therefore less religious because more judicious, I shall claim the privilege of a Freethinker; and take the liberty to inquire into the object, principles, and method of demonstration admitted by the mathematicians of the present age, with the same freedom that you presume to treat the principles and mysteries of Religion; to the end that all men may see what right you have to lead, or what encouragement others have to follow you....

Berkeley wrote *The Analyst* in 1734, not too many years after the deaths of Newton and Leibniz. In his essay Berkeley forcefully made the argument given in Section 5, in criticism of the logic used in infinitesimal calculus. He pointed out quite rightly that the seventeenth century was content to accept arguments that the ancient Greeks would have discarded as inadequate. The implication was that seventeenth-century mathematicians were accepting arguments on faith, not on reason. Berkeley then went on

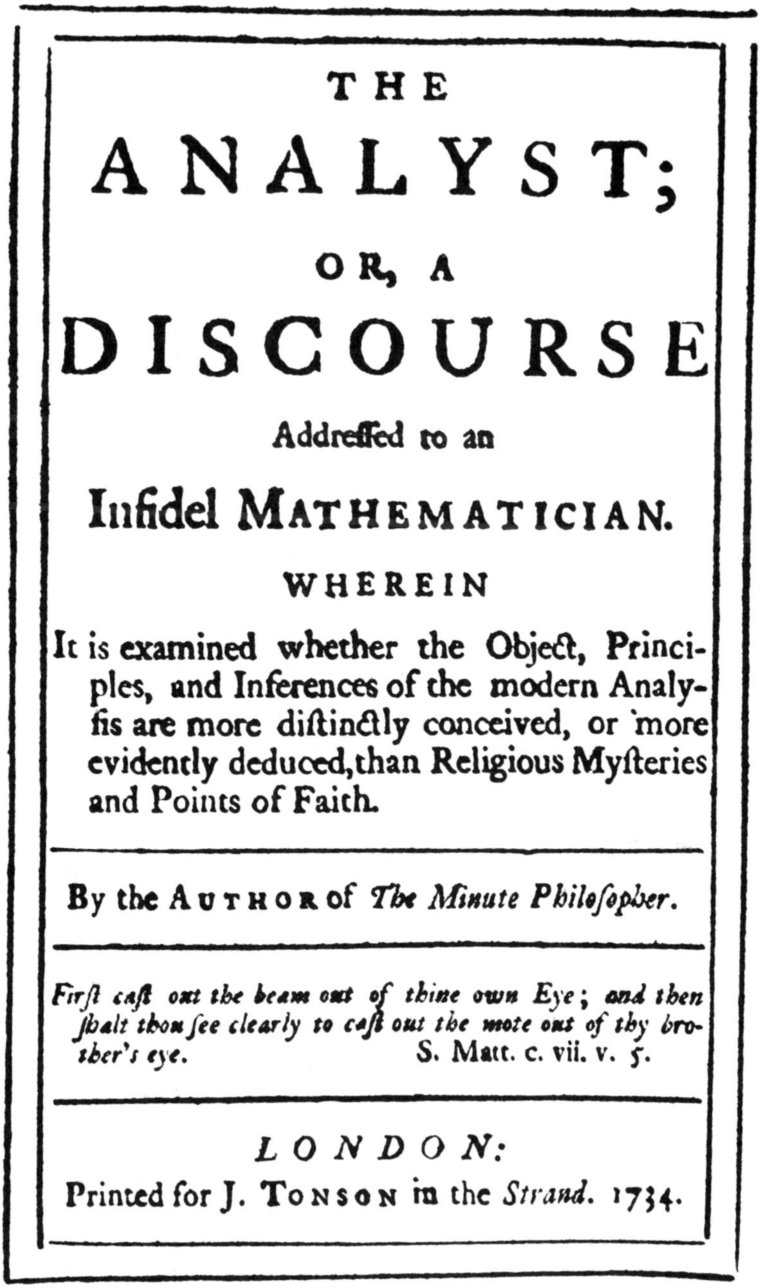

THE ANALYST;

OR, A

DISCOURSE

Addreſſed to an

Infidel MATHEMATICIAN.

WHEREIN

It is examined whether the Object, Principles, and Inferences of the modern Analyſis are more diſtinctly conceived, or more evidently deduced, than Religious Myſteries and Points of Faith.

By the AUTHOR of *The Minute Philoſopher*.

Firſt caſt out the beam out of thine own Eye; and then ſhalt thou ſee clearly to caſt out the mote out of thy brother's eye.   S. Matt. c. vii. v. 5.

LONDON:
Printed for J. TONSON in the *Strand*. 1734.

Figure 5. Title page of Berkeley's "The Analyst".
[H]e who can digest a second or third fluxion . . . need not, methinks, be squeamish about any point in divinity.

drily to inquire whether infinitesimals were not "ghosts of departed quantities", implying that nothing in theology could be more ghostlike than the basic notion of infinitesimal calculus.

At the time it was written, little attention was given by mathematicians to Berkeley's splendid philippic. Today it is admitted by virtually every student of mathematics that some of Berkeley's objections to the calculus were unanswerable until the late nineteenth century, when analysis at last produced a precise definition of the notion of a *limit*.

EXERCISES

7.1. Read one of the following chapters from *Mathematics in Western Culture*, Morris Kline, Oxford University Press, New York, 1953.
   (a) Chapter XVI, "The Newtonian Influence: Science and Philosophy".
   (b) Chapter XVII, "The Newtonian Influence: Religion".
   (c) Chapter XVIII, "The Newtonian Influence: Literature and Aesthetics".

7.2. Read the excerpts and commentary arising from Berkeley's *Analyst* given on pp. 286–293 of *World of Mathematics*, edited by James R. Newman, Simon and Schuster, New York, 1956.

7.3. Read Chapter 13, "From Intuition to Absolute Rigor, 1700–1900", in *The Development of Mathematics*, E. T. Bell, McGraw-Hill, New York, 1945.

§8. Conclusion

It is curious that, despite his fulminations, Bishop Berkeley accepted the calculus on faith. He believed, he said, that correct results in the calculus were the product of some "compensation of errors" in reasoning.

Mathematicians soon became aware of the shaky ground on which the calculus was erected, as indicated by the admonition of d'Alembert (1717–1783): "Go forward, and faith will follow!" In the conflict between faith and reason, mathematics had the potential for use by either side. Among the founders of the calculus both Leibniz and Newton exhibited strong interest in theology. Leibniz made a serious attempt to reunite the Protestant and Catholic churches, and Newton thought of his work as helping to prove the existence of God. The study of things eternal may tend to heighten one's awareness of religion.

The first edition (1771) of the *Encyclopaedia Britannica* quoted with approval the sentiments expressed a century earlier by Issac Barrow. Barrow was Newton's teacher, who resigned from Cambridge in order that Newton be given his professorship. The words could have been written by Plato himself:

> The mathematics ... effectually exercise, not vainly delude, nor vexatiously torment, studious minds with obscure subtilties; but plainly demon-

8. Conclusion

strate every thing within their reach, draw certain conclusions, instruct by profitable rules, and unfold pleasant questions. These disciplines likewise enure and corroborate the mind to a constant diligence in study; they wholly deliver us from a credulous simplicity, most strongly fortify us against the vanity of scepticism, effectually restrain us from a rash presumption, most easily incline us to a due assent, perfectly subject us to the government of right reason. While the mind is abstracted and elevated from sensible matter, distinctly views pure forms, conceives the beauty of ideas, and investigates the harmony of proportions; the manners themselves are sensibly corrected and improved, the affections composed and rectified, the fancy calmed and settled, and the understanding raised and excited to more divine contemplations.*

Nevertheless, mathematics appears generally seen as allied with reason in opposition to faith. Voltaire pointed to the spectacular achievements of "rational" science and mathematics, and demanded the right to examine everything under the authority of reason. Voltaire won out, for a time, and it was little noted that mathematicians of the Enlightenment were using their instinct more than their intellect:

> They defined their terms vaguely and used their methods loosely, and the logic of their arguments was made to fit the dictates of their intuition. In short, they broke all the laws of rigor and of mathematical decorum.
> The veritable orgy which followed the introduction of the infinitesimals ... was but a natural reaction. Intuition had too long been held imprisoned by the severe rigor of the Greeks. Now it broke loose, and there were no Euclids to keep its romantic flight in check.
>
> Tobias Dantzig†

The Enlightenment, emphasizing intellect, was to be washed aside by the Romantic Movement that declared, with Rousseau, the primary nature of instinct. Romanticism can be, and has been, described in a variety of ways. But if it is marked by a reaction against Neoclassicism, an emphasis upon imagination, a disregard for decorum, and a predilection for the seer to the sage, then the Romantic Movement was already rampant in mathematics at the very height of the Age of Reason.

Is this a paradox? Only, perhaps, for those who see mathematics in too narrow a light. Mathematics is romance in reason.

---

* "Mathematics", *Encyclopaedia Britannica*, Vol. III, 1771, pp. 30–31.
† *Number, the Language of Science*, Doubleday Anchor, 1956, p. 130.

Review Problems for Chapters 1–10

1. Given a right triangle of sides a, b, and c, construct three others just like it, and put all four triangles together to help form a large square. Then stare at the large square until you see a new way to prove the Pythagorean theorem.

   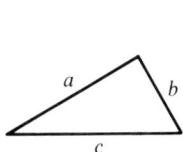
   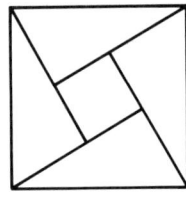

   Hint. Begin by observing that the area of the large square can be expressed in two ways.

2. Below is a right triangle inscribed in a semicircle. The legs of the triangle are diameters of two other semicircles. Prove that the sum of the areas of the two lunes is equal to the area of the right triangle.

   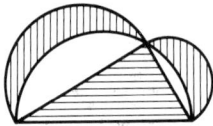

   (Lunes fascinated Leonardo da Vinci. See Coolidge, *The Mathematics of Great Amateurs*, Dover paperback, p. 43 ff.)

3. Find the area of this lune.

   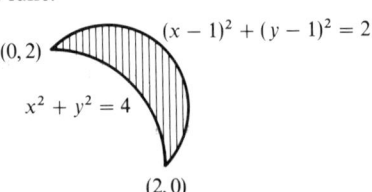

4. Find the indicated areas. *Hint.* In each case, split the area up into a triangle and a pie-shaped sector.

(a)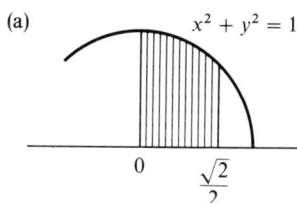

(b)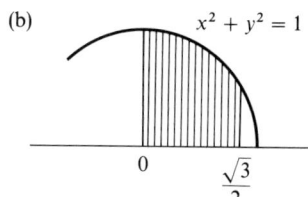

5. Prove the irrationality of each of the following numbers. (a) $\sqrt{2}/2$; (b) $\sqrt{3}/2$.

6. Describe the Pythagorean philosophy. What influence did it have upon Plato?

7. Plato and Aristotle held differing views about the nature and the value of mathematics. Explain these views.

8. Read the catalogue of courses of study offered by a college or university in which you are interested. Separate the offerings into those which are "Greek" and those which are "non-Greek". (*Anthropology*, for example, is a Greek word meaning study of man; *biology* is Greek for study of life, etc.) Does the spirit of Greece still live?

9. (a) Solve the equation $bx + c = 0$.
   (b) Find the limit, as a tends to zero, of $(-b + \sqrt{b^2 - 4ac})/2a$.
   (c) Explain why it should not be surprising that parts (a) and (b) have the same answer.

10. "Limits are as simple as pi." This pun contains a good deal of truth. Explain.

11. A rectangular field bordering a road is to be fenced in. The fence along the road will cost $9 per meter, while each of the other three sides will cost $4 per meter.
    (a) Find the dimensions yielding the cheapest cost if an area of 100 square meters is to be enclosed.
    (b) Find the maximal area that can be enclosed if the total cost of the fence is $1800.

12. The sum of two positive numbers is 60, and the numbers are chosen so that the product of the first by the cube of the second is as large as possible. Find the numbers.

13. A can in the shape of a cylinder is to be placed inside a sphere of radius r inches. Find the largest possible volume the can could have.

14. A cone-shaped paper cup is to be designed whose edge is to be 10 centimeters in length. What should be the height h of the cup, in order to maximize its volume?

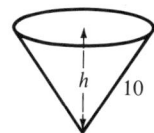

15. A cone is to be inscribed within a sphere of radius r inches. Find the largest possible volume the cone could have.

16. Santa Claus has a problem. Out of 100 square feet of tin a rectangular box with square base and open top is to be constructed. What dimensions should be used to maximize the amount of reindeer food the box can hold?

17. One corner of a long rectangular sheet of paper of width 8 inches is folded over until it just touches the opposite edge. Find the minimum length of the crease.

18. A man is in a boat 100 meters from a straight shoreline. His cabin is 100 meters downshore. If he can walk three times as fast as he can row, where should his boat be docked in order that he get to his cabin as quickly as possible?

19. A ladder is to be carried horizontally around the indicated corner. Find the length of the longest ladder that can pass through.

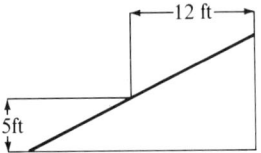

20. A ladder 10 ft long leans against a fence 4 ft high. Find the maximum horizontal distance the ladder can reach beyond the fence.

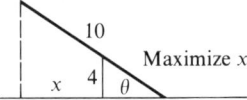

21. A circular lake is 200 meters in diameter. A man standing on one side wishes to get to the point directly opposite. He can run twice as fast as he can swim. What is the quickest way across?

22. A string of length 40 centimeters is to be cut in two, and the pieces are to be folded into a circle and an equilateral triangle respectively. How should the string be cut in order to
    (a) *minimize* the combined area of the circle and triangle?
    (b) *maximize* the combined area?

23. (a) Define the terms *critical point* and *inflection point*.
    (b) Give an example of a curve with no highest point and no lowest point.
    (c) Outline the steps by which one can find the highest and lowest points on a curve, using calculus.

24. The curves drawn in problem 41 at the end of Chapter 4 are given without equations, but the person sketching these curves had certain equations in mind. Can you guess the equations represented by these curves? (The curve in (b), for example, is intended to be the graph of $y = \ln|x|$.)

25. Can you match equations to the curves pictured in problem 25 at the end of Chapter 5?

26. Match equations to the curves pictured in problem 24 at the end of Chapter 6.

27. Consider the equation $C = 10L + (40/L)$.
    (a) Sketch this curve, indicating all critical points.
    (b) Find the range, if the domain is given by $1 < L < 5$.
    (c) Find the range if the domain is given by $L \neq 0$.

28. Consider the function given by $f(x) = 2x^2 + (3/x) - 4$.
    (a) Sketch this curve, indicating all critical points and inflection points. *Hint.* A little work has already been done in exercise 3.1 of Chapter 4.
    (b) Find the range, if the domain is given by $\frac{1}{2} < x \leq 3$.
    (c) Find the area beneath f, from $x = 2$ to $x = 5$.

29. From a tower 124 feet high, a ball is thrown downwards at an initial speed of 40 feet per second. Treating the ball as a freely falling body, answer the following.
    (a) Where is the ball one second after it is released?
    (b) When will the ball hit the ground?
    (c) What speed is the ball approaching at the instant it hits the ground?

30. A car accelerates from 0 to 50 ft/sec in 10 seconds. In this time, assuming constant acceleration, find how far the car travels.

31. A rock thrown straight up reaches a maximum height of 64 feet. What was its initial speed?

32. A balloon, shaped like a sphere, is being blown up at a constant rate so that its volume is increasing at 2 cubic feet per second. When its radius is 3 feet,
    (a) how fast is its radius increasing?
    (b) how fast is its surface area increasing?

33. A paper cup shaped like a cone has height 10 centimeters and has a circular top of radius 5 centimeters. Water is poured in at a constant rate of 3 cubic centimeters per second. Letting h denote the height of water in the cup, find dh/dt when
    (a) $h = 1$ cm.
    (b) $h = 4$ cm.
    (c) $h = 8$ cm.

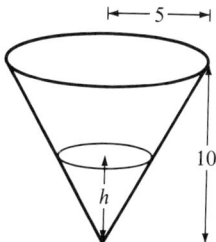

34. An observer stands one mile from the launch site of a rocket. If the speed of the rocket is 300 mi/hr when the angle of inclination θ of the rocket is 45 degrees, find $d\theta/dt$ at this instant.

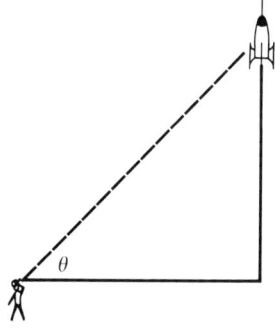

35. A particle is moving along the y-axis in such a way that its position is given by $y = 5\sin t + 12\cos t$.
    (a) Find its position, velocity, and acceleration at the instant when $t = 5\pi/6$.
    (b) Find the range of positions y assumed by the particle between time $t = 0$ and $t = 2\pi$.

36. An observer is 48 feet away from a spot where a ball is thrown straight up with an initial speed of 64 ft/sec. Letting θ be the angle of inclination of the ball from the observer's position, find $d\theta/dt$ and $d^2\theta/dt^2$
    (a) one second after the ball is released.
    (b) two seconds after the ball is released.
    (c) three seconds after the ball is released.

37. In problem 36, let s be the distance between the observer and the ball. Find ds/dt and d^2s/dt^2 at times one, two, and three seconds after the ball is released.

38. Use appropriate formulas for sums to evaluate each of the following.
    (a) $1 + 2 + 3 + \cdots + 1000$.
    (b) $1 + 4 + 9 + \cdots + 1{,}000{,}000$.
    (c) $1 + 8 + 27 + \cdots + 1{,}000{,}000{,}000$.
    (d) $1 + (1/2) + (1/4) + (1/8) + \cdots + (1/1024)$.
    (e) $1 + 3 + 9 + 27 + \cdots + 59{,}049$.

39. Consider the function given by $f(x) = 3x^2 - 2x + 5$.
    (a) Illustrate Fermat's method in finding the slope of the tangent line to the curve f at $(1, 6)$.
    (b) Illustrate Eudoxus' method in finding the integral of f from $x = 0$ to $x = 4$.

40. Evaluate the integral $\int_a^b x^2\, dx$ by Eudoxus' method.

41. Evaluate the integral $\int_0^c x^3\, dx$ by Eudoxus' method.

42. Consider the function given by $f(t) = e^{-t^2}$.
    (a) Find the derivative of f.
    (b) Find the antiderivative of f that takes the value 3 when t is 2.
    (c) Find an algebraic approximation to f that is valid for t near 0.
    (d) Use your answer to part (c) to give a good estimation for the value of the integral $\int_0^1 e^{-t^2}\, dt$.

43. (a) Use trigonometric identities to show $\tan(\theta + \tfrac{1}{2}\pi) = -1/\tan\theta$.
    (b) Use the result of part (a) to show that if a line has slope $b \neq 0$, then a perpendicular line has slope $-1/b$. (Lines are perpendicular if their slopes are negative reciprocals of each other.)

44. Consider the curve $y = x^x$.
    (a) Write an equation of the tangent line at $(4, 256)$.
    (b) The *normal line* to a curve is the line perpendicular to the tangent line. Write an equation of the normal line at $(4, 256)$. *Hint.* Use the result of part (b) of problem 43.

45. The equation $y = \int_0^x e^{-t^2}(t-1)(t-3)\, dt$ describes a curve in the x-y plane.
    (a) Fill in the following table.

x	y	y'	y''
0			

(b) Draw a picture of the curve locally, as it passes through the origin, indicating whether it is rising or falling, concave up or down.
(c) Tell whether the curve is rising or falling, concave up or down, when $x = 2$.
(d) Find all the critical points of the function specified by the equation above, and tell whether they are local minima or local maxima.

46. In Chapter 6 Cavalieri's principle was discussed for figures lying in a plane. Actually, Cavalieri formulated an analogous principle for figures lying in three-dimensional space. Can you? "Two solid figures have the same volume if . . ."?

47. (a) What does it mean to say that a function f is *continuous* at a certain point c?
(b) If $f'(c)$ exists, does it follow that f is continuous at the point c?
(c) If f is continuous at the point c, does it follow that $f'(c)$ exists?

48. A **cycloid** is the curve traced out by a point on a circle, as the circle rolls along a straight line. (Think of the path of the head of a nail embedded in a tire, as the tire rolls.) The cycloid was the subject of many investigations in the seventeenth century. Let the circle have radius r, let the point tracing out the cycloid be initially at the origin, and let the circle roll along the x-axis. Show that, after the circle has rotated through an angle of θ radians, the coordinates of the point tracing out the cycloid are given by $x = r\theta - r\sin\theta$, $y = r - r\cos\theta$. *Hint*. After the tire moves through an angle θ, the situation looks like this. Begin by observing that the circular arc must be equal in length to the distance traveled by the circle.

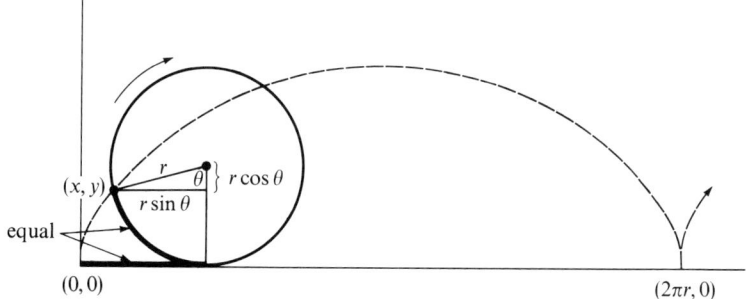

49. Consider the cycloid of problem 48, whose coordinates x and y are parametrized by θ. (The radius r is a constant.)
(a) Find $dx/d\theta$ and $dy/d\theta$.
(b) Find dy/dx with help from part (a) and the chain rule.
(c) The point $(r(\frac{1}{3}\pi - \frac{1}{2}\sqrt{3}), \frac{1}{2}r)$ lies on the cycloid. Find the slope of the tangent line at this point.

50. Let us try to determine the length of arc of the cycloid of problem 48. Let s denote the arclength traversed as a function of θ, beginning at $\theta = 0$ (when the point tracing out the cycloid is located at the origin.)
(a) Use the nonrigorous type of reasoning illustrated in Section 5 of Chapter 10 to make it plausible that $(ds/d\theta)^2 = (dx/d\theta)^2 + (dy/d\theta)^2$. Begin by applying the Pythagorean theorem to an infinitesimal right triangle.

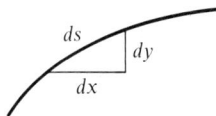

(b) Using the equation of part (a) and the result of part (a) of problem 49, conclude that
$$\frac{ds}{d\theta} = r\sqrt{2 - 2\cos\theta}.$$

(c) Conclude from part (b) that the arclength of the cycloid from the origin to the point $(2\pi r, 0)$ is given by the integral
$$r \int_0^{2\pi} \sqrt{2 - 2\cos\theta}\, d\theta.$$

(d) Evaluate the integral in part (c). *Hint.* To find an antiderivative, use the trick in problem 24 at the end of Chapter 7.

51. (a) Show that $\frac{1}{4}\pi = \arctan\frac{1}{2} + \arctan\frac{1}{3}$. *Hint.* Let $\theta_1 = \arctan\frac{1}{2}$, let $\theta_2 = \arctan\frac{1}{3}$, and use the tangent sum law to show that $\tan(\theta_1 + \theta_2) = 1 = \tan\frac{1}{4}\pi$.
    (b) Use the result established in part (a) to obtain a close approximation to π. *Hint.* The quantities $\arctan\frac{1}{2}$ and $\arctan\frac{1}{3}$ can be found very accurately with little trouble. For example (see problem 26 of Chapter 9),
$$\arctan\frac{1}{2} = \int_0^{1/2} \frac{1}{1 + x^2}\, dx = \int_0^{1/2} (1 - x^2 + x^4 - x^6 + \cdots)\, dx.$$

52. The method of approximating π given in the preceding problem offers a considerable improvement over the method of Leibniz given in problem 26 of Chapter 9. An even quicker method of approximating π was discovered by John Machin in 1706, who established first that
$$\frac{\pi}{4} = 4 \arctan\frac{1}{5} - \arctan\frac{1}{239}.$$

(This can be proved by repeated application of the tangent sum law.) Use Machin's identity to estimate π to six decimal places.

53. Evaluate each of the following limits.
    (a) $\text{Limit}_{x \to 0} (1 - e^x)/(1 - \pi^x)$.
    (b) $\text{Limit}_{x \to 1} (1 - x^e)/(1 - x^\pi)$.
    (c) $\text{Limit}_{x \to 1} \ln((1 - x^3)/(1 - x))$.
    (d) $\text{Limit}_{x \to 2^+} [\ln(x^2 - 4) - \ln(x - 2)]$.

54. Let A and B be unequal positive numbers. Prove that the quantity $(A - B)/(\ln A - \ln B)$ lies between A and B. *Hint.* Apply an appropriate form of the mean-value theorem to the natural log function on the interval from A to B.

55. Integrals were introduced in order to represent areas. Yet there are people having no interest whatever in areas who still place high value on the study of integrals. List at least six reasons why integrals are of interest, aside from their use in measuring areas.

56. Some decomposed carbon has been found in a cave thought to have been inhabited by ancient man. If it is established that one fourth of the original carbon has decomposed, how long has the carbon been in the cave? (The half-life of carbon is 5568 years.)

57. In one month 20 g of radioactive substance X decays to 14 g, while during the same period 50 g of radioactive substance Y decays to 30 g. When will there be the same amount of substances X and Y?

58. Consider the function given by $f(x) = \sqrt[3]{1000 + x}$.
   (a) Find an algebraic approximation agreeing with f to third order at the point $x = 0$.
   (b) Use part (a) to estimate the cube root of 1010 to several decimal places.

59. (a) Find the derivative of te^{-t}.
   (b) Use your answer to part (a) to guess an *antiderivative* of te^{-t}.
   (c) Guess an antiderivative of $t^2 e^{-t}$. *Hint.* First find its derivative.
   (d) Investigate this conjecture: An antiderivative of $f(t)e^{-t}$ is given by $-e^{-t}(f(t) + f'(t) + f''(t) + \cdots)$.
   (e) Find an antiderivative of $(t^2 - 3t + 6)e^{-t}$.
   (f) Find an antiderivative of $(t^2 - 3t + 6)e^{t}$.
   (g) Make your own conjecture about antiderivatives of $f(t)e^{t}$ and attempt to verify it.

60. (a) Give a one-sentence definition of calculus.
   (b) Give a one-sentence definition of mathematics.

61. Mathematics is made up largely of *methods*. We have discussed mainly Fermat's method and the method of Eudoxus, but we have also seen the method of Archimedes, the "Sherlock Holmes method", and the method of Simpson (Simpson's rule). Without referring to the text,
   (a) explain the purpose of each method,
   (b) explain how each method works, and
   (c) evaluate how well each method accomplishes its purpose.

62. Much of our knowledge of Greek mathematics comes from a commentary written on the first book of Euclid's *Elements* by Proclus, who lived in the fifth century A.D. In his commentary Proclus says that mathematics

   > arouses our innate knowledge, awakens our intellect, purges our understanding, brings to light the concepts that belong essentially to us, takes away the forgetfulness and ignorance that we have from birth, sets us free from the bonds of unreason; and all this by the favor of the god [Hermes?] who is truly the patron of this science, who brings our intellectual endowments to light, fills everything with divine reason, moves our souls towards Nous [the highest form of knowledge], awakens us as it were from our heavy slumber, through our searching turns us back upon ourselves, through our birthpangs perfects us, and through the discovery of pure Nous leads us to the blessed life. (*Proclus: A Commentary on the First Book of Euclid's Elements*, translated by Glenn R. Morrow, Princeton University Press, 1970, p. 38)

   (a) Compare this passage with the passage quoted in Chapter 10, Section 8, written some 1200 years later.
   (b) From the quotation above it is obvious whom Proclus thought to be the supreme philosopher. Was it Plato or Aristotle?

63. Name at least five noted philosophers who have also been mathematicians. Is it just an accident that some of the most eminent philosophers have also been mathematicians?

64. The Greek writings of Plato, the French of Pascal, Descartes, and Poincaré, the English of Russell and Whitehead have all been acclaimed as models of prose style by students of literature. Is it just an accident that some of the most eminent writers have also devoted themselves to mathematics?

65. Consider each of the following relatively recent statements. Which of them is virtually a restatement of the principle of continuity? Which of them remind you of Pythagoras? of Plato? of Archimedes? of Leibniz?
   (a) Remote from human passions, remote even from the pitiful facts of nature, the generations have gradually created an ordered cosmos, where pure thought can dwell as in its natural home and where one, at least, of our nobler impulses can escape from the dreary exile of the actual world (Bertrand Russell, twentieth century).
   (b) Since the fabric of the world is the most perfect and was established by the wisest Creator, nothing happens in this world in which some reason of maximum or minimum would not come to light (Leonhard Euler, eighteenth century).
   (c) Apart from a certain smoothness in the nature of things, there can be no knowledge, no useful method, no intelligent purpose (Alfred North Whitehead, twentieth century).
   (d) The great book of Nature lies ever open before our eyes and the true philosophy is written in it . . . But we cannot read it unless we have first learned the language and the characters in which it is written . . . It is written in mathematical language (Galileo, seventeenth century).
   (e) I have never done anything 'useful' . . . . The case for my life . . . is this: that I have added something to knowledge, and helped others to add more; and that these somethings have a value which differs in degree only, and not in kind, from that of the creations of the great mathematicians, or of any of the artists, great or small, who have left some kind of memorial behind them (G. H. Hardy, twentieth century).

66. William Blake (1757–1827) wrote the following poem. What is it about?

> Mock on, Mock on Voltaire, Rousseau:
> Mock on, Mock on: 'tis all in vain!
> You throw the sand against the wind,
> And the wind blows it back again.
>
> And every sand becomes a Gem
> Reflected in the beams divine;
> Blown back they blind the mocking Eye,
> But still in Israel's paths they shine.
>
> The Atoms of Democritus
> And Newton's Particles of light
> Are sands upon the Red sea shore,
> Where Israel's tents do shine so bright.

Appendix 1
Writings "About" Mathematics*

The complaint of some humanists that mathematicians make no attempt to describe to others their function is unjustified. On the contrary, a list of articles and books published with this purpose in mind is extensive, owing to the efforts of a number of writers, many of whom have been distinguished mathematicians.

This appendix calls attention to some of these writings about mathematics, which are by and large non-technical in nature and addressed to the general reader. It also serves as a way of acknowledging, however inadequately, the debt owed by the author of this book to the writings of others. The list of works mentioned has been purposely kept short—so as not to overwhelm the reader—and reflects to some degree, of course, the taste of the author.

A much more extensive list may be found in the useful *Annotated Bibliography of Expository Writings in the Mathematical Sciences*, by Matthew P. Gaffney and Lynn Arthur Steen, published by the Mathematical Association of America, Washington, D.C., 1976.

§1. The Nature of Mathematics

The quotation just inside the title page of this text is taken from Alfred North Whitehead's chapter on mathematics in *Science and the Modern World* [55][†]. This essay, "Mathematics as an Element in the History of

---

* A revision of *A Brief Guide to Writings 'About' Mathematics*, W. M. Priestley, Copyright © 1972 (privately published).

† Numbers in square brackets refer to the listing in the bibliography.

Thought", is a modern classic. Quite a different kind of article, but equally celebrated, is Henri Poincaré's "Mathematical Creation". The ideas expressed here by the great French mathematician are often referred to in discussions of creativity. Poincaré's article may be found in [16], for example, as well as in [34]. It inspired Hadamard's *The Psychology of Invention in the Mathematical Field* [17].

The attractiveness of mathematics as an activity similar to creative endeavors in the fine arts is the theme of an excellent article [18] by Paul Halmos. This theme is given its most eloquent expression, perhaps, in the lines of Russell: "Mathematics, rightly viewed, possesses not only truth, but supreme beauty . . ." (See Chapter 5, Section 1, for the rest.) It is only fair to add here that Russell later [39] partially repudiated some of these sentiments, owing to certain developments in twentieth-century philosophy.

G. H. Hardy's defense [19] of mathematics rests largely on aesthetic grounds. *A Mathematician's Apology* has been reprinted, with a foreword by C. P. Snow. Of this unique book Graham Greene has said, "There is nothing here which the layman cannot understand except possibly one theorem, and I know no writing—except perhaps Henry James's introductory essays—which conveys so clearly and with such an absence of fuss the excitement of the creative artist."

Hardy says that mathematics is not a contemplative but a creative subject. One cannot know the nature of mathematics without doing some mathematics. The books [10], [35], [36], and [48] are designed to help the reader do just that.

§2. About Mathematicians

E. T. Bell's *Men of Mathematics* [1] must surely be the most popular book about mathematicians. In his inimitable, crusty style Bell relates the lives and achievements of thirty-odd mathematicians, the latest being Georg Cantor, who died in 1918.

Few mathematicians write autobiographies, the only one of great literary merit being Bertrand Russell's [40]. Since Russell did little in mathematics after World War I, his is a mathematician's autobiography only as far as Volume I. Norbert Wiener, who received his Ph.D. degree at the age of eighteen and studied under Russell and Hardy, has written an autobiography [56], [57]. Wiener is the father of cybernetics.

Oystein Ore's biography [33] of the great Norwegian mathematician Niels Henrik Abel is well worth reading. Abel died at the age of twenty-six. Leopold Infeld [21] writes movingly of the remarkable French mathematician Evariste Galois. Galois died at twenty.

An absorbing biography [37] of David Hilbert has been written by Constance Reid, who followed it with the life story [38] of Richard Courant.

Hilbert, long associated with Göttingen, helped chart the course of twentieth-century mathematics in a famous address in 1900; and Courant, in moving from Göttingen to New York, was instrumental in raising the level of mathematical research in the United States.

As a rule, only the celebrated mathematician is written about. The ordinary teacher of mathematics, upon whom real inspiration descends grudgingly and fleetingly, is usually thought undeserving of attention. But Donald Weidman has written a sympathetic account [51] of the run-of-the-mill mathematician's fate.

§3. History and Development of Mathematics

For the general reader D. J. Struik's *A Concise History of Mathematics* [46] is most handy. Two good textbooks used by undergraduate courses in history of mathematics are Eves [12] and Boyer [6]. Eves leans toward geometry, Boyer toward analysis. Boyer wrote an earlier book [5] on the development of calculus, but on this subject the real gem is by Toeplitz [49]. For the development of mathematics a variety of works may be consulted. Dantzig's book [11], dealing mainly with analysis, is easy to read. Meschkowski's little book [28] is good, as is Kline's big book [23]. Bell [2] is unique.

If Greek mathematics is your interest, then van der Waerden [50] is for you. Here is a beautiful, exciting description of a glorious epoch in mathematics. Neugebauer [31] discusses Babylonian and Egyptian mathematics.

For samples of little gems by ancient and classical mathematicians, such as Archimedes, Pascal, and Leibniz, see the collections by Coolidge [9] or Meschkowski [27]. Extensive source books are Smith [42] and Struik [47].

The definitive work in the history of mathematics (up to 1800) is the massive tome of Moritz Cantor [7], which Salomon Bochner [4] characterized as "one of those large-scale works by bearded gaslight-Victorians which the 20th century does not quite know how to supersede with whatever it might try to supersede them with."

What about the future of mathematics? See Weil [52].

§4. Philosophy of Mathematics

What is mathematics? Is it created or is it discovered? What are the foundations of mathematics? Textbooks, such as Eves and Newsom [15] and Wilder [58], consider such questions, whose answers may vary according to whether one subscribes to the philosophy of intuitionism, formalism, or logicism.

Hermann Weyl, who was an intuitionist, wrote a deep book [53], *Philosophy of Mathematics and Natural Science*. It demands deliberate reading. Poincaré is most articulate on this subject, as on every other subject collected in his essays [34]. Lately the intuitionist position has been revived and modified, with the coming into prominence of constructive methods. See the review [45] of Bishop's book.

Hilbert was the foremost proponent of the formalist thesis, and passages in [37] attempt to capture the spirit of the early twentieth-century debate over the validity of this thesis. The celebrated theorem proved by Kurt Gödel in 1931, which was such a setback for formalist hopes, has been outlined for the general reader by Nagel and Newman [30].

Russell and Whitehead promulgated the logistic thesis. See Henkin [20] for some comments on its current state.

Many important issues in the philosophy of mathematics turn on one's answer to a seemingly simple question: *What is a real number*? One might think that all mathematicians would agree upon real numbers by now, since numbers have been around for a long time, if not eternally. But see Steen [43].

§5. Collections of Expository Articles

If a guide to writings about mathematics listed only one entry, it would have to be the four-volume set *World of Mathematics* [32], edited by James R. Newman. Some of the articles mentioned above are reprinted (though sometimes in abridged form) in this superb collection. Anyone interested in mathematics is probably already familiar with this work.

A compilation [22] of articles on mathematics that have appeared in *Scientific American* is impressive, as are the collections [3] and [41], all of which attempt to give in a non-technical way something of the flavor of today's mathematical research.

Shortly after World War II there appeared in France an ambitious collection of expository essays about mathematics. Written mainly by French mathematicians, these essays cover a great range of topics, even including the relationship between mathematics and music, aesthetics, philosophic idealism, social change, and Marxism. Though some are dated by now, many retain their original striking quality. It is good to have them [24] available in English.

§6. Miscellaneous Writings

The book [25] by Littlewood, Hardy's great collaborator, surely goes under this heading. Hermann Weyl's *Symmetry* [54] and Hugo Steinhaus's *Mathematical Snapshots* [44] have been widely admired. J. D. Williams has written for the layman a delightful book on game theory, *The Compleat Strategyst*

[59]. Menninger's *Number Words and Number Symbols* [26] must be on the coffee table of every modern Pythagorean. Eves's books [13], [14] help preserve the folklore of mathematics, as does the older collection [29] of Moritz, for those who are interested in "witty, profound, amusing passages about mathematics and mathematicians".

Finally, who is unfamiliar with the work [8] of that dour Oxford professor of mathematics, the Reverend Charles Lutwidge Dodgson?

Bibliography

[1] Bell, E. T. *Men of Mathematics*, Simon and Schuster, New York, 1937.
[2] Bell, E. T. *The Development of Mathematics*, McGraw-Hill, New York, 1945.
[3] Bers, L., ed. *The Mathematical Sciences*, M.I.T. Press, Cambridge, 1969.
[4] Bochner, S. *The Role of Mathematics in the Rise of Science*, Princeton University Press, Princeton, 1966.
[5] Boyer, C. B. *The History of the Calculus and its Conceptual Development*, Dover, New York, 1959.
[6] Boyer, C. B. *A History of Mathematics*, Wiley, New York, 1968.
[7] Cantor, M. *Vorlesungen über Geschichte der Mathematik*, 4 vols., Johnson reprint, New York, 1965.
[8] Carroll, L. *Alice's Adventures in Wonderland*, Centennial Edition, Random House, New York, 1965.
[9] Coolidge, J. L. *The Mathematics of Great Amateurs*, Dover, New York, 1963.
[10] Courant, R. and Robbins, H. *What is Mathematics?* Oxford, New York, 1941.
[11] Dantzig, T. *Number, the Language of Science*, Doubleday Anchor, New York, 1956.
[12] Eves, H. *An Introduction to the History of Mathematics*, Holt, Rinehart, and Winston, New York, 1969.
[13] Eves, H. *In Mathematical Circles*, Prindle, Weber, and Schmidt, Boston, 1969.
[14] Eves, H. *Mathematical Circles Revisited*, Prindle, Weber, and Schmidt, Boston, 1971.
[15] Eves, H. and Newsom, C. *Foundations and Fundamental Concepts of Mathematics*, Holt, Rinehart, and Winston, New York, 1964.
[16] Ghiselin, B., ed. *The Creative Process*, Mentor, New York, 1952.
[17] Hadamard, J. *The Psychology of Invention in the Mathematical Field*, Dover, New York, 1954.
[18] Halmos, P. Mathematics as a creative art, *The American Scientist* **56**, 375–389 (1968).

[19] Hardy, G. H. *A Mathematician's Apology*, Cambridge, New York, 1967.
[20] Henkin, L. Are logic and mathematics identical? *Science* **138**, 788–794 (1962).
[21] Infeld, L. *Whom the Gods Love*, Whittlesey House, New York, 1948.
[22] Kline, M., ed., *Mathematics in the Modern World*, Freeman, San Francisco, 1968.
[23] Kline, M. *Mathematical Thought from Ancient to Modern Times*, Oxford, New York, 1972.
[24] Le Lionnais, F., ed. *Great Currents of Mathematical Thought*, 2 vols., Dover, New York, 1971.
[25] Littlewood, J. E. *A Mathematician's Miscellany*, Methuen, London, 1953.
[26] Menninger, K. *Number Words and Number Symbols*, M.I.T. Press, Cambridge, 1969.
[27] Meschkowski, H. *Ways of Thought of Great Mathematicians*, Holden-Day, San Francisco, 1964.
[28] Meschkowski, H. *Evolution of Mathematical Thought*, Holden-Day, San Francisco, 1965.
[29] Moritz, R. E. *On Mathematics*, Dover, New York, 1958.
[30] Nagel, E. and Newman, J. R. *Gödel's Proof*, N.Y.U. Press, New York, 1958.
[31] Neugebauer, O. *The Exact Sciences in Antiquity*, Dover, New York, 1969.
[32] Newman, J. R., ed. *The World of Mathematics*, 4 vols., Simon and Schuster, New York, 1956.
[33] Ore, O. *Niels Henrik Abel, Mathematician Extraordinary*, U. Minn. Press, Minneapolis, 1957.
[34] Poincaré H. *The Foundations of Science*, The Science Press, New York, 1929.
[35] Pólya, G. *Mathematical Discovery*, Wiley, New York, 1962.
[36] Rademacher, H. and Toeplitz, O. *The Enjoyment of Mathematics*, Princeton University Press, Princeton, 1957.
[37] Reid, C. *Hilbert*, Springer-Verlag, New York, 1970.
[38] Reid, C. *Courant in Göttingen and New York*, Springer-Verlag, New York, 1976.
[39] Russell, B. The retreat from Pythagoras, *The Basic Writings of Bertrand Russell*, Simon and Schuster, 252–256 (1961).
[40] Russell, B. *The Autobiography of Bertrand Russell*, 3 vols., Atlantic-Little Brown, Boston, 1967.
[41] Saaty, T. L. and Weyl, J., eds. *The Spirit and the Uses of the Mathematical Sciences*, McGraw-Hill, New York, 1969.
[42] Smith, D. E. *A Source Book in Mathematics*, 2 vols., Dover, New York, 1959.
[43] Steen, L. New models of the real-number line, *Scientific American* **225**, 92–99 (1971).
[44] Steinhaus, H. *Mathematical Snapshots*, Oxford, New York, 1969.
[45] Stolzenberg, G. Review of Errett Bishop's *Foundations of Constructive Analysis*, *Bulletin of the American Mathematical Society* **76**, 301–323 (1970).
[46] Struik, D. J. *A Concise History of Mathematics*, Dover, New York, 1967.
[47] Struik, D. J. *A Source Book in Mathematics, 1200–1800*, Harvard, Cambridge, 1969.
[48] Tietze, H. *Famous Problems of Mathematics*, Graylock, New York, 1965.
[49] Toeplitz, O. *The Calculus: A Genetic Approach*, University of Chicago Press, Chicago, 1963.
[50] van der Waerden, B. L. *Science Awakening*, Oxford, New York, 1961.
[51] Weidman, D. R. Emotional perils of mathematics, *Science* **149**, 1048 (1965).
[52] Weil, A. The future of mathematics, *American Mathematical Monthly* **57**, 295–306 (1950).
[53] Weyl, H. *Philosophy of Mathematics and Natural Science*, Princeton University Press, Princeton, 1949.

[54] Weyl, H. *Symmetry*, Princeton University Press, Princeton, 1952.
[55] Whitehead, A. N. *Science and the Modern World*, Macmillan, New York, 1925.
[56] Wiener, N. *Ex-prodigy*, M.I.T. Press, Cambridge, 1964.
[57] Wiener, N. *I am a Mathematician*, M.I.T. Press, Cambridge, 1964.
[58] Wilder, R. L. *Introduction to the Foundations of Mathematics*, Wiley, New York, 1952.
[59] Williams, J. D. *The Compleat Strategyst*, McGraw-Hill, New York, 1954.

Appendix 2
Sums and Their Limits

This appendix is intended to supplement the discussion of Eudoxus' method given in Chapter 6 by presenting proofs of several summation formulas and by offering some more examples of summation techniques. What follows may be read immediately after Section 6 of Chapter 6.

Let us begin by discussing a problem which may be no further away than the nearest supermarket. We shall solve it two ways. The first solution makes no use of summation techniques.

Problem A. Oranges are stacked in the form of a pyramid whose base is rectangular, with 6 oranges along one side and 10 oranges along the other. How many oranges are in the pyramid?

Solution 1. The bottom level, of dimensions 6 by 10, has 4 more oranges along one side than along the other. As any experienced stacker of oranges knows, it follows that every level will have 4 more oranges in one dimension than in the other. The size of the top level has to be 1 orange by 5 oranges, the next level must be of size 2 by 6, followed by a level of size 3 by 7, and so on. Adding the oranges in each level—beginning at the top level—we see that the total number of oranges is given by

$$1 \cdot 5 + 2 \cdot 6 + 3 \cdot 7 + 4 \cdot 8 + 5 \cdot 9 + 6 \cdot 10 = 175. \qquad \Box$$

Solution 2. The k-th level (counting from the top level down) will have k oranges along one side and $k + 4$ oranges along the other. Hence the k-th level will contain $k(k + 4) = k^2 + 4k$ oranges. There are obviously 6 levels in all, so the total number of oranges is equal to

$$(*) \qquad \sum_{k=1}^{6} \text{oranges on } k\text{-th level} = \sum_{k=1}^{6} (k^2 + 4k) = \sum_{k=1}^{6} k^2 + 4 \sum_{k=1}^{6} k.$$

Appendix 2 Sums and Their Limits

There are easy formulas for $\sum k$, for $\sum k^2$, and for $\sum k^3$, given as follows:

$$\sum_{k=1}^{n} k = \frac{n(n+1)}{2}, \tag{1}$$

$$\sum_{k=1}^{n} k^2 = \frac{n(n+1)(2n+1)}{6}, \tag{2}$$

$$\sum_{k=1}^{n} k^3 = \frac{n^2(n+1)^2}{4}. \tag{3}$$

(We shall prove these formulas shortly.) When $n = 6$, formulas (1) and (2) become

$$\sum_{k=1}^{6} k = \frac{6(7)}{2} = 21; \qquad \sum_{k=1}^{6} k^2 = \frac{6(7)(13)}{6} = 91.$$

These, when put together with equation (*) above, show that the total number of oranges is equal to $91 + 4(21) = 175$. □

In Problem A we needed to add together only six quantities. With such a small number of summands the use of summation techniques saves little time. Solution 2 will probably consume as much time as Solution 1. With a large number of summands, however, the use of summation techniques is almost indispensable. The reader will find it helpful to memorize formulas (1), (2), and (3).

Problem A⁺. Baseballs are stacked in the form of a huge pyramid, with a rectangular base of 60 balls by 50 balls. How many baseballs does the pyramid contain?

Solution 1. Every level will have 10 more balls in one dimension than in the other, and there will be 50 levels in all. Counting from the top level down, we see that the total number of balls is given by

$$1 \cdot 11 + 2 \cdot 12 + 3 \cdot 13 + \cdots + 50 \cdot 60 = ??! \qquad \square$$

Solution 2. The k-th level from the top will have k balls along one side and $k + 10$ balls along the other. Hence the k-th level contains $k(k + 10) = k^2 + 10k$ balls, and the total number of balls in the entire pyramid is given by

$$\sum_{k=1}^{50} (k^2 + 10k) = \sum_{k=1}^{50} k^2 + 10 \cdot \sum_{k=1}^{50} k$$

$$= \frac{50(51)(101)}{6} + 10 \cdot \frac{50 \cdot 51}{2} \quad \text{[by (1) and (2)]}$$

$$= 55{,}675. \qquad \square$$

§1. Collapsing Sums; Proofs of Formulas (1), (2) and (3)

Here are some examples of a simple but important type of summation. The sum depends only upon the first and last terms, since the intermediate terms cancel. The sum "collapses", making it quite easy to add up. The following equalities are obvious.

$$\left(1 - \frac{1}{2}\right) + \left(\frac{1}{2} - \frac{1}{3}\right) + \left(\frac{1}{3} - \frac{1}{4}\right) + \cdots + \left(\frac{1}{n} - \frac{1}{n+1}\right) = 1 - \frac{1}{n+1}, \quad (4)$$

$$[1^2 - 0^2] + [2^2 - 1^2] + [3^2 - 2^2] + \cdots + [n^2 - (n-1)^2] = n^2, \quad (5)$$

$$[1^3 - 0^3] + [2^3 - 1^3] + [3^3 - 2^3] + \cdots + [n^3 - (n-1)^3] = n^3. \quad (6)$$

In summation notation, formulas (4), (5), and (6) are expressed as follows.

$$\sum_{k=1}^{n} \left(\frac{1}{k} - \frac{1}{k+1}\right) = 1 - \frac{1}{n+1}; \quad (4')$$

$$\sum_{k=1}^{n} (k^2 - (k-1)^2) = n^2; \quad (5')$$

$$\sum_{k=1}^{n} (k^3 - (k-1)^3) = n^3. \quad (6')$$

What is this good for? It seems too obvious to lead to anything interesting. Yet interesting results are immediately at hand. Since $(1/k) - (1/(k+1)) = 1/(k^2 + k)$, the obvious formula (4') immediately yields the interesting summation formula

$$\sum_{k=1}^{n} 1/(k^2 + k) = n/(n+1),$$

a result which may not be obvious. And the obvious formulas (5') and (6') lead immediately to proofs of formulas (1) and (2).

PROOF OF FORMULA (1). First note that

$$k^2 - (k-1)^2 = k^2 - (k^2 - 2k + 1) = 2k - 1,$$

so that formula (5') becomes

$$\sum_{k=1}^{n} (2k - 1) = n^2.$$

Therefore,

$$2 \cdot \sum_{k=1}^{n} k - \sum_{k=1}^{n} 1 = n^2,$$

$$2 \cdot \sum_{k=1}^{n} k - n = n^2,$$

$$2 \cdot \sum_{k=1}^{n} k = n^2 + n,$$

$$\sum_{k=1}^{n} k = \frac{n^2 + n}{2}. \qquad \square$$

1. Collapsing Sums; Proofs of Formulas (1), (2) and (3)

PROOF OF FORMULA (2). First note that
$$k^3 - (k-1)^3 = k^3 - (k^3 - 3k^2 + 3k - 1) = 3k^2 - 3k + 1,$$
so that formula (6′) becomes
$$\sum_{k=1}^{n}(3k^2 - 3k + 1) = n^3.$$
Therefore,
$$3 \cdot \sum_{k=1}^{n} k^2 - 3 \cdot \sum_{k=1}^{n} k + \sum_{k=1}^{n} 1 = n^3,$$
$$3 \cdot \sum_{k=1}^{n} k^2 - 3\frac{n(n+1)}{2} + n = n^3.$$

We want to derive formula (2) by solving this equation for $\sum k^2$. This is easier to do if first we multiply through by 2:
$$6 \cdot \sum_{k=1}^{n} k^2 - 3n(n+1) + 2n = 2n^3.$$
Therefore (the reader is asked to supply the missing steps),
$$6 \cdot \sum_{k=1}^{n} k^2 = 2n^3 + 3n^2 + n$$
$$= n(2n^2 + 3n + 1)$$
$$= n(n+1)(2n+1),$$
and formula (2) is obtained upon dividing by 6. □

PROOF OF FORMULA (3). Since the idea of these proofs should be familiar by now, only the main steps are given. The reader is asked to fill in the details. Beginning with the collapsing sum
$$\sum_{k=1}^{n}(k^4 - (k-1)^4) = n^4,$$
and noting that $k^4 - (k-1)^4 = 4k^3 - 6k^2 + 4k - 1$, we get
$$4 \cdot \sum k^3 - 6 \cdot \sum k^2 + 4 \cdot \sum k - \sum 1 = n^4,$$
where all the summations run from $k = 1$ to $k = n$. Using formulas already derived for $\sum k^2, \sum k,$ and $\sum 1$, we obtain the equation
$$4 \cdot \sum k^3 - (2n^3 + 3n^2 + n) + (2n^2 + 2n) - n = n^4.$$
Therefore,
$$4 \cdot \sum k^3 = n^4 + 2n^3 + n^2$$
$$= n^2(n^2 + 2n + 1)$$
$$= n^2(n+1)^2,$$
and formula (3) is obtained upon dividing by 4. □

The summation formulas (1), (2), and (3) were known to the Greeks, but the proofs presented here are modern. The modern approach, using summation notation, has the advantage of applying equally well to the determination of formulas for $\sum k^4$, $\sum k^5$, etc.—sums which the Greeks apparently did not consider. The determination of such formulas is left to the reader as an exercise.

§2. Integrals of Quadratics and Cubics

Many examples of Eudoxus' method of calculating integrals are given in Chapter 6, but most of them deal with relatively simple linear functions. Here are some slightly more complicated applications of the method.

EXAMPLE 1. Calculate the integral $\int_0^1 x^2\,dx$ directly from its definition as a limit of sums.

Solution. We are asked to calculate $\int_a^b f(x)\,dx$, where $a = 0$, $b = 1$, and $f(x) = x^2$. Here we have $\Delta x = 1/n$, $x_k = k/n$, and $f(x_k) = k^2/n^2$. An approximating sum for the desired integral is then given by

$$S_n = \sum_{k=1}^n f(x_k)\,\Delta x$$

$$= \sum_{k=1}^n \left(\frac{k^2}{n^2}\right)\left(\frac{1}{n}\right)$$

$$= \frac{1}{n^3}\sum_{k=1}^n k^2$$

$$= \frac{n(n+1)(2n+1)}{6n^3} \quad [\text{by (2)}]$$

$$= \frac{1}{6}\left(1 + \frac{1}{n}\right)\left(2 + \frac{1}{n}\right).$$

Therefore,

$$\int_0^1 x^2\,dx = \text{Limit } S_n = \text{Limit}\,\frac{1}{6}\left(1 + \frac{1}{n}\right)\left(2 + \frac{1}{n}\right) = \frac{1}{6}(1)(2) = \frac{1}{3}. \qquad \square$$

EXAMPLE 2. Calculate the integral $\int_0^\pi ax^2\,dx$ directly from its definition as a limit of sums.

Solution. This is only a slight modification of the preceding example. Here we have $\Delta x = \pi/n$, $x_k = k\pi/n$, and $f(x_k) = ak^2\pi^2/n^2$.

$$S_n = \sum_{k=1}^n \left(\frac{ak^2\pi^2}{n^2}\right)\left(\frac{\pi}{n}\right) = \frac{a\pi^3}{n^3}\sum_{k=1}^n k^2 = \frac{a\pi^3 n(n+1)(2n+1)}{6n^3}.$$

2. Integrals of Quadratics and Cubics

Therefore,

$$\int_0^\pi ax^2\,dx = \text{Limit } S_n = \text{Limit } \frac{a\pi^3}{6}\left(1+\frac{1}{n}\right)\left(2+\frac{1}{n}\right) = \frac{a\pi^3}{3}. \qquad \square$$

EXAMPLE 3. Calculate the integral $\int_0^t ax^2\,dx$ by Eudoxus' method.

Solution. This is treated just like the preceding example except that we have t in place of π. It is therefore obvious that the bottom line will read

$$\int_0^t ax^2\,dx = \text{Limit } S_n = \text{Limit } \frac{at^3}{6}\left(1+\frac{1}{n}\right)\left(2+\frac{1}{n}\right) = \frac{at^3}{3}. \qquad \square$$

EXAMPLE 4. Calculate the integral $\int_0^1 x^3\,dx$ by Eudoxus' method.

Solution. Here we have

$$S_n = \sum_{k=1}^n \left(\frac{k^3}{n^3}\right)\left(\frac{1}{n}\right)$$

$$= \frac{1}{n^4}\sum_{k=1}^n k^3$$

$$= \frac{n^2(n+1)^2}{4n^4} \quad [\text{by (3)}]$$

$$= \frac{1}{4}\left(1+\frac{1}{n}\right)^2.$$

Therefore,

$$\int_0^1 x^3\,dx = \text{Limit } S_n = \text{Limit } \frac{1}{4}\left(1+\frac{1}{n}\right)^2 = \frac{1}{4}(1) = \frac{1}{4}. \qquad \square$$

EXAMPLE 5. Calculate the integral $\int_a^b x^3\,dx$ by Eudoxus' method.

Solution. Here are the main steps. The details are left to the reader. To save space the index k is suppressed in the summations below. The summations are understood to run from $k=1$ to $k=n$.

Since $x_k = a + k\,\Delta x$, we have

$$S_n = \sum x_k^3\,\Delta x$$
$$= \sum (a + k\,\Delta x)^3\,\Delta x$$
$$= \sum [a^3 + 3a^2 k(\Delta x) + 3ak^2(\Delta x)^2 + k^3(\Delta x)^3]\,\Delta x$$
$$= a^3(\Delta x)\sum 1 + 3a^2(\Delta x)^2\sum k + 3a(\Delta x)^3\sum k^2 + (\Delta x)^4\sum k^3.$$

Using the fact that $\Delta x = (b-a)/n$ and using the summation formulas for $\sum 1, \sum k, \sum k^2$, and $\sum k^3$, we get

$$S_n = a^3(b-a) + \frac{3}{2}a^2(b-a)^2\left(1+\frac{1}{n}\right) + \frac{1}{2}a(b-a)^3\left(1+\frac{1}{n}\right)\left(2+\frac{1}{n}\right)$$
$$+ \frac{1}{4}(b-a)^4\left(1+\frac{1}{n}\right)^2.$$

Taking the limit of S_n as n increases without bound, we see that

$$\int_a^b x^3 \, dx = a^3(b-a) + \frac{3}{2}a^2(b-a)^2 + a(b-a)^3 + \frac{1}{4}(b-a)^4$$

$$= \frac{b^4}{4} - \frac{a^4}{4}. \qquad \square$$

§3. Geometric Series and Applications

The equation $(1 + x)(1 - x) = 1 - x^2$ is such a simple algebraic identity that the most interesting thing about it is rarely noticed. The interesting thing is that it comes from a collapsing sum.

$$(1 + x)(1 - x) = (1)(1 - x) + (x)(1 - x) = (1 - x) + (x - x^2) = 1 - x^2.$$

This is about the simplest possible example of a collapsing sum. We should ask whether this example generalizes readily and whether the generalization is even more interesting. The answer is *yes* to both questions.

The immediate generalization is this:

$$(1 + x + x^2)(1 - x) = (1 - x) + (x - x^2) + (x^2 - x^3) = 1 - x^3.$$

And the far-reaching generalization is one of the most important identities in mathematics:

$$(1 + x + x^2 + \cdots + x^n)(1 - x) = (1 - x) + (x - x^2) + \cdots + (x^n - x^{n+1})$$
$$= 1 - x^{n+1}.$$

The identity is important because from it we get the following summation formula (by dividing both sides of the identity by $1 - x$):

$$1 + x + x^2 + \cdots + x^n = \frac{1 - x^{n+1}}{1 - x}, \quad \text{if } x \neq 1. \tag{7}$$

The series on the left is called a *geometric* series and its sum is given in equation (7). The result just obtained is useful enough to be called a theorem.

Theorem on Geometric Series. *For a geometric series the following summation formula is valid, provided $x \neq 1$:*

$$\sum_{k=0}^{n} x^k = \frac{1 - x^{n+1}}{1 - x}$$

(where x^0 is understood to be 1).

The reader should note the difference between the type of series now being considered and the type that was considered in Section 1. In Section 1 we

3. Geometric Series and Applications

found, for example, that

$$\sum_{k=1}^{n} k^2 = \frac{n(n+1)(2n+1)}{6};$$

whereas by the theorem just proved (with $x = 2$) we have

$$\sum_{k=0}^{n} 2^k = \frac{1 - 2^{n+1}}{1 - 2} = 2^{n+1} - 1.$$

The reader should be careful not to confuse $\sum 2^k$ with $\sum k^2$. One is a geometric series while the other is not. Note also that the formula for the geometric series above is for the sum beginning with index $k = 0$. By subtracting 1 from both sides of equation (7) we get an analogous formula where the index k begins at 1:

$$\sum_{k=1}^{n} x^k = \frac{x - x^{n+1}}{1 - x}, \quad x \neq 1.$$

EXAMPLE 6. Evaluate the sum

$$1 + \frac{1}{4} + \left(\frac{1}{4}\right)^2 + \left(\frac{1}{4}\right)^3 + \cdots + \left(\frac{1}{4}\right)^n.$$

Solution. The sum in question comes from the geometric series

$$\sum_{k=0}^{n} \left(\frac{1}{4}\right)^k.$$

By (7) its sum is

$$\frac{1 - (\frac{1}{4})^{n+1}}{1 - \frac{1}{4}} = \frac{4}{3} - \frac{1}{3}\left(\frac{1}{4}\right)^n \approx \frac{4}{3}$$

if n is a large positive integer. □

EXAMPLE 7. Consider the function given by $f(x) = 1/(1-x)$. Find an approximation of this function given by nonnegative powers of x.

Solution. Equation (7) says

$$1 + x + x^2 + x^3 + \cdots + x^n = \frac{1}{1-x} - \frac{x^{n+1}}{1-x} \approx \frac{1}{1-x}$$

if $x \approx 0$. Therefore,

$$\frac{1}{1-x} \approx \sum_{k=0}^{n} x^k, \quad \text{if } x \approx 0.$$

This answer should be compared to the result of problem 24 at the end of Chapter 9. □

Here is a dazzling application of the use of a geometric series. It is due to Fermat. While the theorem is easy to prove using the fundamental theorem of calculus, Fermat was able to prove it before the fundamental theorem was known. The proof demands careful attention, as some details of it are left to the reader.

Theorem (Fermat). *Let A be the area beneath the graph of the curve $y = t^n$ (where n is a positive integer) between $t = 0$ and $t = b$. Then $A = b^{n+1}/(n + 1)$.*

PROOF. Let x be a positive number just less than 1 and consider the (infinite) sequence of numbers

$$b, bx, bx^2, bx^3, \ldots, bx^k, \ldots$$

which subdivide the interval from $t = 0$ to $t = b$ into infinitely many subintervals. For fixed x, let A_x be the area beneath the staircase built upon these subintervals.

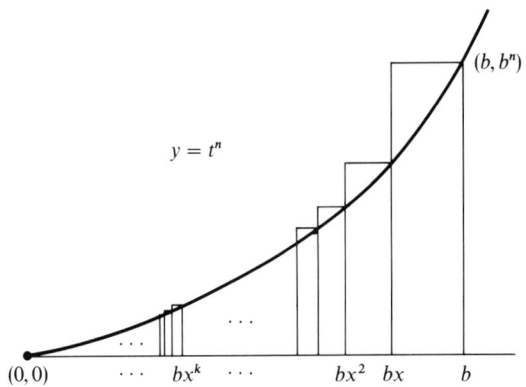

As $x \to 1^-$ the staircase approximates the curve ever more closely. Hence the area A beneath the curve is given by

$$A = \operatorname*{Limit}_{x \to 1^-} A_x.$$

We shall first calculate A_x. To do this it is convenient to start at the top step and go down, with the top step counted as the 0-th step. Then the k-th step looks like this.

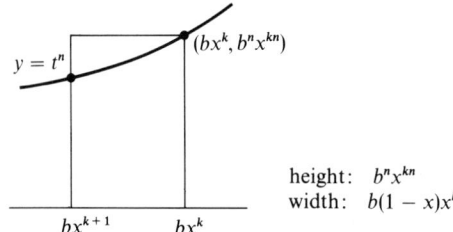

The area beneath the k-th step is then $b^{n+1}(1-x)(x^{n+1})^k$, and the total area A_x (where $x < 1$) is given by summing, beginning at $k = 0$:

$$A_x = b^{n+1}(1-x)[1 + x^{n+1} + (x^{n+1})^2 + (x^{n+1})^3 + \cdots]$$

$$= b^{n+1}(1-x)\left[\frac{1}{1-x^{n+1}}\right] \quad (\text{why?})$$

$$= b^{n+1}\left[\frac{1-x}{1-x^{n+1}}\right]$$

$$= b^{n+1}/(1 + x + x^2 + \cdots + x^n) \quad [\text{by (7)}].$$

Therefore,

$$A = \underset{x \to 1^-}{\text{Limit }} A_x = \underset{x \to 1^-}{\text{Limit }} b^{n+1}/(1 + x + x^2 + \cdots + x^n) = b^{n+1}/(n+1). \quad \square$$

PROBLEMS (OPTIONAL)

1. Baseballs are stacked in the form of a pyramid with a rectangular base of 40 balls by 36 balls. How many balls are in the pyramid?

2. Evaluate each of the following sums.
   (a) $1 \cdot 8 + 2 \cdot 9 + 3 \cdot 10 + 4 \cdot 11 + \cdots + 50 \cdot 57$.
   (b) $(1/1 \cdot 2) + (1/2 \cdot 3) + (1/3 \cdot 4) + (1/4 \cdot 5) + \cdots + (1/99 \cdot 100)$.
   (c) $\sum_{k=1}^{100} k^3$.
   (d) $\sum_{k=1}^{100} 3^k$.
   (e) $\sum_{k=0}^{100} (\frac{1}{2})^k$.

3. (a) Prove that $\sum_{k=1}^{n} k^4 = n(n+1)(2n+1)(3n^2 + 3n - 1)/30$.
   (b) Use the result of part (a) to calculate $\int_0^1 x^4 \, dx$ by Eudoxus' method.

4. Find a formula for $\sum k^5$ and use it to calculate $\int_0^1 x^5 \, dx$. *Caution*: Keep a cool head. This problem can cause nervous breakdowns.

5. Apply Eudoxus' method to calculate each of the following integrals.
   (a) $\int_1^4 (3x^3 - 2x^2) \, dx$.
   (b) $\int_{-2}^2 (2x^3 - 7x) \, dx$.
   (c) $\int_{-1}^5 (x^2 - x + 4) \, dx$.

6. (*For those who think they understand infinity*) This book has avoided mentioning the symbol for infinity until now, because the symbol is so easily misunderstood. Test whether you understand it or not, by explaining why it is natural to write

$$\sum_{k=0}^{\infty} x^k = 1/(1-x), \quad \text{if } -1 < x < 1;$$

and yet at the same time to write

$$\sum_{k=0}^{\infty} x^k \neq 1/(1-x), \quad \text{if } x < -1 \text{ or } x > 1.$$

7. (a) When did Fermat die?
   (b) When did Newton discover the fundamental theorem of calculus?
   (c) Look again at the theorem of Fermat's proved in the last section. Give a one-line proof of this theorem by making use of the fundamental theorem of calculus.

8. (*An ambitious project*) Geometric series continue to find many surprising applications, even to the present day. Yet none could be more charming than the application made by Archimedes to effect a quadrature of the parabola. He proved that the area of a segment of a parabola is equal to four-thirds the area of its largest inscribed triangle. (The factor 4/3 comes from the fact that $\sum_{k=0}^{\infty} (\frac{1}{4})^k = 4/3$, as seen in Example 6.) In modern terminology we may state this result as follows:

   Theorem (Archimedes). *Let A be the area enclosed between the graph of a linear function and the graph of a quadratic function, and let T be the area of the largest triangle that can be inscribed in A. Then $A = (4/3)T$.*

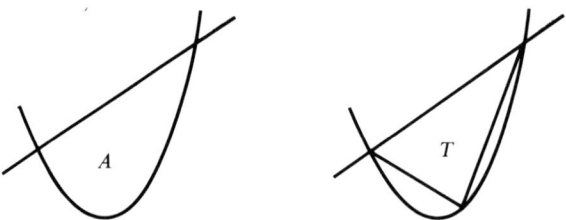

   Either (a) prove this result through the use of calculus; or (b) look up Archimedes' original proof and write a paper on it, sketching the main points.

Appendix 3
Archimedes

The mind of Archimedes is modern. Though he was born about 287 B.C., one may expect to have difficulty understanding his work unless one knows something of the developments in mathematics that took place two thousand years later. It may be well to read much of Chapters 1–6, together with Chapter 10, before expecting to understand everything in the small sample of Archimedes' work that is outlined here in this appendix.

In his published papers Archimedes characteristically put together his ideas with such tight logic that the adjective *archimedean* came to refer to any logical demonstration meeting the very highest standards of rigor. The reader interested in seeing truly archimedean demonstrations is invited to consult T. L. Heath, *The Works of Archimedes*, Cambridge, 1897 (also available in paperback by Dover Publications). This appendix outlines only a few of his ideas, and these are given presentations that may be described as casual if compared with archimedean standards.

§1. Archimedes and the Classical Problems

The three so-called "classical problems of antiquity" are as follows.

(*The Trisection Problem*) Given an angle, devise a method for constructing another angle one-third as large.

(*The Quadrature of the Circle*) Given a circle, devise a method for constructing a square having the same area.

(*The Duplication of the Cube*) Given a cube, devise a method for constructing another cube whose volume is twice as large.

It is probably safe to say that every Greek mathematician worked seriously on at least one of these famous problems. It ought to be surprising, therefore, that Euclid's *Elements* gives no account of them.

Why did Euclid not discuss these easily stated, natural problems? The reason is simple. *Euclid did not know how to do them. Nor did anyone else.* The construction of the required trisection, quadrature, or duplication eluded the efforts of the greatest mathematicians.

It is important to understand what Euclid meant by a "construction". In Euclidean geometry a construction may use a ruler and a compass, *but nothing else*. And the "ruler" can have no distance markings on it, its use being only as a straightedge to draw straight lines through points already constructed. Using Euclidean constructions, no one was able to solve any of the three problems above.

Archimedes somehow recognized the futility of Euclidean methods of attacking these problems, and reacted in a thoroughly modern way. If traditional theory proves inadequate to handle the type of thing for which it was designed, then something new is needed.

In a sense, it is "obvious" that each of the problems above has a solution. For example, it is obvious that there exists a trisection of a sixty-degree angle. (An angle of twenty degrees, of course, does the trick.) The whole problem is in *constructing* an angle of twenty degrees from a given angle of sixty degrees *through the use of ruler and compass alone*. Archimedes devised the following construction of striking simplicity.

Given an angle, we may construct a circle whose center O lies at the angle's vertex:

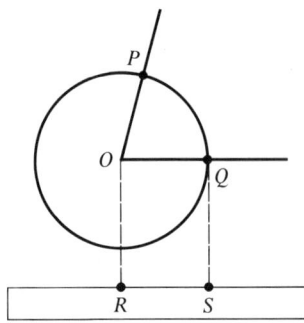

On a ruler, or straightedge, mark off two points R and S the distance between which is equal to the radius OQ. Now perform the following trick with the straightedge. Keeping the point R on the line through OQ and keeping the point S on the circle, manipulate the straightedge until it touches the point P:

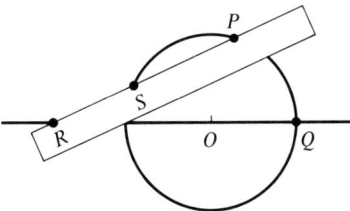

The angle PRQ is the required trisection of the given angle POQ. The proof of this is easy and is left to the reader. *Hint.* Begin by drawing the triangles RSO and SOP and note that they are isosceles triangles.

As Archimedes pointed out, the construction just outlined is done with ruler and compass, but it is not a Euclidean construction. Why is it not a Euclidean construction?

Archimedes' answer to the problem of squaring the circle resulted in one of the most important papers in mathematics. Instead of finding a square of the same area as the circle, Archimedes found a triangle, which is just as good.

Archimedes' Quadrature of the Circle. *A circle has the same area as a triangle whose base is equal to the circumference of the circle and whose height is equal to the radius.*

PROOF. Let A be the area of the circle of radius r and let B be the area of the right triangle with legs of lengths r and C.

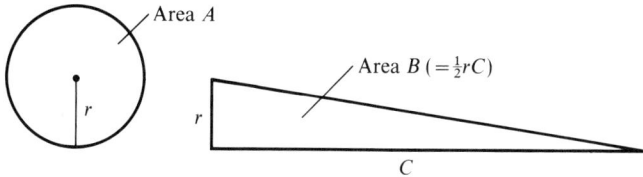

There are clearly three logical possibilities:

(a) $A > B$,
(b) $A < B$,
(c) $A = B$.

To prove (c) Archimedes used the principle of elimination. He proved that neither possibility (a) nor possibility (b) could be true. This leaves (c) as the only case not eliminated.

Proof that possibility (a) *is false.* We use the method of *reductio ad absurdum.* Suppose $A > B$. Then the number B is not equal to A but is only an approximation. How can we get a better approximation? That is easy. We can approximate the circle as closely as we please by a regular polygon inscribed inside, and therefore there is such a polygon whose area P is a better approximation to A. Then we have

$$A > P > B. \qquad (1)$$

But this leads quickly to a contradiction. Let p denote the length of the polygon's perimeter and let r' denote the polygon's "radius" (see the figure below).

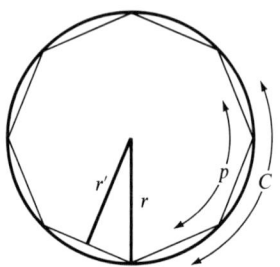

It follows that

$$P = \tfrac{1}{2}r'p \quad \text{(by problem 14, Chapter 2)}$$
$$< \tfrac{1}{2}rC \quad \text{(since } r' < r \text{ and } p < C\text{)}.$$

Therefore, $P < \tfrac{1}{2}rC$. But $\tfrac{1}{2}rC = B$, so

$$P < B. \qquad (2)$$

Statements (1) and (2) contradict each other. This contradiction arises from the supposition that $A > B$. This supposition is therefore false. Possibility (a) has been proved false.

Proof that possibility (b) *is false.* This proof follows closely the lines of the proof above, except that a *circumscribed* polygon approximating the circle is brought into play. Suppose that $A < B$. Then there is a circumscribed regular polygon satisfying condition (1) *but with the inequalities reversed.* This leads quickly to a contradiction (the demonstration of which is left to the reader), which shows that the supposition $A < B$ must be false.

By "double elimination" it follows that $A = B$. □

In the theorem above, Archimedes tells us how to construct a triangle equal in area to a given circle. However, the construction given is not a Euclidean construction. (Why not?)

1. Archimedes and the Classical Problems

In fact, this theorem occupies only a small (though essential) place in Archimedes' celebrated paper on the quadrature of the circle. The main body of the paper is concerned with estimating the numerical value of the ratio of the circumference of a circle to its diameter. (Today this ratio is always denoted by the Greek letter π—the first letter in the Greek word for *perimeter*—but this notation was introduced only in the eighteenth century by Euler.) Archimedes found this ratio to be between $3\frac{10}{71}$ and $3\frac{1}{7}$. The establishment of such a close approximation for π required the display of an awesome technique of calculation that involve delicate estimates made from inscribed and circumscribed polygons of 96 sides.

Actually, there existed before Archimedes other successful (but not Euclidean) methods of trisecting angles and squaring circles. In fact, a single curve—aptly called the quadratrix—could be employed in solving both problems. Hippias and Dinostratus had shown how to do this, but at the expense of a considerable departure from traditional methods.

Like Hippias and Dinostratus, Archimedes did not hesitate to break with tradition when tradition prevented him from attending his calling. But when he broke, he evidently did not like to go further away than he had to. When Archimedes found Euclidean constructions inadequate he tried to develop adequate constructions that were almost Euclidean. Happily he found them, as we have seen above, even though he also found a single curve—the spiral—that could be used to do the same job as the quadratrix.

In the case of the third of the classical problems, the duplication of the cube, Archimedes offered no new solution. Solutions (using non-Euclidean constructions) had already been given by Archytas, Eudoxus, Eratosthenes, Apollonius, and others. In modern terms the problem can be stated as follows. Given a cube whose sides have length s (yielding a volume of s^3), construct a cube with sides x whose volume is $2s^3$, or twice as large. This means one must construct a length x satisfying the equation

$$x^3 = 2s^3, \tag{3}$$

where s is given. There are many (non-Euclidean) ways of doing this.

Archimedes did something very much harder. Instead of posing for himself the problem of solving the simple cubic equation (3), Archimedes tackled the analysis of cubic equations in general. Since the Greeks couched all their algebra in geometric terms, and since they did not consider negative numbers, it would not be said today that Archimedes gave the first complete analysis of the general cubic equation. But if it were said, it would not be far wrong.

As we have seen, Archimedes failed in his attempts to solve the three classical problems of antiquity through the exclusive use of Euclidean methods. These are among his few failures, but we now know that they are nothing to be ashamed of. No Euclidean method, no matter how ingenious, will solve any of these problems. The inadequacy of Euclidean methods in this regard was conclusively demonstrated in the nineteenth century. This demonstration may be found in many undergraduate texts on modern algebra, but it is beyond the scope of the present text.

§2. Archimedes' Method

When does a seesaw balance? *Answer*: when the moments on each side are equal, the *moment* being defined as the product of a weight with its distance from the lever's fulcrum. This principle is a cornerstone of *statics*, a branch of physics that studies conditions of equilibrium.

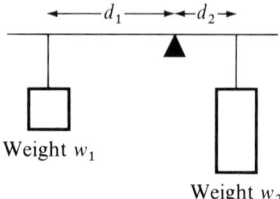

Law of the lever:
The lever is in equilibrium if $w_1 d_1 = w_2 d_2$

This principle was known to the Greeks before Archimedes was born. Yet Archimedes was the one to see how this tool could be used to open the way toward mathematical physics. He postulated simple axioms about statics, from which he proceeded to deduce the law of the lever and much, much more. He began investigating, with great success, the problem of finding the centroid, or center of gravity, of a solid figure. When he incorporated into all this his famous *principle of buoyancy* (the upward force on an object submerged in water is exactly equal to the weight of the water displaced), he invented the science of *hydrostatics*. Though Archimedes is said to have deplored "the whole trade of engineering", he could not have failed to know that his work would have practical applications to engineering. Everything from the design of more efficient compound pulleys to the design of more stable floating vessels is connected with it.

As impressive as all this might be, there is yet another application of the law of the lever that is even more surprising. Archimedes perceived—in what must be described as a flash of genius—that the lever can be brought into play with problems of pure mathematics. While a physical principle cannot, of course, be admitted into an archimedean demonstration of pure mathematics, a physical principle (or anything else, for that matter) can certainly be used to make guesses. And an archimedean guess precedes an archimedean demonstration.

The law of the lever applies only to the physical world, of course. But it occurred to Archimedes that there ought to be an analogous law in the realm of geometry! What should such a law say? Archimedes began to play

2. Archimedes' Method

with the idea of balancing geometric objects against each other. He reasoned, for example, that equilibrium would hold in the following situation.

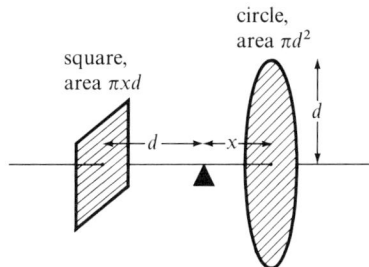

Since $\pi x d \cdot d = \pi d^2 \cdot x$, equilibrium obtains

Since a body behaves as if all its weight is concentrated at its center of gravity, the configuration below is essentially the same as the one above, and is therefore in equilibrium. (The combined area of the two circles below is equal to the area of the square in the picture above.)

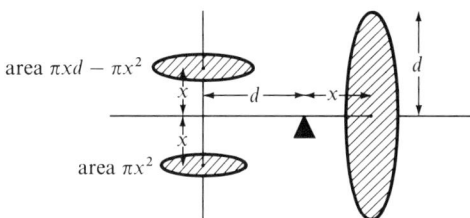

Reasoning somewhat as Cavalieri was to do centuries later (see Chapter 6), Archimedes concluded that we must have equilibrium in the following figure—for each vertical slice through the cylinder is exactly balanced by a corresponding pair of horizontal slices in the sphere and cone.

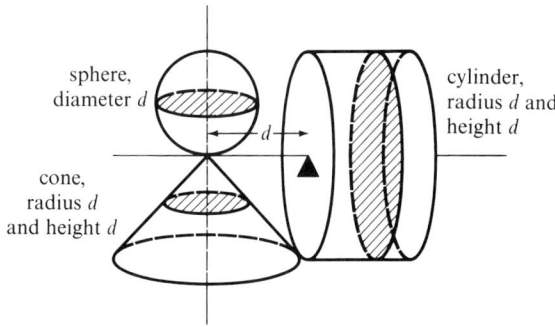

What is all this good for? Archimedes used it to guess the volume of a sphere. The volumes of cones and cylinders were well known already. Since the configuration above balances, the law of the lever says that

$$\text{(volume of sphere and cone)} \cdot d = \text{(volume of cylinder)} \cdot \frac{d}{2}.$$

(The length $d/2$ is the moment arm of the center of gravity of the cylinder.) Dividing by d in this equation shows that

$$\text{volume of sphere} + \text{volume of cone} = \tfrac{1}{2}(\text{volume of cylinder}).$$

Let V denote the volume of the sphere of diameter d. Then by known formulas for the volumes of cones and cylinders the above equation becomes

$$V + \frac{1}{3}\pi d^3 = \frac{1}{2}\pi d^3,$$

from which it follows that

$$V = \frac{1}{6}\pi d^3 = \frac{4}{3}\pi r^3, \tag{4}$$

where $r(=\tfrac{1}{2}d)$ is the radius. The volume of a sphere is then given by equation (4).

This is one of several extraordinary balancing acts that Archimedes was able to perform. They are all examples of his so-called "method", described in his famous letter to Eratosthenes. He emphasized that his method was used only to make guesses at what seemed to be plausible. Once he knew the likely truth he could prove it by rigorous means, such as the principle of double elimination illustrated in Section 1.

Let us look at just one more example of what a genius can see. In his letter to Eratosthenes, Archimedes says

> ... judging from the fact that any circle is equal to a triangle with base equal to the circumference and height equal to the radius of the circle, I apprehended that, in like manner, any sphere is equal to a cone with base equal to the surface of the sphere and height equal to the radius.*

Archimedes has guessed this *cubature* of the sphere:

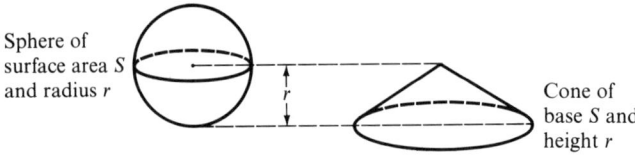

Sphere of surface area S and radius r

Cone of base S and height r

---

* From *The Method of Archimedes*, pp. 20–21 of the supplement to *The Works of Archimedes*, edited by T. L. Heath, Cambridge, 1912.

If this guess is correct it follows that

$$\frac{4}{3}\pi r^3 = \frac{1}{3} Sr$$

from the formulas for the volumes of spheres and cones. Solving this equation for the surface area S yields

$$S = 4\pi r^2. \tag{5}$$

In this way Archimedes guessed the correct formula (5) for the surface area of a sphere. The surface area is exactly four times as large as any great circle in the sphere, according to Archimedes. Having guessed the right answer he then proved it by completely different means, giving a rigorous demonstration to meet his standards.

As Archimedes once noted on a different occasion, a light touch—if properly applied—can move the earth.

Problems (Optional)

1. Prove that Archimedes' trisection technique actually works, as follows: in the figure of Section 1 illustrating this technique, let α = angle PRQ, let β = angle POQ, and prove that $\beta = 3\alpha$.

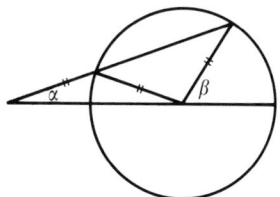

2. (a) Given lengths x and y, outline a Euclidean construction that produces the length $\sqrt{xy}$. *Hint*. Ponder the figure below. What is the length PQ?

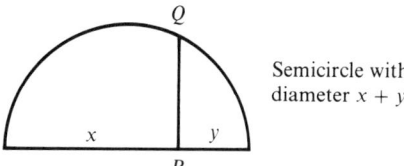

Semicircle with diameter $x + y$

(b) Devise a way of effecting a quadrature of a rectangle, using only Euclidean methods. That is, given a rectangle of sides x and y, construct a square having the same area as the rectangle. *Hint*. Use the result of part (a).

(c) Devise a way of effecting a quadrature of a triangle, using only Euclidean methods. *Hint*. First construct by Euclidean methods a rectangle having the same area as the given triangle. Then use the result of part (b).

3. Many references are listed at the end of Appendix 1. Several of them discuss the quadratrix of Hippias and Dinostratus.
   (a) Find a book that discusses the quadratrix (also called the trisectrix).
   (b) Be prepared to illustrate in class how the quadratrix can be employed to trisect an angle and to square a circle.
4. Find a book that discusses Archimedes' spiral and be prepared to illustrate in class how the spiral can be employed to trisect an angle and to square a circle.
5. It is impossible to duplicate the cube using only Euclidean constructions. Find a way to do it that uses constructions that are "almost" Euclidean. If you need to, find and use a book that discusses the Greek attempts to solve the *Delian problem* (as the problem of duplicating the cube was known).
6. We have already noted (in exercise 10.4 of Chapter 6) that Archimedes found the ratio of the volume of a cylinder to the volume of an inscribed sphere. Archimedes also found the ratio of the surface area of the cylinder (including its base and top) to the surface area of the inscribed sphere. What is this ratio?
7. Archimedes' "balancing act" described in Section 2 works not only for a sphere but also (as Archimedes pointed out) for a segment of a sphere. Only a slight modification of the method described in Section 2 is needed to determine the volume of this segment of a sphere:

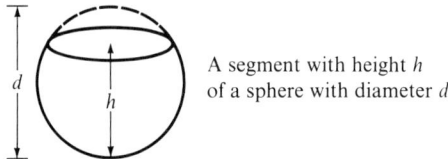

A segment with height h of a sphere with diameter d

Guess what this volume is by using the method of Archimedes. Then verify your result by calculus. *Hint*. Let V denote the volume of the segment of height h pictured above. By Archimedes' method of balancing, derive the relation

$$\left(V + \frac{1}{3}\pi h^3\right)d = (\pi d^2 h)\frac{h}{2}.$$

Then solve for V. Check your answer to see that it is the same as the integral

$$\int_{-r}^{-r+h} \pi(r^2 - x^2)\,dx,$$

which gives V as the volume of a solid of revolution.

8. The proof given in Section 1 of Archimedes' quadrature of the circle is left unfinished. The proof that possibility (b) is false is left to the reader. Write out this proof in detail.
9. Write a short essay either defending or attacking Voltaire's assertion that Archimedes is superior in imagination to Homer.

Appendix 4
Clean Writing in Mathematics

Style is like good manners. Its lambent presence is barely noticeable, but its absence is conspicuous. Taken in a broad sense, style can be discerned almost everywhere. One can speak of style (or its absence) in playing tennis, in hosting a dinner party, in presiding over a meeting, in teaching a class, or even—the subject of this appendix—in writing out the solution to a problem in calculus.

In such activities style is characterized by the light touch that draws harmony out of imminent disorder and makes difficult things seem easy. Everyone hates the burden of unnecessary fuss and bother; the grace that comes from easing this burden is the hallmark of style. In any purposeful activity it is style that eases the way.

Style must be natural because it cannot be affected. Affectation will draw attention only to itself, while style would draw attention straightway to the goal at hand.

Style is an outgrowth of education, not a product of it, for style cannot be readily taught or learned. It is acquired almost incidentally, like good manners, by those who want to please. Yet the final aim of education may be the cultivation of a sense for style.

> Finally [out of education], there should grow the most austere of all mental qualities; I mean the sense for style. It is an aesthetic sense, based on admiration for the direct attainment of a foreseen end, simply and without waste. Style in art, style in literature, style in logic, style in practical execution have fundamentally the same aesthetic qualities, namely, attainment and restraint. The love of a subject in itself and for itself, where it is not the sleepy pleasure of pacing a mental quarter-deck, is the love of style as manifested in that study.
>
> ... Style, in its finest sense, is the last acquirement of the educated mind; it is also the most useful. It pervades the whole being. The administrator with

a sense for style hates waste; the engineer with a sense for style economises his material; the artisan with a sense for style prefers good work. Style is the ultimate morality of mind.

But above style, and above knowledge, there is something, a vague shape like fate above the Greek gods. That something is Power. Style is the fashioning of power, the restraining of power. But, after all, the power of attainment of the desired end is fundamental. The first thing to do is to get there. Do not bother about your style, but solve your problem, justify the ways of God to man, administer your province, or do whatever else is set before you.

Where, then, does style help? In this, with style the end is attained without side issues, without raising undesirable inflammations. With style you attain your end and nothing but your end. With style the effect of your activity is calculable, and foresight is the last gift of gods to men.

Alfred North Whitehead*

§1. What to Do After Solving a Problem

Much of this text aims at aiding the reader to acquire the power to solve problems. This appendix is not about solving problems, but about what to do afterwards. Unless a problem is so easy that its answer is virtually apparent at the outset, one should not be content with merely finding the answer. One ought to develop a style of justifying what one believes to be true.

The tone of that justification should be geared to the expectations of those to whom it is addressed. Archimedes aimed at satisfying the highest expectations of his most critical fellow mathematicians.

> [Archimedes' deliberate style] suggests the tactics of some great strategist who foresees everything, eliminates everything not immediately conducive to the execution of his plan, masters every position in its order, and then suddenly (when the very elaboration of the scheme has almost obscured, in the mind of the spectator, its ultimate object) strikes the final blow. Thus we read in Archimedes proposition after proposition the bearing of which is not immediately obvious but which we find infallibly used later on; and we are led by such easy stages that the difficulty of the original problem, as presented at the outset, is scarcely appreciated.
>
> T. L. Heath†

Plutarch must have been right in suggesting that it was only by means of the greatest labor that Archimedes' works appear so unlabored. Archimedes was willing to put forth any amount of time and effort in his work. He was baffled for years, he tells us, before he was able to write some of his papers.

No one (including the instructor) in an introductory calculus course should be expected to meet archimedean standards. But some standards of clean

---

* Presidential address to the Mathematical Association of England, 1916. (Reprinted in *The Aims of Education*, by A. N. Whitehead, Macmillan, 1929, p. 24.)

† Preface to *The Works of Archimedes*, Cambridge, 1897, p. vi.

1. What to Do After Solving a Problem

exposition can be developed and maintained. Here are a few rules that might be considered.

(1) Do not slavishly follow any set of rules, even these.
(2) State clearly what information has been given at the outset, and make each succeeding step in your reasoning follow from what has gone before.
(3) If you introduce a symbol such as "x", be sure to indicate what it stands for. Your reader may not guess.
(4) Say exactly what you mean. Do not, for example, put an "equals" sign between unequal quantities.
(5) Write complete sentences and punctuate them correctly. Remember that an equation is (usually) a sentence.
(6) By being as concise and as natural as you can, disguise whatever effort it may have cost you to attain your goal. Be serious but not solemn.
(7) When you have completed your argument and have led your reader to the end, state your full conclusion in a complete sentence. Then stop writing.
(8) Review what you have written and delete anything irrelevant.

All of these rules may be condensed into one short Latin phrase:

Respice finem!*

It takes thought and time to produce a clear and concise piece of writing. The story is told that Pascal—a master of French prose—once apologized at the end of a long letter, saying that he simply had not had time to write a short letter. The great mathematician C. F. Gauss told a friend:

> You know that I write slowly. This is chiefly because I am never satisfied until I have said as much as possible in a few words, and writing briefly takes far more time than writing at length.†

But one can write too little just as easily as one can write too much. A proper balance must be struck.

EXAMPLE. Consider the function given by $f(x) = x^2 - 4x + 2$. Find the coordinates of the highest point on the graph of f if the domain of f is specified by the inequality $0 \leq x \leq 3$.

"Solution" by Student D.

$$y = x^2 - 4x + 2 = 2x - 4 = 0$$
$$2x = 4$$
$$x = 2$$

---

* Literally, "Respect your goal!" or "Have a high regard for the final result!" The phrase is often understood in its broadest sense, where it expresses a philosophy of life.

† From a letter by Gauss, as quoted in *Ways of Thought of Great Mathematicians*, by Herbert Meschkowski, Holden-Day, 1964, p. 62.

Remark. Student D appears to be slavishly following rule 1, for he has broken rules 4, 5, and 7. His statement
$$x^2 - 4x + 2 = 2x - 4$$
adds a touch of algebraic humor to this brief comedy of errors.

"Solution" by Student C. If $y = x^2 - 4x + 2$, then $y' = 2x - 4$. The derivative is then equal to zero when $2x - 4 = 0$, or $x = 2$. The point $x = 2$ is then the highest point on the graph of f.

Remark. Although Student C demonstrates knowledge of a nonalgebraic language—and thus appears to be better educated than Student D—his attempted solution is still inadequate. For one thing, the point $x = 2$ is on the x-axis and not on the graph of f.

"Solution" by Student B. To find the highest point, we set the derivative $2x - 4$ equal to zero. We get $x = 2$. Since $f(2) = -2$ we have a horizontal tangent line to the graph of f at the point $(2, -2)$. The highest point is therefore $(2, -2)$.

Remark. Student B has favored us with four informative sentences indicating much knowledge of calculus. But the fourth sentence does not follow from the third, and this breaks rule 2.

Solution by Student A. The only critical point occurs when $x = 2$. Since the largest value attained by a continuous function must occur at a critical point or at an endpoint, we need only glance at the following table to see that $(0, 2)$ is the highest point on the graph of f.

x	y
0	2
2	-2
3	-1

Remark. Student A has style.

Solution by Student A^+. Since the second derivative (given by $f''(x) = 2$) is always positive, the curve f is always concave upwards. Every such curve, like every smile, reaches its highest point at one end, and the endpoints here are $(0, 2)$ and $(3, -1)$. The highest point on the curve is then $(0, 2)$.

§2. Rewriting

"This first thing to do is to get there. Do not bother about your style, but solve your problem..." Whitehead's point is well taken. Virtually any means of solving a problem is legitimate, whether by a calculated method, by eliminat-

2. Rewriting

ing wrong answers, or by pure guesswork. If you are like the author of this book, you will make a big mess. You will fill up pages with hastily scrawled, illegible handwriting (half of which will be crossed out, being irrelevant), you will sketch badly drawn figures (which will not be improved when you spill coffee on them), and you will lose your pencil (the one that still had a good eraser). You will begin to believe those who say that scientific research is the purest example of an essentially comic activity.

But you learn, after all, through play; comic activity serves a serious purpose. Almost miraculously, your playful attempts may begin to give form to something new, however dimly conceived. Then your work is really cut out for you. What is becoming clear to you must be shown related to things familiar to all. It is here, with your end already in mind, that you begin to worry about style.

The chances are that you must rethink your whole project. First you must decide for whom you are writing. Are you addressing your instructor and classmates, or some wider circle? It is well to keep in mind some real or imaginary audience.

What is your goal? Is it to impress your reader with your knowledge, or is it to lead your reader to that knowledge? Or do you see your task as offering the most direct possible justification of some assertion? Your goal will determine your style.

EXAMPLE. A Norman window is in the shape of a rectangle surmounted by a semicircle. If the perimeter of the window is 16 feet, find the dimensions maximizing the area.

Comic Activity.

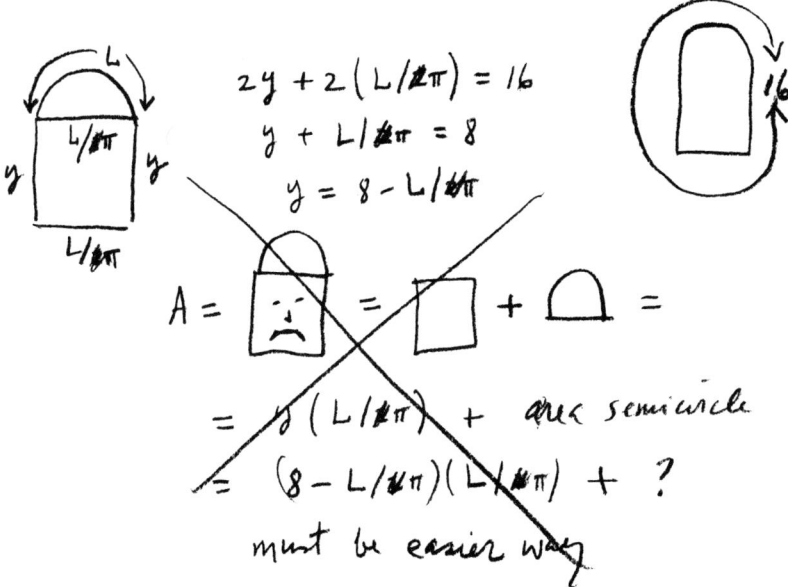

[Handwritten scratch work:]

NORMAN

$$2y + \pi r + 2r = 16$$

$$y = \frac{16 - \pi r - 2r}{2}$$

$$A = \tfrac{1}{2}\pi r^2 + 2ry$$

$$= \tfrac{1}{2}\pi r^2 + 2r\left(\frac{16 - \pi r - 2r}{2}\right)$$

$$= \tfrac{1}{2}\pi r^2 + 16r - \pi r^2 - 2r^2$$

$$A' = \pi r + 16 - 2\pi r - 4r = 0 \text{ when}$$

$$-\pi r - 4r = -16$$

$$(-\pi - 4)r = -16$$

$$r = \frac{16}{4 + \pi} \approx 2 \quad \text{EUREKA}$$

$$A'' = \pi - 2\pi - 4 = -\pi - 4 < 0$$

So frown → Max

Solution A. Let r be the radius of the semicircle making up the top of a Norman window whose perimeter is 16 feet. It follows easily that the rectangular portion of the window must be of dimensions $2r$ by $(16 - \pi r - 2r)/2$ feet. The area A of the window is the sum of the areas of the semicircle and the rectangle:

$$A = \frac{1}{2}\pi r^2 + 2r(16 - \pi r - 2r)/2$$

$$= \left(-2 - \frac{1}{2}\pi\right)r^2 + 16r.$$

This is just a simple quadratic function whose leading term is negative, so it attains a maximum at its critical point. To find the critical point we set the derivative A' equal to zero:

$$2\left(-2 - \frac{1}{2}\pi\right)r + 16 = 0,$$

$$(4 + \pi)r = 16,$$

$$r = 16/(4 + \pi)$$

$$\approx 2.24 \text{ feet.}$$

To maximize the area, the rectangular portion of the window must be of dimensions 4.48 by 2.24 feet, approximately.

Solution A^+. It is no harder to consider the more general problem where a fixed perimeter P is specified, and to prove the following theorem.

Norman Window Theorem. *Let P be the perimeter of a Norman window, made up of a rectangle surmounted by a semicircle. Then the area of the window is maximized if the rectangle has base $2P/(4 + \pi)$ and height $P/(4 + \pi)$.*

PROOF. If r is the radius of the semicircle, it follows by easy algebra that the rectangle has these dimensions:

(*) $$\text{base} = 2r,$$

(**) $$\text{height} = \frac{P - \pi r - 2r}{2}.$$

The area A of the window is the sum of the areas of the semicircle and the rectangle. When these are calculated and combined we get

$$A = -\left(\frac{4 + \pi}{2}\right)r^2 + Pr,$$

$$A' = -(4 + \pi)r + P,$$

$$A'' = -(4 + \pi).$$

Setting A' equal to zero immediately yields $r = P/(4 + \pi)$. This gives the only critical point, which is a maximum since A'' is negative. Substituting this value of r into equations (*) and (**) shows that the area A is maximized when the rectangle is of dimensions $2P/(4 + \pi)$ by $P/(4 + \pi)$. □

§3. Summary

Like virtually every course taught in the liberal arts, a course in mathematics is in part a course in writing. A student cannot learn to think like a mathematician without learning to write like a mathematician.

Style in writing is of little use, however, unless you first have something to say. To find something new you must strike out on your own, with a willingness to make mistakes, to learn from them, and to laugh at yourself. By playing the fool in a comedy of errors, you may find the means to climb up to a more serious level.

What has been discussed in this appendix is the classical sense of style that is so well depicted by Whitehead. This sense derives to some extent from classical mathematics. In modern times other notions about style have arisen, the most notable being that style is virtually synonymous with self-expression. Since everyone agrees that is is fatal to imitate the style of another, many believe that style is acquired only by writing away after one's own fashion with a proud indifference to any discipline imposed from outside.

Mathematics was born, however, in a less (or perhaps more) sophisticated time when self-expression was not so important. The purpose of education was not then to learn to express yourself; it was to learn to tell the truth. And even today, when writing in the discipline of mathematics, you may find yourself most squarely standing between the reader and the truth. In this austere place there is little room for self-expression. You must get yourself out of the way. It is only good manners to bow.

PROBLEMS (OPTIONAL)

1. Consider the function given by $f(x) = 3x^2 + 10x + 7$. Find the coordinates of the highest point on the graph of f if the domain of f is specified by the inequality $0 \leq x \leq 3$.

2. A Norman window is in the shape of a rectangle surmounted by a triangle. If the perimeter of the window is 16 feet, find the dimensions maximizing the area of the *rectangular portion* of the window.

3. An athletic field is to be built roughly in the shape of an oval, with a 400-meter track as its perimeter. The field is to consist of a rectangle with a semicircle at each end. Find the dimensions of the field maximizing the area of the rectangular portion.

4. In problem 3, find the dimensions of the field maximizing its total area.

5. What do you think is the final aim of education? What, if anything, does the study of mathematics contribute to the attainment of this final aim?

Natural Logarithms of Numbers

N	0	1	2	3	4	5	6	7	8	9
1.0	0.0 0000	0995	1980	2956	3922	4879	5827	6766	7696	8618
1.1	0.0 9531	*0436	*1333	*2222	*3103	*3976	*4842	*5700	*6551	*7395
1.2	0.1 8232	9062	9885	*0701	*1511	*2314	*3111	*3902	*4686	*5464
1.3	0.2 6236	7003	7763	8518	9267	*0010	*0748	*1481	*2208	*2930
1.4	0.3 3647	4359	5066	5767	6464	7156	7844	8526	9204	9878
1.5	0.4 0547	1211	1871	2527	3178	3825	4469	5108	5742	6373
1.6	0.4 7000	7623	8243	8858	9470	*0078	*0682	*1282	*1879	*2473
1.7	0.5 3063	3649	4232	4812	5389	5962	6531	7098	7661	8222
1.8	0.5 8779	9333	9884	*0432	*0977	*1519	*2078	*2594	*3127	*3658
1.9	0.6 4185	4710	5233	5752	6269	6783	7294	7803	8310	8813
2.0	0.6 9315	9813	*0310	*0804	*1295	*1784	*2271	*2755	*3237	*3716
2.1	0.7 4194	4669	5142	5612	6081	6547	7011	7473	7932	8390
2.2	0.7 8846	9299	9751	*0200	*0648	*1093	*1536	*1978	*2418	*2855
2.3	0.8 3291	3725	4157	4587	5015	5442	5866	6289	6710	7129
2.4	0.8 7547	7963	8377	8789	9200	9609	*0016	*0422	*0826	*1228
2.5	0.9 1629	2028	2426	2822	3216	3609	4001	4391	4779	5166
2.6	0.9 5551	5935	6317	6698	7078	7456	7833	8208	8582	8954
2.7	0.9 9325	9695	*0063	*0430	*0796	*1160	*1523	*1885	*2245	*2604
2.8	1.0 2962	3318	3674	4028	4380	4732	5082	5431	5779	6126
2.9	1.0 6471	6815	7158	7500	7841	1881	8519	8856	9192	9527
3.0	1.0 9861	*0194	*0526	*0856	*1186	*1514	*1841	*2168	*2493	*2817
3.1	1.1 3140	3462	3783	4103	4422	4740	5057	5373	5688	6002
3.2	1.1 6315	6627	6938	7248	7557	7865	8173	8479	8784	9089
3.3	1.1 9392	9695	9996	*0297	*0597	*0896	*1194	*1491	*1788	*2083
3.4	1.2 2378	2671	2964	3256	3547	3837	4127	4415	4703	4990
3.5	1.2 5276	5562	5846	6130	6413	6695	6976	7257	7536	7815
3.6	1.2 8093	8371	8647	8923	9198	9473	9746	*0019	*0291	*0563
3.7	1.3 0833	1103	1372	1641	1909	2176	2442	2708	2972	3237
3.8	1.3 3500	3763	4025	4286	4547	4807	5067	5325	5584	5841
3.9	1.3 6098	6354	6609	6864	7118	7372	7624	7877	8128	8379
4.0	1.3 8629	8879	9128	9377	9624	9872	*0118	*0364	*0610	*0854
N	0	1	2	3	4	5	6	7	8	9

Natural Logarithms of Numbers (*Cont.*)

N	0	1	2	3	4	5	6	7	8	9
4.0	1.3 8629	8879	9128	9377	9624	9872	*0118	*0364	*0610	*0854
4.1	1.4 1099	1342	1585	1828	2070	2311	2552	2792	3031	3270
4.2	1.4 3508	3746	3984	4220	4456	4692	4927	5161	5395	5629
4.3	1.4 5862	6094	6326	6557	6787	7018	7247	7476	7705	7933
4.4	1.4 8160	8387	8614	8840	9065	9290	9515	9739	9962	*0185
4.5	1.5 0408	0630	0851	1072	1293	1513	1732	1951	2170	2388
4.6	1.5 2606	2823	3039	3256	3471	3687	3902	4116	4330	4543
4.7	1.5 4756	4969	5181	5393	5604	5814	6025	6235	6444	6653
4.8	1.5 6862	7070	7277	7485	7691	7898	8104	8309	8515	8719
4.9	1.5 8924	9127	9331	9534	9737	9939	*0141	*0342	*0543	*0744
5.0	1.6 0944	1144	1343	1542	1741	1939	2137	2334	2531	2728
5.1	1.6 2924	3120	3315	3511	3705	3900	4094	4287	4481	4673
5.2	1.6 4866	5058	5250	5441	5632	5823	6013	6203	6393	6582
5.3	1.6 6771	6959	7147	7335	7523	7710	7896	8083	8269	8455
5.4	1.6 8640	8825	9010	9194	9378	9562	9745	9928	*0111	*0293
5.5	1.7 0475	0656	0838	1019	1199	1380	1560	1740	1919	2098
5.6	1.7 2277	2455	2633	2811	2988	3166	3342	3519	3695	3871
5.7	1.7 4047	4222	4397	4572	4746	4920	5094	5267	5440	5613
5.8	1.7 5786	5958	6130	6302	6473	6644	6815	6985	7156	7326
5.9	1.7 7495	7665	7843	8002	8171	8339	8507	8675	8842	9009
6.0	1.7 9176	9342	9509	9675	9840	*0006	*0171	*0336	*0500	*0665
6.1	1.8 0829	0993	1156	1319	1482	1645	1808	1970	2132	2294
6.2	1.8 2455	2616	2777	2938	3098	3258	3418	3578	3737	3896
6.3	1.8 4055	4214	4372	4530	4688	4845	5003	5160	5317	5473
6.4	1.8 5630	5786	5942	6097	6253	6408	6563	6718	6872	7026
6.5	1.8 7180	7334	7487	7641	7794	7947	8099	8251	8403	8555
6.6	1.8 8707	8858	9010	9160	9311	9462	9612	9762	9912	*0061
6.7	1.9 0211	0360	0509	0658	0806	0954	1102	1250	1398	1545
6.8	1.9 1692	1839	1986	2132	2279	2425	2571	2716	2862	3007
6.9	1.9 3152	3297	3442	3586	3730	3874	4018	4162	4305	4448
7.0	1.9 4591	4734	4876	5019	5161	5303	5445	5586	5727	5869
N	0	1	2	3	4	5	6	7	8	9

Natural Logarithms of Numbers (*Cont.*)

N	0	1	2	3	4	5	6	7	8	9
7.0	1.9 4591	4734	4876	5019	5161	5303	5445	5586	5727	5869
7.1	1.9 6009	6150	6291	6431	6571	6711	6851	6991	7130	7269
7.2	1.9 7408	7547	7685	7824	7962	8100	8238	8376	8513	8650
7.3	1.9 8787	8924	9061	9198	9334	9470	9606	9742	9877	*0013
7.4	2.0 0148	0283	0418	0553	0687	0821	0956	1089	1223	1357
7.5	2.0 1490	1624	1757	1890	2022	2155	2287	2419	2551	2683
7.6	2.0 2815	2946	3078	3209	3340	3471	3601	3732	3862	3992
7.7	2.0 4122	4252	4381	4511	4640	4769	4898	5027	5156	5284
7.8	2.0 5412	5540	5668	5796	5924	6051	6179	6306	6433	6560
7.9	2.0 6686	6813	6939	7065	7191	7317	7443	7568	7694	7819
8.0	2.0 7944	8069	8194	8318	8443	8567	8691	8815	8939	9063
8.1	2.0 9186	9310	9433	9556	9679	9802	9924	*0047	*0169	*0291
8.2	2.1 0413	0535	0657	0779	0900	1021	1142	1263	1384	1505
8.3	2.1 1626	1746	1866	1986	2106	2226	2346	2465	2585	2704
8.4	2.1 2823	2942	3061	3180	3298	3417	3535	3653	3771	3889
8.5	2.1 4007	4124	4242	4359	4476	4593	4710	4827	4943	5060
8.6	2.1 5176	5292	5409	5524	5640	5756	5871	5987	6102	6217
8.7	2.1 6332	6447	6562	6677	6791	6905	7020	7134	7248	7361
8.8	2.1 7475	7589	7702	7816	7929	8042	8155	8267	8380	8493
8.9	2.1 8605	8717	8830	8942	9054	9165	9277	9389	9500	9611
9.0	2.1 9722	9834	9944	*0055	*0166	*0276	*0387	*0497	*0607	*0717
9.1	2.2 0827	0937	1047	1157	1266	1375	1485	1594	1703	1812
9.2	2.2 1920	2029	2138	2246	2354	2462	2570	2678	2786	2894
9.3	2.2 3001	3109	3216	3324	3431	3538	3645	3751	3858	3965
9.4	2.2 4071	4177	4284	4390	4496	4601	4707	4813	4918	5024
9.5	2.2 5129	5234	5339	5444	5549	5654	5759	5863	5968	6072
9.6	2.2 6176	6280	6384	6488	6592	6696	6799	6903	7006	7109
9.7	2.2 7213	7316	7419	7521	7624	7727	7839	7932	8034	8136
9.8	2.2 8238	8340	8442	8544	8646	8747	8849	8950	9051	9152
9.9	2.2 9253	9354	9455	9556	9657	9757	9858	9958	*0058	*0158
10.0	2.3 0259	0358	0458	0558	0658	0757	0857	0956	1055	1154
N	0	1	2	3	4	5	6	7	8	9

Four-Place Logarithms to Base 10

10	0	1	2	3	4	5	6	7	8	9
10	0000	0043	0086	0128	0170	0212	0253	0294	0334	0374
11	0414	0453	0492	0531	0569	0607	0645	0682	0179	0755
12	0792	0828	0864	0899	0934	0969	1004	1038	1072	1106
13	1139	1173	1206	1239	1271	1303	1335	1367	1399	1430
14	1461	1492	1523	1553	1584	1614	1644	1673	1703	1732
15	1761	1790	1818	1847	1875	1903	1931	1959	1987	2014
16	2041	2068	2095	2122	2148	2175	2201	2227	2253	2279
17	2304	2330	2355	2380	2405	2430	2455	2480	2504	2529
18	2553	2577	2601	2625	2648	2672	2695	2718	2742	2765
19	2788	2810	2833	2856	2878	2900	2923	2945	2967	2989
20	3010	3032	3054	3075	3096	3448	3139	3160	3181	3201
21	3222	3243	3263	3284	3304	3324	3345	3365	3385	3404
22	3424	3444	3464	3483	3502	3522	3541	3560	3579	3598
23	3617	3636	3655	3674	3692	3711	3729	3747	3766	3784
24	3802	3820	3838	3856	3874	3892	3909	3927	3945	3962
25	3979	3997	4014	4031	4048	4065	4082	4099	4416	4133
26	4150	4466	4483	4200	4216	4232	4249	4265	4281	4298
27	4314	4330	4346	4362	4378	4393	4409	4425	4440	4456
28	4472	4487	4502	4518	4533	4548	4564	4579	4594	4609
29	4624	4639	4654	4669	4683	4698	4713	4728	4742	4757
30	4771	4786	4800	4814	4829	4843	4857	4871	4886	4900
31	4914	4928	4942	4955	4969	4983	4997	5011	5024	5038
32	5051	5065	5079	5092	5105	5119	5132	5145	5159	5172
33	5185	5198	5211	5224	5237	5250	5263	5276	5289	5302
34	5315	5328	5340	5353	5366	5378	5391	5403	5446	5428
35	5441	5453	5465	5478	5490	5502	5514	5527	5539	5551
36	5563	5575	5587	5599	5611	5623	5635	5647	5658	5670
37	5682	5694	5705	5717	5729	5740	5752	5763	5775	5786
38	5798	5809	5821	5832	5843	5855	5866	5877	5888	5899
39	5911	5922	5933	5944	5955	5966	5977	5988	5999	6010
40	6021	6031	6042	6053	6064	6075	6085	6096	6107	6117
41	6128	6138	6149	6160	6170	6180	6191	6201	6212	6222
42	6232	6243	6253	6263	6274	6284	6294	6304	6314	6325
43	6335	6345	6355	6365	7375	6385	6395	6405	6415	6425
44	6135	6444	6154	6464	6474	6484	6493	6503	6513	6522
45	6532	6542	6551	6561	6571	6580	6590	6599	6609	6618
46	6628	6637	6646	6656	6665	6675	6684	6693	6702	6712
47	6721	6730	6739	6749	6758	6767	6776	6785	6794	6803
48	6812	6821	6830	6839	6848	6857	6866	6875	6884	6893
49	6902	6911	6920	6928	6937	6946	6955	6964	6972	6981
50	6990	6998	7007	7016	7024	7033	7042	7050	7059	7067
51	7076	7084	7093	7101	7110	7118	7126	7135	7443	7452
52	7160	7168	7177	7185	7193	7202	7210	7218	7226	7235
53	7243	7251	7259	7267	7275	7284	7292	7300	7308	7316
54	7324	7332	7340	7348	7356	7364	7372	7380	7388	7396

Four-Place Logarithms to Base 10 (*Cont.*)

55	0	1	2	3	4	5	6	7	8	9
55	7404	7412	7419	7427	7435	7443	7451	7459	7466	7474
56	7482	7490	7497	7505	7513	7520	7528	7536	7543	7551
57	7559	7566	7574	7582	7589	7597	7604	7612	7619	7627
58	7634	7642	7649	7657	7664	7672	7679	7686	7694	7701
59	7709	7716	7723	7731	7738	7745	7752	7760	7767	7774
60	7782	7789	7796	7803	7810	7818	7825	7832	7839	7846
61	7853	7860	7868	7875	7882	7889	7896	7903	7910	7917
62	7924	7931	7938	7945	7952	7959	7966	7973	7980	7987
63	7993	8000	8007	8014	8021	8028	8035	8041	8048	8055
64	8062	8069	8075	8082	8089	8096	8102	8109	8116	8122
65	8129	8136	8142	8149	8156	8162	8169	8176	8182	8189
66	8195	8202	8209	8215	8222	8228	8235	8241	8248	8254
67	8261	8267	8274	8280	8287	8293	8299	8306	8312	8319
68	8325	8331	8338	8344	8351	8357	8363	8370	8376	8382
69	8388	8395	8401	8407	8414	8420	8426	8432	8439	8445
70	8451	8457	8463	8470	8476	8482	8488	8494	8500	8506
71	8513	8519	8525	8531	8537	8543	8549	8555	8561	8567
72	8573	8579	8585	8591	8597	8603	8609	8615	8621	8627
73	8633	8639	8645	8651	8657	8663	8669	8675	8681	8686
74	8692	8698	8704	8710	8716	8722	8727	8733	8739	8745
75	8751	8756	8762	8768	8774	8779	8785	8791	8797	8802
76	8808	8814	8820	8825	8831	8837	8842	8848	8851	8859
77	8865	8871	8876	8882	8887	8893	8899	8904	8910	8915
78	8921	8927	8932	8938	8943	8949	8954	8960	8965	8971
79	8976	8982	8987	8993	8998	9004	9009	9015	9020	9025
80	9031	9036	9042	9047	9053	9058	9063	9069	9074	9079
81	9085	9090	9096	9101	9106	9112	9117	9122	9128	9133
82	9138	9143	9149	9154	9159	9165	9170	9175	9180	9186
83	9191	9196	9201	9206	9212	9217	9222	9227	9232	9238
84	9243	9248	9253	9258	9263	9269	9274	9279	9284	9289
85	9294	9299	9304	9309	9315	9320	9325	9330	9335	9340
86	9345	9350	9355	9360	9365	9370	9375	9380	9385	9390
87	9395	9400	9405	9440	9445	9420	9425	9430	9435	9140
88	9445	9450	9455	9460	9465	9469	9474	9479	9484	9489
89	9494	9499	9504	9509	9513	9518	9523	9528	9533	9538
90	9542	9547	9552	9557	9562	9566	9571	9576	9581	9586
91	9590	9595	9600	9605	9609	9614	9619	9621	9628	9633
92	9638	9643	9647	9652	9657	9661	9666	9671	9675	9680
93	9685	9689	9694	9699	9703	9708	9713	9717	9722	9727
94	9734	9736	9741	9745	9750	9754	9759	9763	9768	9773
95	9777	9782	9786	9791	9795	9800	9805	9809	9844	9818
96	9823	9827	9832	9836	9811	9815	9850	9854	9859	9863
97	9868	9872	9877	9881	9886	9890	9891	9899	9903	9908
98	9912	9917	9921	9926	9930	9934	9939	9943	9948	9952
99	9956	9961	9965	9969	9974	9978	9983	9987	9991	9996

Answers to Selected Problems

CHAPTER 1

1. (a) $C = (84/W) + 4W$. (b) $0 < W$.
3. (a) $A = 1200w - 2w^2$. (b) $0 < w < 600$.
9. (a) domain F is

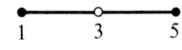

   or $1 \leq x \leq 5$, $x \neq 3$. range F is

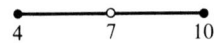

   or $4 \leq y \leq 10$, $y \neq 7$.
   (b) domain f is

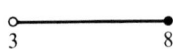

   or $3 < x \leq 8$. range f is

   or $y = 4$ or 7.
11. (a) No. (b) No.
13. (a) range F is

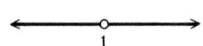

   or $y \neq 1$.

17. (b) $A = 20x - (1/12)x^2$, $0 < x < 240$.

19. (c) $0 < L < \sqrt{120}$.

CHAPTER 2

1. (f) 2500.

3. *Eureka!*

5. (d) center (0,0), radius $\sqrt{2}$.

7. (a) if $\sqrt{8} = 2\sqrt{2}$ were rational, it would be equal to some ratio m/n of integers, which leads to a contradiction, as follows: $2\sqrt{2} = m/n$ implies that $\sqrt{2} = m/2n = $ a rational number, contradicting the fact that $\sqrt{2}$ is irrational.

9. domain $0 < r$; range $0 < A$.

11. *Eureka!*

15. Statement (c) is true. (See the appendix on Archimedes.)

21. (c) (For the origin and meaning of the word *mathematics*, see S. Bochner, *The Role of Mathematics in the Rise of Science*, Princeton University Press, 1966, pp. 22–28.)

22. *No Roman lost his life because he was absorbed in the contemplation of a mathematical diagram.*—A. N. Whitehead. *In mathematics all roads lead back to Greece.*—Tobias Dantzig.

CHAPTER 3

3. (a) -6.
   (b) falling.
   (c) $x < \tfrac{3}{2}$.
   (d)

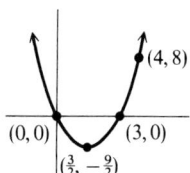

   (e) $-4.5 \leq y \leq 8$.
   (f) $-4 < y < 0$.
   (g) $-4.5 \leq y$.

5. (a) $-1 \leq y \leq 8$. (b) $-7 < y < 14$.

7. (4,2).

9. (This problem is discussed in detail in Appendix 4.)

11. The wire should be cut so as to make the first part $500\pi/(4 + \pi) \approx 220$ centimeters long in order to minimize the combined areas.

13. $f'(x) = \text{Limit}_{h \to 0}((f(x + h) - f(x))/h) = \text{Limit}_{h \to 0}((7 - 7)/h) = \text{Limit}_{h \to 0} 0 = 0$.

15. 1200 square meters.

17. (a) $y - 1 = 3(x - 1)$, or $y = 3x - 2$. (b) Yes. (c) Yes. (d) No, the tangent line of slope 3 through (1,1) would be eliminated from consideration in "Holmes's method", because it cuts the curve twice.

18. The derivative of (a) is pictured in (d); the derivative of (b) is (d); of (c) is (g); of (d) is (g); of (e) is (h); of (f) is (b); of (g) is (k); of (h) is (j); of (i) is (b); of (j) is (k); of (k) is (k).

CHAPTER 4

1. $92.95.

3. $y - 5 = -10(x - 1)$.

5. (a) rising. (b) to the right.

7. (a) 3. (b) to the left. (c) lowest.

9. range f is

11. (a)

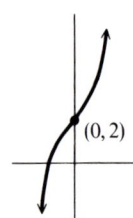

(b) (0,2).
(c) $-2 \leq y \leq 78$.

13.
t	y	y'	y''
0	$-\frac{3}{2}$	$\frac{3}{4}$	$\frac{1}{4}$

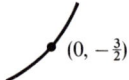

15. The first number should be $(-1 + \sqrt{61})/3 \approx 2.27$, and the second should be ≈ 7.73.

17. 3.

19. The sides of the cut-out square should be of length $(14 - \sqrt{76})/6 \approx 0.88$ meters, in order to maximize the volume.

Answers to Selected Problems

21. (d) Distance PQ should be $\sqrt{4/5} \approx 0.8944$ miles.

23. (a) min C occurs when $L = \sqrt[3]{16} \approx 2.52$ meters.

25. $\sqrt[3]{3}$ by $\sqrt[3]{3}$ by $\sqrt[3]{3}$ meters.

27. (a) approximately 7.30 by 7.30 by 3.65 feet.
    (b) approximately 5.16 by 5.16 by 5.16 feet.
    (c) both radius and height should be $\sqrt{160/3\pi} \approx 4.12$ feet.
    (d) radius should be approximately 2.91 feet and height 5.82 feet.

29. (b) The first part should be $\pi A/(4 + \pi)$ centimeters long.

31. (a) $2(x/(x - 6))((x - 6 - x)/(x - 6)^2)$. (f) $(1/2\sqrt{x^5})(5x^4)$.

41. (Partial answer) the derivative of (a) is (d); the derivative of (j) is (a).

CHAPTER 5

1. (a) 100 miles. (b) 50 mi/hr. (c) 40 mi/hr. (d) accelerating. (e) decelerating.

3. (a) $\Delta A = x(\Delta y) + y(\Delta x) + (\Delta x)(\Delta y)$. (b) Δx and Δy must tend to zero since x and y are differentiable, hence continuous, functions of t.

5. (a) $4(x^2 + 7x)^3(2x + 7)$. (c) $3((x - 2)/(x + 2))^2(4/(x + 2)^2)$.

7. (a) $\frac{15}{4}$ ft/sec. (b) $\frac{20}{3}$ ft/sec.

9. $\frac{15}{7}$ ft/sec.

11. $(dr/dt)|_{r=2} = 3/4\pi$ in/sec.

13. (b) $F(t) = 1 - (1/t)$. (This curve F is pictured in (d) of problem 25.) The answer is not unique, because the function pictured in figure (f) of problem 25 is also an antiderivative of $1/t^2$ that takes the value 0 when $t = 1$.

15. (a) 80 ft/sec. (b) 80 ft/sec. (c) 62 ft/sec.

17. The upward speed at impact is $\text{Limit}_{t \to 2.3^-}(-32t - 50) = -123.6$ ft/sec.

19. $g(3) = 3$.

23. (b) 208 ft.

CHAPTER 6

1. 49 km.

3. $dA/dt = 1/(t + 2)$.

5.

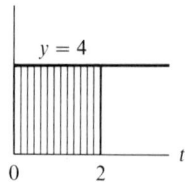

The area is $2 \cdot 4 = 8$.

7. $8 - \pi$ square units. (We do not yet know an antiderivative of $\sqrt{4-t^2}$.)

9. (a) $\int_0^2 4x\,dx = \text{Limit } S_n = \text{Limit } \sum_{k=1}^{n} 4(2k/n)(2/n) = \text{Limit}(16/n^2)(n(n+1)/2) = 8$.

11. (b) 2.

13. (a) $16\pi/15$ cubic units. (b) $4\pi/3$ cubic units.

15. $280\pi/3$ cubic units.

17. (b) Slice 2.

23. $(-1/2, 15/4)$.

CHAPTER 7

3. (a) 2. (b) 0. (c) 1. (d) $\frac{1}{6}$.

5. (a) $x\cos x + \sin x$. (b) $-\sin^2 x + \cos^2 x$. (c) $(-x\sin x + \cos x)\cos(x\cos x)$.
   (d) $\pi\cos\pi x$. (e) $\frac{1}{2}\cos(\frac{1}{2}x)$. (f) $2\sec^2 x \tan x$. (g) $\sec^3 x + \sec x \tan^2 x$.

7.

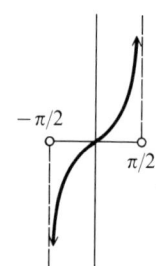

9.

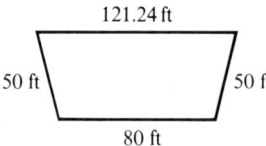

11. (Show, using techniques of optimization, that the minimum time of travel is half an hour.)

Answers to Selected Problems

13. (a) $\sqrt{3}/2$ square units. (b) $\frac{5}{2}$ square units. (c) $\sqrt{3}$ square units.
15. TTFTTFFFFTTFFTTFTF.
17. (From problem 16 we have $\cos 3x = 4\cos^3 x - 3\cos x$. Differentiate both sides and solve for the expression $\sin 3x$. Get it in terms of sin alone, without using cos.)
19. $A_1 = 7\pi/12$ square units, $A_2 = \pi/4$ square units, $A_3 = \pi/3$ square units.
21. To maximize the angle, let x equal $\sqrt{700} \approx 26.46$.
23. (a) 0. (b) $\frac{2}{5}$. (c) π. (d) $\pi/2$. (e) $-\frac{2}{5}$. (f) 0. (g) 0. (h) $5\sqrt{2}/42$.
25. (Partial answer) The inverse of (a) is (e); the inverse of (e) is (a).
33. (a) $-\frac{5}{12}$. (b) $\frac{5}{12}$. (c) 2. (d) $-\frac{1}{4}$.

Chapter 8

3. (Partial answer) (f) No. (g) Yes, $\int_0^2 f(x)\,dx = 10$.
5. (Partial answer) The average of the bounds will be slightly larger than the integral $\int_1^5 (1/t)\,dt$, since the curve $y = 1/t$ is concave up when $1 \le t \le 5$.
7. $76/45 \approx 1.69$.
9. (a) 161.67. (b) 161.67.
11. $8T^2/3$ feet.
13. (a) 1. (b) $\pi/3$. (c) 2. (d) 0.
15. (a) 4. (b) $\cos^3 2$. (c) $-\frac{1}{2}$. (d) 0.
17. (a) $\int_0^x |t|\,dt$.
19. (a) $5/3, -0.099\ldots, 2 + \sin 1$. (b) $43, 1 + \ln(5/3), 2 + \sin 5$.
21. (a) $1/x$. (b) $1/x$.
23. (a) $\pi/4$ ft. (b) $\sqrt{3}$ min.
29. (a) $16{,}380\pi$ ft-lbs.
31. $\left(9x + \dfrac{4}{x}\right)\cos(3x^3 + 4x) - \left(\dfrac{1}{1+x^2} + \dfrac{\arctan x}{x}\right)\cos(x \arctan x)$
$+ \dfrac{1}{x^2}(\sin(x \arctan x) - \sin(3x^3 + 4x))$.

Chapter 9

1. $52.5 + 48\ln 4 \approx 119.042$ square units.
3. (a) 1. (b) 1. (c) $\ln 10$. (d) $1/\ln 10$. (e) $\frac{1}{2}$. (f) $\ln 5$. (g) $-1/(2\ln 2)$. (h) $1/(2\ln 2)$.

7. The only critical point is at $x = -1$, and the only inflection point is at $x = -2$. As x runs through large negative values, e^x tends to zero. So does xe^x.

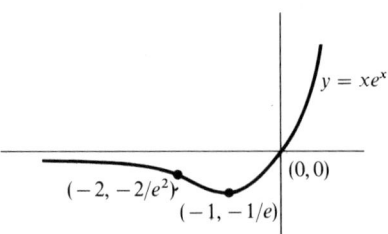

11. (a) $(-20/7)(\ln 2)(\frac{1}{2})^{t/35}$.   (c) $5(3 + \cos t)^{5t}(\ln(3 + \cos t) - (t \sin t/(3 + \cos t)))$.

13. (a) $\frac{1}{2}\pi(e^2 - 1)$.   (b) $\frac{1}{2}\pi(e^2 - 4e + 5)$.   (c) $\pi \ln 2$.   (d) $99\pi/(2 \ln 10)$.

15. (a) $2013.75.   (b) $1448.21.   (c) In 15.69 years.

17. 707.

19. after 11.4 days.

21. 144.6 grams.

23. (a) $\sqrt{25 + x} \approx 5 + \frac{1}{10}x - \frac{1}{1000}x^2 + \frac{1}{50,000}x^3$.
    (b) $\sqrt{28} \approx 5.29154$. (Correct value is $5.29150\ldots$)

25. Your answer should be the same as that obtained from (74) by differentiation.

27. (a) $dy/dt = k(y - 20)$, where y is the temperature in degrees Celsius of the metal ball, and k is the negative (why?) constant of proportionality.
    (b) $y = 20 + 160e^{kt}$.   (d) $50.625°C$.   (e) when $t \approx 8.39$.

Chapter 10

1. The area of the large square is equal to the sum of the areas of four congruent right triangles and the area of a small square whose sides have length $a - b$. Therefore, $4(\frac{1}{2}ab) + (a - b)^2 =$ area of large square $= c^2$, from which it follows by simple algebra that $a^2 + b^2 = c^2$.

3. 2 square units.

9. (a) $x = -c/b$.

11. (a) The fence along the road should be $\sqrt{800/13} \approx 7.845$ m. and the depth of the fenced-in area should be ≈ 12.748 m.   (b) approximately 7788 square meters.

13. $4\pi r^3/3\sqrt{3}$ cubic units.

15. $32\pi r^3/81$ cubic units.

17. $6\sqrt{3}$ inches.

19. 26 ft.

21. Swimming directly across is the quickest way.

Answers to Selected Problems

25. Yes. (For help, see problem 28 at the end of Chapter 7.)
27. (b) $40 \leq C < 58$.
29. (a) 68 ft. from the ground. (b) approximately 1.802 seconds after it is released. (c) 97.65 ft/sec downwards, approximately.
31. 64 ft/sec.
33. (a) $12/\pi$ cm/sec. (b) $3/4\pi$ cm/sec. (c) $3/16\pi$ cm/sec.
35. (b) $-13 \leq y \leq 13$.

37.

t	s	ds/dt	d^2s/dt^2
1		$16\sqrt{2}$	$-32\sqrt{2}/3$
2		0	-25.6
3		$-16\sqrt{2}$	$-32\sqrt{2}/3$

41. $\int_0^t x^3\, dx = \text{Limit} \sum_{k=1}^n (kt/n)^3(t/n) = \text{Limit}(t^4/n^4) \sum_{k=1}^n k^3 = \text{Limit}(t^4/n^4)(n^2(n+1)^2/4) = t^4/4$

45.

x	y	y'	y''	
0	0	3	-4	(rising, concave down)
1		0	$-2/e$	(local maximum)
2		$-1/e^4$	$4/e^4$	(falling, concave up)
3		0	$2/e^9$	(local minimum)

47. (b) Yes. (c) No.
49. (c) $\sqrt{3}$.
53. (a) $1/\ln \pi$. (b) e/π. (c) $\ln 3$. (d) $\ln 4$.
57. in about 5.94 months.
63. *It is one of the rarest gifts to be able to hold a view with conviction and detachment at the same time. Philosophers and scientists more than other men strive to train themselves to achieve it, though in the end they are usually no more successful than the layman. Mathematics is admirably suited to foster this kind of attitude. It is by no means accidental that many great philosophers were also mathematicians.*—Bertrand Russell.

Index

Abel, Niels Henrik   386
Absolute value   242
Acceleration   115
Action   14
  versus purpose   17–19
Age of Reason   352, 360, 375
Algebra   352
Algebraic approximations   337–342
Algebraic functions   7, 24
  convention about domain of   10
Analysis   96, 157, 179, 345, 371
Analytic geometry   33, 74
Angle:
  of inclination   240
  radian measure of   196
Antiderivative   127
  and area   140, 156–157
  and distance   135–137
  and integral   178, 282
  uniqueness of   131
Apollonius   33–34, 43, 352, 407
Arabs   352
Arccos t, arccosine   225
Archimedes   28, 42–44, 45, 48, 138, 148, 153, 165, 186, 189, 248, 251, 259, 271, 274, 352, 353, 355, 356, 358, 370, 384, 387, 402, 403–412, 414
  method of   408–411
  principle of buoyancy of   408

Archytas   407
Arcsin t, arcsine   222
Arctan t, arctangent   224
Area   152–154
  and antiderivatives   148–151, 154–157
  between two curves   159–163
  of circle   45, 138
  of ellipse   192
  of parabolic segment   402
  principle   160
  of sector   198
Aristotle   38–39, 358, 359, 360
Arrow paradox   356
Atomism   355, 371
Average   258
  of function   261, 264
  speed   112–113
  weighted   265
Axiom, axiomatic method   38–39

Babylonians   28, 194, 195, 196, 387
Barrow, Isaac   149, 374
Base of a logarithm   300
Bell, Eric Temple   39, 48, 361, 372, 374, 386, 387
Bending right, bending left   85–88
Berkeley, George   372–374
Bernoulli, Johann   277

435

Bishop, Errett 371–372, 388
Blake, William 384
Bochner, Salomon 40, 387
Boyer, C.B. 387

Calculus 1
  differentialis 127
  integralis 127
Cantor, Georg 386
Cantor, Moritz 387
Cavalieri, Buonaventura 164, 409
  principle of 164–165, 193, 381
Chain rule 121, 297–298
Change 108
  and continuity 117–118
Chord, chordal length 205
Cicero, Marcus Tullius 43
Circle:
  area of 45, 138
  circumference of 48, 138
  equation of 32
  functions 199–200
  unit 199
Collapsing sum 394, 398
Comic activity 417–418
Compound interest 324
Concave upward 104
Concavity 85–88, 254
Cone, volume of 181–183
Connected domain 130–131
Continuity 14, 17–19, 24
  and attaining extreme values 96
  in Δ-notation 117
  of a differentiable function 105
  and existence of integral 176
  Leibniz's principle of 116
Continuous:
  function 18, 19
  interest 325
  rate of growth 326
  versus discrete 361
Coolidge, J. L. 387
Cosh t 348
Cos t, cosine 195, 197, 200
Cot t, cotangent 216
Courant, Richard 353, 386–387
Critical point 71, 95–97, 99–100
Csc t, cosecant 216
Cycloid 381

d, Δ 110
d'Alembert, Jean le Rond 374
Dante Alighieri 194
Dantzig, Tobias 375, 387, 427
Decimal expansion 35
Degree measure 195
Delian problem 412
Delta-notation 109
Democritus 355–357, 384
Derivative 69
  chain rule 121, 297–298
  of implicit function 234
  and integral 285
  of inverse function 229
  second 84
  shortcut rules 95
  sign of 63
  sign of second 85–88
  See also rate of change, speed, tangent line
Derivative of:
  a^x 315
  arccos x 232
  arcsin x 230–231
  arctan x 232
  c f(x) 82
  constant function 105
  cos x 216
  COS x 243
  cosh x 348
  cot x 216
  csc x 216
  e^x 308
  f(x) + g(x) 83
  $(f(x))^2$ 90
  f(x) g(x) 91
  f(x)/g(x) 94
  $\sqrt{g(x)}$ 92
  inverse function 229
  ln x 303
  1/x 79
  1/g(x) 81
  powers 105, 314
  quadratics 69
  sec x 134
  sin x 212–214
  SIN x 243
  sinh x 348
  squaring function 61
  tan x 216

x^n 146
x^t 314
Descartes, René 33, 74, 141, 357, 358, 360, 383
Difference quotient 78
Differential 362
Differential calculus 74, 177
Differential equation 284
Differentiation 127
　implicit 234
　logarithmic 320
Dinostratus 407, 412
Discrete versus continuous 361
Discriminant 72
Distance formula 31
Dodgson, C. L. 389
Domain and range 8–12, 96–99, 101
Double-angle formulas 208, 211
Dryden, John 43
Dummy variable 158, 257
Duplicating cube 404, 407

e 309
　approximations to 311, 340, 345
e^x 309
　series for 340
Egyptians 28, 194, 387
Elimination principle 55–59, 405–406
Ellipse 192
Ellipsoid 193
Enlightenment 352, 353, 375
Equations and curves 33
Eratosthenes 40–41, 43, 237, 356–357, 407, 410
Euclid 38–40, 352, 358, 371, 404
Eudoxus 28, 36–37, 148, 153, 165, 183, 246, 248, 356, 407
　method of 164–173, 177
　See also Appendix 2
Euler, Leonhard 367, 384, 407
Evaluation theorem 246
Even function 263
Eves, Howard 387, 389
Existence theorem 176, 246, 254
Exponential:
　decay 335
　function 307
　growth 327

Exp x 307
Extreme values of continuous function 96, 101

F1 246, 248
F2 246
F3 246
F4 285
Faith versus reason 372–374
Fermat, Pierre de 24, 33, 59, 74, 143, 153, 187, 246, 399–400, 402
　method of 59–62, 69, 80, 165, 171, 177–178
Fluents, fluxions 362
Freely falling body 115, 131–133
Function 3, 4, 5, 24
　algebraic 7, 24
　continuous 18, 19
　domain of 8–10
　even 263
　exponential 307, 313
　hyperbolic 348
　identity 241
　implicit 234
　inverse 225
　linear 52
　logarithmic 302
　odd 262
　quadratic 55, 70–73
　range of 9, 10
　reciprocal 69, 79
　square root 69
　transcendental 245
　trigonometric 194
Functional existence 15
Fundamental theorem(s) of calculus 140, 148, 177–179, 246, 248, 256, 285, 363, 402

g, long-suffering 15
Galilei, Galileo 384
Galois, Évariste 386
Gaffney, Matthew 385
Gauss, Carl Friedrich 354–355, 415
Geometric series 398
Geometry 40
Global (versus local) 87, 89
Gödel, Kurt 388

Graph   5
Gravity   115, 131–132, 187, 287
Greeks, *see individual listings*
Greene, Graham   386
Growth law   329
Guessing   64–65, 408

Hadamard, Jacques   386
Half-angle formulas   208, 211
Half-life   334
Halley, Edmund   372
Halmos, Paul   386
Hardy, Godfrey Harold   384, 386, 388
Heath, T. L.   403, 414
Henkin, Leon   388
Heraclitus   15, 19, 138
Hilbert, David   386–387, 388
Hippias   407, 412
Hippocrates   46, 47, 274
Holmes, Sherlock, principle of   55–59, 76–77
Homer   28, 412
Homomorphism   302, 308
Hydrostatics   42, 408
Hyperbolic functions   348

Identities   203
  trigonometric   204, 208, 209
Implicit differentiation   234
Indeterminate form   276
Index of summation   169
Infeld, Leopold   386
Infinitesimal calculus   363–370
Infinitesimals   362
Infinity   401
Inflection point   88
Instantaneous rate of change   108–110, 112–113
integ-   164
Integer   2, 9, 164
Integral   167
  and antiderivative   178, 282, 285
  as average   261
  definite   250
  as distance   291
  error in approximating   253
  existence of   176, 254–255
  as limit of sums   168–173, 249–250
  mean-value theorem for   281
  properties of   174–176
  sign ($\int$)   167
  as volume   181–186
  as work   289
Integral calculus   130, 164, 177
Integrating factor   330
Interest:
  compound   324
  continuous   325
Inverse function   225
Irrational numbers   34–37, 49
Irrationality of $\sqrt{2}$ 35–36

James I, king of England   360
Jefferson, Thomas   39

Kepler, Johannes   187
Kline, Morris   374, 387

Leibniz, Gottfried Wilhelm von   24, 74, 108–111, 117, 119, 127, 140, 141–143, 148, 149, 155, 159, 164, 165, 177, 189, 362, 370, 372, 374, 384, 387
  approximation for π of   350
  notation of   108, 111, 125
  principle of continuity of   116–117, 145
Leonardo da Vinci   376
Lever, law of   408
L'Hôpital's rule   277, 279
Lim   14–19
Limit   14, 19–24, 42, 48, 153, 165, 345, 361, 370, 371
  from one side   17, 22, 114
Lincoln, Abraham   39
Line:
  equation of   54
  rising, falling   50
  slope of   51
Linear function   52–54
Littlewood, J. R.   388
ln x   302
Local maximum, minimum   89

Logarithm 300, 318
 See also tables 421–425
Logarithmic differentiation 320
Logarithmic function 146, 302
Lunes of Hippocrates 46, 376

Marcellus 42
Maximum, minimum:
 applications of 64–67, 96, 97,
  100–101, 220–221, 223–224,
  234–235, 415–416, 417–419
 on a closed interval 96
 steps for finding 99
Mean-value theorem 278, 281
Menninger, Karl 389
Meschkowski, Herbert 370, 387
Millay, Edna St. Vincent 39
Minerva 6, 13, 15
Model 133, 329
Moment 408
Moritz, R. E. 389
Moser, Jürgen 141

n! (n factorial) 340
Nagel, Ernest 388
Natural logarithm 302
Nebuchadnezzar 196
Neoclassicism 352, 375
Neugebauer, Otto 387
Newman, J. R. 37, 374, 388
Newsom, C. V. 387
Newton, Isaac 74, 108, 140, 148, 149,
 155, 159, 187–189, 354–355, 360,
 361–362, 370, 372, 374, 384, 402
 law of cooling of 350
 law of gravity of 187, 287
Non-standard analysis 371
Number 29, 34–36, 248–250, 312–313
Number line 4

Odd function 262
Omar Khayyám 352
Optimization 5, 13, 64, 69, 95–101,
 220–221, 223–224, 234–235,
 415–416, 417–419
Ore, Oystein 386

π 45, 407
 approximations to 249, 255, 271, 294,
  350, 382
Pan, Peter 56
Parabola, quadrature of 402
Pascal, Blaise 74, 149, 354, 383, 387,
 415
Plato 37–38, 194, 358, 359, 360, 383,
 384
Plutarch 42–43, 414
Poetry and mathematics 110–111
Poincaré, Henri 383, 386, 388
Pólya, George 354
Polygon 47
Pooh, Winnie the 56
Pope, Alexander 187
Population growth 329–332
Position function 135
Principia 187–188
Proclus 383
Product rule 91, 106, 144, 348
Purpose versus action 17–19
Pythagoras, Pythagoreans 28–31,
 33–35, 37–38, 44, 46, 169, 189,
 248–249, 273, 311, 349, 384, 389
Pythagorean theorem 29, 44, 69, 125,
 198, 376, 381

Q. E. D. 30, 35, 40
Quadrant 202
Quadratic:
 approximation 338
 formula 72–73
 functions 55, 70–73
Quadrature 273–274
 of circle 48
  See also Appendix 3
 of parabola 402
 of rectangle 411
 of triangle 411
Quotient rule 94

Radian measure 196
Radioactive decay 334
Range and domain 8–12, 96–99, 101
Rate of change:
 average 113
 instantaneous 108–110

Rational number   34−37
Real number system   248−249, 388
Reciprocal rule   81
Reductio ad absurdum   35−36, 45
Reid, Constance   386
Related rates   123−125, 233−234
Renaissance   352
Respice finem   415
Rising, falling   50−52
Robbins, Herbert   353
Robinson, Abraham   371−372
Roman mathematics   XLIII, CDXXVII
Romantic Movement   352, 375
Rousseau, Jean Jacques   375, 384
Rule for squares   90
Russell, Bertrand   111, 368, 383, 384, 386, 388, 433

Sage   358
Santa Claus   377
Secant line   58
Second derivative   84
Sec t, secant   216
Sector of a circle   198
Seer   358
Serendipity   173, 311
Series:
  geometric   398
  Taylor   340−342, 351
Series for:
  cos x   343
  e 340
  e^x 340
  e^{-x} 341
  ln(1 + x)   341
  π 350
  sin x   341
Shakespeare, William   110
Siegel, C. L.   141
Sigma-notation   166−169
  See also Appendix 2
Signature of a function   338
Simpson's rule   265−271
Sinh x   348
Sin t, sine   195, 197, 200
Slope   51
Slope-intercept form   54
Smile   440
  and positive second derivative   87

Smith, D. E.   387
Snow, C. P.   386
Socrates   38
Solid of revolution   182
Speed   112−114
Square root rule   92
Squaring function   55
Steen, Lynn Arthur   385, 388
Steinhaus, Hugo   388
Struik, D. J.   387
Style, *see* Appendix 4
Sumerians   28
Summation formulas, *see* Appendix 2
Surface area of sphere   411

Tangent line   51, 56−64
Tan t, tangent   195, 197, 201
Taylor's theorem   351
Thales   28−29, 38
Tigger   56
Toeplitz, Otto   387
Tolstoy, Leo   371
Transcendental functions   245
Trigonometric functions, *see individual listings*
Trigonometric identities   204, 208, 209
Trigonometry   194
Trisection problem   403, 404−405, 411

Uniqueness theorem   246, 248
Unit circle   199

Van der Waerden, B. L.   387
Variable   3, 24
Velocity, *see* speed
Voltaire   141, 375, 384, 412
Volume   182
  of a cone   181−183
  of a cylinder   181
  of a pyramid   356, 357
  of a solid of revolution   184−185, 365−366
  of a sphere   185−186, 410

Weidman, Donald   387
Weil, André   360, 387

Weyl, Hermann   387, 388
Whewell, William   360
Whitehead, Alfred North   49, 358, 361, 383–384, 385, 388, 413–414, 416, 420
Wiener, Norbert   386
Wilder, R. L.   387
Williams, J. D.   388
Winnie the Pooh, *see* Pooh
Wordsworth, William   187
Work   287, 289

Xenophanes   14, 15, 19, 138

Yoknapatawpha County   56

Zeno   165, 356
Zero   251

Undergraduate Texts in Mathematics

Apostol: Introduction to Analytic Number Theory.
1976. xii, 370 pages. 24 illus.

Childs: A Concrete Introduction to Higher Algebra.
1979. xiv, 338 pages. 8 illus.

Chung: Elementary Probability Theory with Stochastic Processes. Third Edition
1979. 336 pages.

Croom: Basic Concepts of Algebraic Topology.
1978. x, 177 pages. 46 illus.

Fleming: Functions of Several Variables. Second edition.
1977. xi, 411 pages. 96 illus.

Halmos: Finite-Dimensional Vector Spaces. Second edition.
1974. viii, 200 pages.

Halmos: Naive Set Theory.
1974. vii, 104 pages.

Kemeny/Snell: Finite Markov Chains.
1976. ix, 210 pages.

Lax/Burstein/Lax: Calculus with Applications and Computing, Volume 1.
1976. xi, 513 pages. 170 illus.

LeCuyer: College Mathematics with A Programming Language.
1978. xii, 420 pages. 126 illus. 64 diagrams.

Malitz: Introduction to Mathematical Logic.
Set Theory - Computable Functions - Model Theory.
1979. 255 pages. 2 illus.

Prenowitz/Jantosciak: The Theory of Join Spaces.
A Contemporary Approach to Convex Sets and Linear Geometry.
1979. Approx. 350 pages. 404 illus.

Priestley: Calculus: An Historical Approach.
1979. 400 pages. 300 illus.

Protter/Morrey: A First Course in Real Analysis.
1977. xii, 507 pages. 135 illus.

Sigler: Algebra.
1976. xi, 419 pages. 32 illus.

Singer/Thorpe: Lecture Notes on Elementary Topology and Geometry.
1976. viii, 232 pages. 109 illus.

Smith: Linear Algebra
1978. vii, 280 pages. 21 illus.

Thorpe: Elementary Topics in Differential Geometry.
1979. 256 pages. Approx. 115 illus.

Whyburn/Duda: Dynamic Topology.
1979. Approx. 175 pages. Approx. 20 illus.

Wilson: Much Ado About Calculus.
A Modern Treatment with Applications Prepared for Use with the Computer.
1979. Approx. 500 pages. Approx. 145 illus.

LIBRARY OF DAVIDSON COLLEGE